STUDIES ON HEPATITIS VIRUSES

STUDIES ON HEPATITIS VIRUSES

Life Cycle, Structure, Functions, and Inhibition

SATYA P. GUPTA
Meerut Institute of Engineering and Technology,
Meerut, India

ELSEVIER

ACADEMIC PRESS
An imprint of Elsevier

Academic Press is an imprint of Elsevier
125 London Wall, London EC2Y 5AS, United Kingdom
525 B Street, Suite 1650, San Diego, CA 92101-4495, United States
50 Hampshire Street, 5th Floor, Cambridge, MA 02139, United States
The Boulevard, Langford Lane, Kidlington, Oxford OX5 1GB, United Kingdom

Library of Congress Cataloging-in-Publication Data
A catalog record for this book is available from the Library of Congress

British Library Cataloguing-in-Publication Data
A catalogue record for this book is available from the British Library

ISBN: 978-0-12-813330-9

For information on all Academic Press publications visit our website at https://www.elsevier.com/books-and-journals

Publisher: John Fedor
Acquisitions Editor: Linda Versteeg-buschman
Editorial Project Manager: Tracy I. Tufaga
Production Project Manager: Poulouse Joseph
Designer: Matthew Limbert

Typeset by TNQ Technologies

To
My wife, Kanak, whose love immunized me
against all odds of life

CONTENTS

ABOUT THE AUTHOR

Satya P. Gupta is presently a professor emeritus at Meerut Institute of Engineering and Technology (MIET), Meerut, India, after retiring as professor in the Department of Applied Sciences at National Institute of Technical Teachers' Training and Research (NITTTR), Bhopal. Earlier, he served Tata Institute of Fundamental Research (TIFR), Mumbai, a world-renowned Institution of India, Birla Institute of Technology and Science (BITS), Pilani, and then (MIET) as its director-cum professor of eminence in the Department of Pharmacy. Professor Gupta has a very long standing of teaching quantum chemistry, pharmaceutical chemistry, biophysics, and drug design. He had obtained his M. Sc and D. Phil degrees from University of Allahabad, Allahabad, in 1967 and 1971, respectively. He is a world-renowned scientist. For his work in drug design, he has bagged several honors and awards. To his credit, Gupta has more than 200 research publications in highly reputed national and international journals and several reviews in such highly prestigious periodicals as *Chemical Reviews* (American Chemical Society), *Progress in Drug Research* (Birkhäuser Verlag Basel, Switzerland), and *Current Medicinal Chemistry* (Bentham Science, Netherlands), and has been on editorial boards of several international journals. Dr. Gupta himself has been editor of several scientific journals and books of international repute published by Springer Verlag, Heidelberg Germany; CRC Press, Taylor and Francis Group; Nova Science Publishers, NY, USA; and Elsevier (Academic Press), USA. He has delivered several lectures on drug design across the world. His scientific activity has brought him to the forefront of the scientific community of the world.

PREFACE

Viral hepatitis is a systemic infection affecting predominantly the liver and causing its inflammation, which may be acute or chronic. Compared to better-publicized infectious diseases such as HIV or malaria, viral hepatitis is often referred to as a silent killer. Viral hepatitis is caused by infection with one of the five known hepatitis viruses, A–E, which are quite divergent in their structure, etiology, epidemiology, and pathophysiology. Of these five hepatitis viruses, hepatitis B, C, and D viruses (HBV, HCV, and HDV) cause chronic liver disease leading to cirrhosis, liver failure, and hepatocellular carcinoma. Liver cirrhosis is an important cause of mortality globally and hepatocellular carcinoma, the most common form of liver cancer, is the third-leading cause of cancer deaths, claiming more than 500,000 lives each year. Thus hepatitis viruses have been the subject of intense study in the last 20 years, and a wealth of information has accumulated related to their life cycle, structure, functions, and inhibition. However, the study still seems to be in its youth. To further accelerate the study on these viruses, all the information available so far on them needed to be put in one place. The book in hand has attempted to serve this purpose. The book may provide further guidelines to work on hepatitis viruses. It may be useful to all those working in the area of virology and medicinal chemistry.

The book overall contains eight chapters. Chapter 1 is a kind of introductory chapter that covers the etiology, epidemiology, and pathophysiology of hepatitis infection. The word etiology refers to the manner of causation of a disease or condition, epidemiology refers to the study of cause of a disease, its patterns, effects on health, its outbreak, and the disease condition in the defined population, and pathophysiology refers to the physiology of abnormal states, specifically the functional changes that accompany a particular syndrome or disease. The chapter describes all these aspects of all viruses in detail. Chapter 2 presents structure and replication of all hepatitis viruses, where structure refers to morphology, genome, classification, and antigens present in them. Under replication, it is discussed how the viruses after infecting the host multiply within the host. To control the viral hepatitis, it is therefore essential that we know the important targets in their life cycle that can be hit to cease their replication. Chapter 3 therefore presents the details of proteases and polymerases that are crucial for the replication of different hepatitis viruses and that can be exploited for the development of antihepatitis drugs. Based on this strategy, several attempts have been made to discover antihepatitis drugs. The present status of such drugs has been presented in detail in Chapter 4. In the design and development of potential, therapeutically useful antiviral drugs, the role of medicinal chemistry has been rewarding, as it covers all aspects of drug design such as recognition of important drug targets, computational chemistry, optimization of drug activity based on their structure-activity relationship, finding the three-dimensional structures of compounds by X-ray crystallography, NMR, and molecular dynamics, and then synthesizing the drugs and evaluating their activity. Chapters 5–7 are therefore devoted to discussing the medicinal chemistry of development of drugs against all hepatitis viruses. Chapter 5 presents such studies only on anti-HAV, anti-HDV, and anti-HEV drugs, as they are not so extensively explored, but Chapter 5 gives an exhaustive account of such study on anti-HBV drugs, and so does Chapter 7 for anti-HCV drugs. This separation was essential to avoid making a single chapter too voluminous to read. Finally, Chapter 8, briefly mentions the present status of viral hepatitis and describes the strategy planned to prevent the hepatitis virus infection. The strategy primarily encompasses vision, goal, targets, guiding principles, directions, and priority actions and outlines key elements of its implementation, including strategic partnerships, monitoring and evaluation, and costing. I hope the readers will enjoy reading this book and the researchers will find it useful to making further advancement in antihepatitis chemotherapy.

Satya P. Gupta
Meerut, India

CHAPTER 1

Viral Hepatitis: Historical Perspective, Etiology, Epidemiology, and Pathophysiology

1. INTRODUCTION

Hepatitis refers to inflammation of the liver, which is primarily caused by viruses called hepatitis viruses. So far, there are five known hepatitis viruses, called hepatitis A virus (HAV), hepatitis B virus (HBV), hepatitis C virus (HCV), hepatitis D virus (HDV), and hepatitis E virus (HEV). In United States, the first three viruses, i.e., HAV, HBV, and HCV, are more prevalent. The symptoms of their infection are usually nausea, abdominal pain, fatigue, malaise, and jaundice (Wasley et al., 2008). Hepatitis may be temporary (acute) or long term (chronic), depending on whether it lasts for less or more than 6 months. Acute hepatitis can sometimes resolve on its own or progress to chronic hepatitis, which over time may progress to liver failure or liver cancer. HBV and HCV may lead to chronic infection, causing cirrhosis and hepatocellular carcinoma (HCC) (Wasley et al., 2008). The most common effect of HAV and HEV is the sudden onset of fever and systemic symptoms, followed a few days later by jaundice. HDV may also cause chronic infection, but it affects only in presence of HBV infection; hence an individual protected against HBV is hardly affected by HDV. Three viruses—HBV, HCV, and HDV—are transmitted in the body parenterally, while the remaining two—HAV and HEV—can be transmitted enterally.

Compared to human immunodeficiency virus (HIV), hepatitis viruses are less publicized. They are recognized as silent killers. According to World Health Organization (WHO), about 400 million people worldwide are suffering with chronic hepatitis viral infection. More than 1 million people die of liver cirrhosis caused by hepatitis viruses, and HCC is the third leading cause of cancer deaths, claiming more than 500,000 lives each year. In spite of the prevalence of hepatitis, no symptoms of chronic viral hepatitis are recognized until patients develop complications of cirrhosis or liver cancer, and once symptoms occur, currently available remedies are often

Studies on Hepatitis Viruses
ISBN 978-0-12-813330-9
https://doi.org/10.1016/B978-0-12-813330-9.00001-6

ineffective and/or expensive. Given the large burden of viral hepatitis, governments around the world have made various efforts to mitigate its impact. WHO has set a goal to eliminate viral hepatitis as a major public health threat by 2030.

In addition to these five hepatitis viruses, other viruses that can also cause liver inflammation are cytomegalovirus, Epstein–Barr virus, yellow fever, and herpes simplex virus (Kaufman et al., 1997). Additionally, two more hepatitis viruses were recognized: hepatitis F virus (HFV) and hepatitis G virus (HGV), where the former is a hypothetical virus linked to hepatitis. None of the HFV candidates that emerged in the 1990s were substantiated (Uchida, 1993; Fagan, 1994; Bowden, 2001). HGV, thought to cause hepatitis, is in fact an orphan virus with no casual links to any human disease (Lefrère et al., 2008). It was initially identified as GB virus C.

2. ETIOLOGY OF INFECTION

The word etiology refers to the manner of causation of a disease or condition. The etiology of hepatitis infection by various hepatitis viruses is described next.

2.1 Hepatitis A Virus

HAV is a member of Picornaviridae family. It is an RNA virus with a size of 7.5 kb and a diameter of 27 nm. It has one serotype and multiple genotypes. HAV is mostly transmitted through consumption of contaminated water or food, but certain sex practices can also be responsible for HAV transmission. HAV infection can be severe and life threatening, but in most of the cases, it is mild and fully recoverable, creating immunity from further HAV infection. It infects people living in areas that have poor sanitation, so the risk is highest in developing countries. Safe and effective vaccines have been developed to prevent HAV infection.

2.2 Hepatitis B Virus

HBV is a member of Hepadnaviridae family. It is a partially double-stranded DNA virus with a size of 3.2 kb. It has an incomplete positive strand, but the complete negative strand has four overlapping genes, as mentioned subsequently.

1. HBsAg is also known as the Australia antigen. It is the surface antigen of HBV virus that indicates current hepatitis B infection.

2. HBcAg is a core antigen of HBV. It indicates active viral replication, meaning that the HBV-infected person is infectious.
3. Gene *P* codes for a DNA polymerase that has reverse transcriptase activity.
4. Gene *X* codes for the X protein that has transcription-regulating activity.

HBV transmission can take place through exposure to infectious blood, semen, and other body fluids. This is termed *horizontal transmission*. Infected mothers can transmit to infants at the time of birth or an infected family member to an infant in early childhood. This is termed *perinatal transmission* (sometimes also called *vertical transmission*). Transfusions of HBV-contaminated blood and blood products and contaminated injections during medical procedures may also be responsible for HBV transmission.

2.3 Hepatitis C Virus

HCV, with a size of 9.4 kb and diameter 55 nm, is an RNA virus and belongs to Flaviviridae family. It can be transmitted through transfusion of HCV-contaminated blood and blood products or contaminated injections. Sexual transmission is also a possibility. There is no vaccine for it, as its genetic variability hinders all efforts in this direction. It has 1 serotype, 6 major genotypes, and more than 80 subtypes.

2.4 Hepatitis D Virus

HDV, which can be transmitted sexually as well as by exposure to infected blood and blood products, belongs to *Deltavirus* genus. It is a 1.7-kb single-stranded RNA virus with a diameter of 36 nm. It is a single species in *Deltavirus* genus. It contains hepatitis D antigen and uses HBsAg as its envelope protein, so its release requires HBV infection (Previsani and Lavanchy, 2001). A serious disease can occur only with dual infection of HDV and HBV, but an effective HBV vaccine can protect from HDV infection.

2.5 Hepatitis E Virus

Like HDV, HEV is also a single species in *Hepevirus* genus with a single-stranded RNA. It has a size of 7.5 kb and a dimeter of 32–34 nm. It is transmitted through contaminated water or food (Previsani and Lavanchy, 2001; Adhami and Levinthal, 2002). HEV infection is common in developing countries. There are vaccines to prevent HEV infection, but they are not easily available.

3. EPIDEMIOLOGY

Epidemiology refers to the study of cause of a disease, its patterns, effects on health, its outbreak, and the disease condition in a defined population. This study cautions about the public health and alerts the public healthcare departments to shape policy decisions and evidence-based practices to stop the spread of the disease. Epidemiology also helps develop methodologies useful for clinical research, public health studies, and related basic research in biologic sciences. Epidemiology, in fact, has a vast umbrella, which can cover the study of disease causation, transmission, outbreak investigation, biomonitoring, and comparisons of treatment effects such as in clinical trials. Epidemiology is also concerned with the assessment of new cases in a given time period, usually 1 year, and the number of cases of a disease in the population at a given time. Following is the epidemiology of hepatitis caused by hepatitis viruses A–E.

3.1 Hepatitis A

Hepatitis A is prevalent in almost 100% of developing countries, but in the United States, it is just above 40%. Hepatitis accounts for about 40% of all acute viral hepatitis. Residents and travelers in underdeveloped countries, homosexual men, and workers and children in daycare are at risk. The early symptoms of hepatitis A are fever, nausea, vomiting, fatigue, diarrhea, and anorexia, which at later stages may lead to right-upper abdominal pain, dark urine, and jaundice. Having lived in the same household with a patient with hepatitis accounts for about 24% of risk, fecal-oral spread of virus in the homosexual act may account for about 11% of risk, and close contact with young children of daycare centers shares about 18% of the risk. Studies have suggested that preschool daycare centers may at times be important foci for transmission within the United States (Hadler et al., 1980). Most cases of hepatitis A were attributed to fecal-oral transmission of the virus. In recent years, illicit use of parenteral drugs has been reported as a risk factor by only 2% of patients, but occasional association of hepatitis A with intravenous drug use was found to be interesting. The early symptoms of hepatitis A include fever, nausea, vomiting, fatigue, diarrhea, and anorexia, which are expressed in later stages as right-upper abdominal pain, dark urine, and jaundice. However, hepatitis A never causes chronic disease and rarely leads to fulminant liver failure. On the other hand, a person when recovered from HAV infection develops lifelong immunity from further infection. However, no effective treatment of HAV infection exists: prevention is the only remedy. Prevention includes washing hands after using the toilets, after changing diapers, and before preparing foods or eating. HAV vaccines can be used to have long-term protection from HCV, high-risk persons, chronic liver disease, and transplant recipients. Vaccination has quite a high success rate (>95%).

3.2 Hepatitis B

HBV infection eventually leads to chronic infection, and around 240 to 350 million people have been estimated to be chronically infected with HBV (Ott et al., 2012; Lavanchy, 2004), with more than 2 billion people globally ever having been infected. Global Burden of Disease Study 2010 had estimated 786,000 deaths due to HBV infection, where 341,000 were because of liver cancer and 312,000 due to cirrhosis (Lozano et al., 2012). However, because of the universal vaccination program, the epidemiology of HBV infection globally is changing (Chang et al., 1997; Liang et al., 2009; Ott et al., 2012; Zoulim and Durantel, 2015; Jennifer and Benjamin, 2015). This change has also been affected by migration between high- and low-prevalence populations (Hahne et al., 2004; Marschall et al., 2008; Kowdley et al., 2012; MacLachlan et al., 2013).

In a recent communication, MacLachlan et al. (2015) have inferred that HBV infection epidemiology is geographically diverse and depends upon population prevalence, age, and mode of occupation, and sensitivity to chronic infection, all of which are mutually interdependent. In the countries where chronic infection is high, individuals are supposed to be sensitive to HBV infection at birth or childhood when the risk of development of chronicity is high (Lavanchy, 2004). It has been estimated that about half of the population of the world lives in areas of high prevalence (prevalence is a measurement of all individuals affected by the disease at a particular time), such as the Asian Pacific and sub-Saharan African regions (Mahoney, 1999). The low HBV prevalence regions include Australia, Asia, Northern and Western Europe, Japan, North America, and some countries in South America. People living in these areas make up the minority of the global population. In these areas, the incidence of vertical and horizontal transmission in childhood is low, with most incident infections occurring in adolescence and adulthood through sexual contact, injecting drug use, and other blood-related exposures, including historically in healthcare settings (MacLachlan et al., 2015).

There are regions that have intermediate HBV prevalence (2%–7%). Such regions include North Africa and the Middle East and parts of Eastern Europe, Southern Europe, Latin America, and South Asia. In these regions, the transmission can be either perinatal or horizontal (Lavanchy, 2004), but as compared to high prevalence countries the former is less common in these regions. In low-prevalence areas, neither the perinatal nor horizontal transmission is very common; rather, it occurs through sexual contact, injected drug use, and other blood-related exposure (MacLachlan et al., 2015).

Since HBV leads to HCC and since there is wide variation in the prevalence of HCV globally, the cancers related to HBV vary considerably according the regions. In developed countries, about a quarter of liver cancer is attributed to HBV, but in developing countries, it is 60% (Jemal et al., 2011). In China, the liver cancer due to HBV has been estimated to be 90% (Gust, 1996).

3.3 Hepatitis C

Regarding the epidemiology of HCV, a systematic study was performed by Petruzziello et al. (2016) using the comprehensive studies reported for 138 countries (about 90% of global population) from 2000 to 2015. In this study, they found that global HCV prevalence was around 2.5% (177.5 million of HCV-infected adults), in which the greatest share was from Africa and the lowest from the Americas. HCV genotype 1 is the most prevalent worldwide (49.1%) and genotype 3 the least (11.0%). Genotypes 1 and 3 are common worldwide and most prevalent globally, but the largest proportion of genotypes 4 and 5 is in lower-income countries. The global distribution of HCV genotypes is, however, influenced by historical events, e.g., migration trends (Markov et al., 2012). HCV infection is one of the main global burdens (Razavi et al., 2013, 2014; Lozano et al., 2012; Deuffic-Burban et al., 2012; Gane et al., 2015; Hatzakis et al., 2015), whose paradigm changes region to region. The HCV disease burden can be reduced by country-specific policy, provided robust epidemiological survey is available. A correct knowledge of the HCV distribution is crucial to contain this global burden disease, but unfortunately, half of the countries in the world do not have robust studies of HCV-infected populations (Petruzziello et al., 2016).

Chronic HCV infection often leads to liver cirrhosis, hepatocellular cancer, liver failure, and death (Lauer and Walker, 2001), especially in HIV-positive patients during active antiretroviral therapy. However, it is expected that HCV infection could be eliminated in the next 15–20 years with focused therapeutic strategies (Wedemeyer et al., 2014; Martin et al., 2013) and good understanding of HCV infections to prevent new infections.

3.4 Hepatitis D

Alfaiate et al. (2015) reported that approximately 15–20 million people are coinfected with HDV and HBV worldwide. In the United States as well as in Mediterranean Basin, HDV infection was more commonly observed in patients who used intravenous drugs. However, the highest HDV prevalence was reported to be in southern Italy, North Africa, the Middle East, and the Amazon basin (Braga et al., 2012). A low prevalence of HDV was observed in the American South Pacific islands of Samoa, Hauru, and Hiue. China, Japan, and Myanmar also have a low prevalence of HDV but with a high prevalence of HBV infection (Heidrich et al., 2009; Makuwa et al., 2009). In a recent study, Ordieres et al. (2017) reported through a survey conducted during 1983–2012 that there could be 8.2% prevalence of anti-HDV among those infected with HBV. HBV particles are required for HDV to replicate and infect hepatocytes (Rizzetto and Verme, 1985; Bean, 2002; Xiridou et al., 2009; Abbas and Afzal, 2013). HDV can infect only 5% of those already infected with HBV and leads to fulminant liver failure in 1% of patients. HBV-HDV superinfection is the most aggressive form of the viral hepatitis (Mailet et al., 2017; Alfaiate et al., 2016). HDV infection is more common in adults than in children, but children in underdeveloped countries are more prone to contract it through breaks in the skin and skin lesions.

3.5 Hepatitis E

HEV infection has been prevalent in the Indian subcontinent, Africa, and southeast and central Asia, threatening the public, especially pregnant women. Once infection occurs in pregnant women, the outcome significantly varies by pregnancy status, stage of pregnancy, and preexisting liver disease. The prevalence of HEV infection greatly varies by geographic location, but it depends upon HEV genotypes (Teshale and Hu, 2011; Kamar et al., 2012; Aggarwal, 2011). It is highest in the regions that lack water, as fecal contamination of drinking water is a major route of transmission (Rein et al., 2012). In such regions, the HEV genotypes 1 and 2 are dominant. In the regions where uncooked or undercooked meat is taken, zoonotic transmission, particularly by genotype 3, is common

(Kamar et al., 2012). HEV genotype 4 causes disease mainly in China and Taiwan, and mode of transmission is mainly zoonotic. Sporadic cases of genotype infection have also been reported in Europe. The outbreaks of HEV genotype 1 or 2 have been reported in 30 countries from three subcontinents and most frequently occur in the Indian subcontinent (Labrique et al., 2010). It has been reported that HEV genotypes 1 and 2 exclusively infect humans. In fact, epidemiologic study of HEV is very limited, and information on disease occurrence and distribution is available only from a few European countries (Hepatitis E Vaccine Working Group, 2014).

4. PATHOPHYSIOLOGY

Pathophysiology refers to the physiology of abnormal states, specifically the functional changes that accompany a particular syndrome or disease. Following is the pathophysiology of each hepatitis.

4.1 Hepatitis A

The HAV infection is marked by several weeks of malaise, anorexia, nausea, vomiting, and elevated aminotransferase levels. Cholestatic hepatitis, marked by the development of an elevated alkaline phosphatase level, is also experienced by some patients, while several relapses during a course can also be observed in some patients. Jaundice develops in more severe cases, but HAV infection does not persist and does not lead to chronic hepatitis. The manifestation of symptoms of disease is related to age; in developing nations, it can occur before age 2, while in Western countries in persons, it occurs in those aged 5–17 years.

4.2 Hepatitis B

The HBV infection may lead to acute as well as a chronic condition of hepatitis. An acute condition is severe and sudden in onset such as an asthma attack, but a chronic condition is a long-developing syndrome, e.g., asthma. Acute conditions, if untreated, may lead to the development of chronic syndrome. HBV may be directly cytopathic to hepatocytes. In a cytopathic effect, the virus causes lysis of the host cell or brings about the structural changes in the host cell, so it becomes unable to reproduce. HBV is a stealth virus that avoids detection and evades the immune system (Chisari et al., 2010). According to Nebbia et al. (2012), HBV causes inflammation and progressive fibrosis in the infected liver by triggering the immune system to attack the hepatocytes. In the chronic condition of hepatitis B, no symptoms are observed, but sometimes fatigue, aches and pains, fever, loss of appetite, nausea, and abdominal pain are observed. In acute conditions also, a majority of cases do not show any symptoms, but around 30% of adults will complain of jaundice, fatigue, poor appetite, weight loss, nausea, vomiting, abdominal pain, pyrexia, dark urine, and light stools (Aspinall et al., 2011).

4.3 Hepatitis C

Chronic HCV infection occurs in 50%–80% of cases, leading to cirrhosis and HCC. In the chronic condition, HCV does not appear to be cytopathic. Cirrhosis is facilitated by external factors such as chronic alcohol consumption, and patients with cirrhosis become sensitive to developing HCC. Initial symptoms of HCV infection are often extrahepatic, most commonly involving the joints, muscle, and skin. Advanced or decomposed liver disease is manifested by synthetic dysfunction and portal hypertension, such as mental status changes (encephalopathy), ankle edema and abdominal distension, and hematemesis or melena (variceal bleeding). Acute HCV infection is usually asymptomatic. The chronic state may or may not be symptomatic, but in this case, fatigue is the predominant system. Twenty percent of the patients with chronic HCV infection have been estimated to move to cirrhosis, and patients with HCV-induced cirrhosis become prone to develop HCC. In the United States, HCC arises in 3%–5% of patients with HCV-induced cirrhosis each year.

4.4 Hepatitis D

HDV infection leads to any pathologic changes only in the liver, the only organ in which HDV can replicate. HDV can have any effect only in the presence of HBV, and together, they lead to more severe liver injury than HBV alone. Chronic HBV and HDV disease tends to progress more rapidly to cirrhosis than chronic HBV infection alone. Immune-mediated liver damage is attributed to HDV infection (Samuel et al., 1995; Smedile et. al., 1998; Niro and Smedile, 2012). It is has been speculated that the infected hepatocytes (cells of the main parenchymal tissue of the liver) in acute HDV infection undergo degenerative changes, consistent with cytopathic hepatocellular damage. This speculation was verified in human studies (Lefkowitch et al., 1987; Popper et al., 1983). HDV superinfection may result in fulminant hepatic failure, i.e., acute liver failure that is complicated by hepatic encephalopathy. A 28-year study of the course of HDV infection had concluded that HDV infection may be a risk factor for cirrhosis and HCC (Romeo et al., 2009).

4.5 Hepatitis E

Acute HEV infection is symptomized by jaundice, fatigue, and nausea. The symptomatic phase coincides with elevated hepatic aminotransferase levels (Hoofnagle et al., 2012). While HEV leads usually to acute disease, it may also cause chronic infection, causing liver fibrosis and cirrhosis (Bonnet et al., 2012). Traditionally, HEV was not believed to cause chronic liver disease; however, several reports have described that organ transplant recipients may have chronic hepatitis due to HEV (Kamar et al., 2008).

5. CONCLUSIONS

There are mainly five types of hepatitis virus, namely HAV, HBV, HCV, HDV, and HEV. The etiology, epidemiology, and pathophysiology of each one has been discussed. Under the etiologic study of each one, which refers to manner of causation of a disease or condition, it has been found that HAV infection occurs through consumption of contaminated water or food, but certain sex practices can also be responsible for it. HBV infection can occur through exposure to infectious blood, semen, and other body fluids, while HCV can be transmitted through transfusion of HCV-contaminated blood and blood products or contaminated injections. HDV can be transmitted sexually as well as by exposure to infected blood and blood products, but it shows its effect only when the person is already infected with HBV. HEV is transmitted through contaminated water or food. In epidemiologic study, which refers to the study of cause of a disease, its patterns, effects on health, its outbreak, and the disease condition in a defined population, it has been reported that HAV is prevalent in almost 100% of developing countries, but in the United States, it is just above 40%. In most cases, it has been attributed to fecal-oral transmission of the virus. Regarding HBV, it has been estimated that about half of the population of the world lives in an area of high prevalence of HBV, and HBV transmission in most of the cases occurs through sexual contact, injected drug use, and other blood-related exposures. Global HCV prevalence was around 2.5%, in which the greatest share was from Africa and the lowest from the Americas. However, half of the countries in the world do not have robust studies of the HCV-infected population. Chronic HCV infection often leads to liver cirrhosis, HCC, liver failure, and death.

Approximately 15–20 million people have been found to be coinfected with HDV and HBV worldwide. In the United States, HDV infection was more commonly observed in patients who used intravenous drugs and those residing in the Mediterranean basin. HDV infection is more common in adults than in children. Epidemiologic study of HEV is very limited, and information on disease occurrence and distribution is available only from a few European countries. However, HEV infection has been reported to be prevalent in the Indian subcontinent, Africa, and southeast and central Asia, threatening the public, especially pregnant women.

In pathophysiologic study, which refers to the physiology of abnormal states, specifically the functional changes that accompany a particular syndrome or disease, the HAV infection has been found to lead to malaise for several weeks, anorexia, nausea, vomiting, and elevated aminotransferase levels, but HAV infection does not persist and does not lead to chronic hepatitis, while the HBV infection leads to acute as well as chronic hepatitis. Chronic HCV infection leads to cirrhosis and HCC. In the United States, HCC arises in 3%–5% of patients with HCV-induced cirrhosis each year. Regarding HDV infection, it has been confirmed that it can have any effect only in the presence of HBV, and together, they lead to more severe liver injury than HBV alone. However, HDV infection may be a risk factor for cirrhosis and HCC. The acute HEV infection is symptomized by jaundice, fatigue, and nausea, and HEV is not believed to lead to any chronic liver disease, but it has been reported that organ transplant recipients may have chronic hepatitis due to HEV.

REFERENCES

Abbas, Z., Afzal, R., 2013. Life cycle and pathogenesis of hepatitis D virus: a review. World J. Hepatol. 5, 666–675.

Adhami, T., Levinthal, G., 2002. Hepatitis E and Hepatitis g/GBV-C. The Cleveland Clinic Disease Management Project.

Aggarwal, R., 2011. Hepatitis E: historical, contemporary, and future perspectives. J. Gastroenterol. Hepatol. 26 (Suppl. 1), 72–82.

Alfaiate, D., Lucifora, J., Abeywickrama-Samarakoon, N., Michelet, M., Testoni, B., Cortay, J.C., Sureau, C., Zoulim, F., Deny, P., Durantel, D., 2016. Hdv RNA replication is associated with HBV repression and interferon-stimulated genes induction in super-infected hepatocytes. Antiviral Res. 136, 19–31.

Alfaiate, D., Deny, P., Durantel, D., 2015. Hepatitis delta virus: from biological and medical aspects to current and investigational therapeutic options. Antiviral Res. 122, 112–129.

Aspinall, E.J., Hawkins, G., Fraser, A., Hutchinson, S.J., Goldberg, D., 2011. Hepatitis B prevention, diagnosis, treatment and care: a review. Occup. Med. 61, 531–540.

Bean, P., 2002. Latest discoveries on the infection and coinfection with hepatitis D virus. Am. Clin. Lab. 2, 25–27.

Bonnet, D., Kamar, N., Izopet, J., Alric, L., 2012. L'hépatite virale E: Une maladie émergente. La Revue de Médecine Interne 33, 328–334.
Bowden, S., 2001. New hepatitis viruses: contenders and pretenders. J. Gastroenterol. Hepatol. 16, 124–131.
Braga, W.S., Castilho, M.C., Borges, F.G., Leão, J.R., Martinho, A.C., Rodrigues, I.S., Azevedo, E.P., Barros Júnior, G.M., Paraná, R., 2012. Hepatitis D virus infection in the Western Brazilian Amazon - far from a vanishing disease. Rev. Soc. Bras. Med. Trop. 45, 691–695.
Chang, M.H., Chen, C.J., Lai, M.S., Hsu, H.M., Wu, T.C., Kong, M.S., Liang, D.C., Shau, W.Y., Chen, D.S., 1997. Universal hepatitis B vaccination in Taiwan and the incidence of hepatocellular carcinoma in children. Taiwan childhood hepatoma, study Group. N. Engl. J. Med. 336, 1855–1859.
Chisari, F.V., Isogawa, M., Stefan, F.W., 2010. Pathogenesis of hepatitis B virus infection. Pathol. Biol. 58, 258–266.
Deuffic-Burban, S., Deltenre, P., Buti, M., Stroffolini, T., Parkes, J., Mühlberger, N., Siebert, U., Moreno, C., Hatzakis, A., Rosenberg, W., Zeuzem, S., Mathurin, P., 2012. Predicted effects of treatment for HCV infection vary among European countries. Gastroenterology 143, 974–985.
Fagan, E.A., 1994. Acute liver failure of unknown pathogenesis: the hidden agenda. Hepatology 19, 1307–1312.
Gane, E., Kershenobich, D., Seguin-Devaux, C., Kristian, P., Aho, I., Dalgard, O., Shestakova, I., et al., 2015. Strategies to manage hepatitis C virus (HCV) infection disease burden – volume 2. J. Viral Hepat. 22 (Suppl. 1), 46–73.
Gust, I.D., 1996. Epidemiology of hepatitis B infection in the Western Pacific and South East Asia. Gut 38, S18–S23.
Hadler, S.C., Webster, H.M., Erben, J.J., Swanson, J.E., Maynard, J.E., 1980. Hepatitis A in day-care centers. A community-wide assessment. N. Engl. J. Med. 302, 1222–1227.
Hahne, S., Ramsay, M., Balogun, K., Edmunds, W.J., Mortimer, P., 2004. Incidence and routes of transmission of hepatitis B virus in England and Wales, 1995–2000: implications for immunisation policy. J. Clin. Virol. 29, 211–220.
Hatzakis, A., Chulanov, V., Gadano, A.C., Bergin, C., Ben-Ari, Z., Mossong, J., Schréter, I., et al., 2015. The present and future disease burden of hepatitis C virus (HCV) infections with today's treatment paradigm- volume 2. J. Viral Hepat. 22 (Suppl, 1), 26–45.
Hepatitis, E., Vaccine Working Group, 2014. Hepatitis E: Epidemiology and Disease Burden. A Document Prepared for Strategic Advisory Group of Experts on Immunization (SAGE).
Heidrich, B., Deterding, K., Tillmann, H.L., Raupach, R., Manns, M.P., Wedemeyer, H., 2009. Virological and clinical characteristics of delta hepatitis in Central Europe. J. Viral Hepat. 16, 883–894.
Hoofnagle, J.H., Nelson, K.E., Purcell, R.H., 2012. Hepatitis E. N. Engl. J. Med. 367, 1237–1244.
Jennifer, H.M., Benjamin, C.C., 2015. Hepatitis B virus epidemiology. Cold Spring Harb. Perspect. Med. 5, a021410.
Jemal, A., Bray, F., Center, M.M., Ferlay, J., Ward, E., Forman, D., 2011. Global cancer statistics. CA Cancer J. Clin. 61, 69–90.
Kamar, N., Bendall, R., Legrand-Abravanel, F., Xia, N.S., Ijaz, S., Izopet, J., Dalton, H.R., 2012. Hepatitis E. Lancet 379, 2477–2488.
Kamar, N., Selves, J., Mansuy, J.M., Ouezzani, L., Pèron, J.M., Guitard, J., Cointault, O., Esposito, L., Abravanel, F., Danjoux, M., Vinel, J.P., Izopet, J., Rostainq, L., 2008. Hepatitis E virus and chronic hepatitis in organ-transplant recipients. N. Engl. J. Med. 358, 811–817.

Kaufman, B., Gandhi, S.A., Louie, E., Rizzi, R., Illei, P., 1997. Herpes simplex virus hepatitis: case report and review. Clin. Infect. Dis. 24, 334–338.
Kowdley, K.V., Wang, C.C., Welch, S., Roberts, H., Brosgart, C.L., 2012. Prevalence of chronic hepatitis B among foreignborn persons living in the United States by country of origin. Hepatology 56, 422–433.
Labrique, A.B., Zaman, K., Hossain, Z., Saha, P., Hossain, A., Ticehurst, J.R., Nelson, K.E., 2010. Epidemiology and risk factors of incident hepatitis E virus infections in rural Bangladesh. Am. J. Epidemiol. 172, 952–961.
Lauer, G.M., Walker, B.D., 2001. Hepatitis C virus infection. N. Engl. J. Med. 345, 41–52.
Lavanchy, D., 2004. Hepatitis B virus epidemiology, disease burden, treatment, and current and emerging prevention and control measures. J. Viral Hepat. 11, 97–107.
Lefrère, J.J., Laperche, S., Roudot-Thoraval, F., 2008. Hepatitis G virus: a suitable marker of in vivo efficacy for pathogen inactivation. Vox Sang. 95, 76–78.
Lefkowitch, J.H., Goldstein, H., Yatto, R., Gerber, M.A., 1987. Cytopathic liver injury in acute delta virus hepatitis. Gastroenterology 92, 1262–1266.
Liang, X., Bi, S., Yang, W., Wang, L., Cui, G., Cui, F., Zhang, Y., Liu, J., Gong, X., Chen, Y., et al., 2009. Epidemiological serosurvey of hepatitis B in China—declining HBV prevalence due to hepatitis B vaccination. Vaccine 27, 6550–6557.
Lozano, R., Naghavi, M., Foreman, K., Lim, S., Shibuya, K., Aboyans, V., Abraham, J., Adair, T., Aggarwal, R., Ahn, S.Y., et al., 2012. Global and regional mortality from235 causes of death for 20 age groups in 1990 and 2010: a systematic analysis for the Global Burden of Disease Study 2010. Lancet 380, 2095–2128.
MacLachlan, J.H., Allard, N., Towell, V., Cowie, B.C., 2013. The burden of chronic hepatitis B virus infection in Australia, 2011. Aust. NZ J Public Health 37, 416–422.
MacLachlan, J.H., Locarnini, S., Cowie, B.C., 2015. Estimating the global prevalence of hepatitis B. Lancet 386 (10003), 1515–1517.
Mailet, V., Hamed, K., Schwarzinger, M., 2017. Prognosis of patients with chronic hepatitis B in France (2008-2013). A nationwide observational and hospital-based study. J. Hepatol. 66, 514–520.
Martin, N.K., Hickman, M., Hutchinson, S.J., Goldberg, D.J., Vickerman, P., 2013. Combination interventions to prevent HCV transmission among people who inject drugs: modeling the impact of antiviral treatment, needle and syringe programs, and opiate substitution therapy. Clin. Infect. Dis. 57 (Suppl. 2), S39–S45.
Markov, P.V., Van de Laar, T.J., Thomas, X.V., Aronson, S.J., Weegink, C.J., Van den Berk, G.E., Prins, M., Pybus, O.G., Schinkel, J., 2012. Colonial history and contemporary transmission shape the genetic diversity of hepatitis C virus genotype 2 in Amsterdam. J. Virol. 86, 7677–7687.
Marschall, T., Kretzschmar, M., Mangen, M.J., Schalm, S., 2008. High impact of migration on the prevalence of chronic hepatitis B in The Netherlands. Eur. J. Gastroenterol. Hepatol. 20, 1214–1225.
Mahoney, F.J., 1999. Update on diagnosis, management, and prevention of hepatitis B virus infection. Clin. Microbiol. Rev. 12, 351–366.
Makuwa, M., Mintsa-Ndong, A., Souquiere, S., Nkoghe, D., Leroy, E.M., Kazanji, M., 2009. Prevalence and molecular diversity of hepatitis B virus and hepatitis delta virus in urban and rural populations in northern Gabon in central Africa. J. Clin. Microbiol. 47, 2265–2268.
Nebbia, G., Peppa, D., Maini, M.K., 2012. Hepatitis B infection: current concepts and future challenges. Quat. J. Med. 105, 109–113.
Niro, G.A., Smedile, A., 2012. Current concept in the pathophysiology of hepatitis delta infection. Curr. Infect. Dis. Rep. 14, 9–14.

Ordieres, C., Navascues, C.A., Gonzalez-Dieguez, M.L., Rodriguez, M., Cadahia, V., Varela, M., Rodrigo, L., Rodriguez, M., 2017. Prevalence and epidemiology of hepatitis D among patients with chronic hepatitis B virus infection: a report from Northern Spain. Eur. J. Gastroenterol. Hepatol. 29, 277–283.

Ott, J.J., Stevens, G.A., Groeger, J., Wiersma, S.T., 2012. Global epidemiology of hepatitis B infection: new estimates of age-specific HBsAg seroprevalence and endemicity. Vaccine 30, 2212–2219.

Petruzziello, A., Marigliano, S., Loquercio, G., Cozzolino, A., Cacciapuoti, C., 2016. Global epidemiology of hepatitis C virus infection: an up-date of the distribution and circulation of hepatitis C virus genotypes. World J. Gastroenterol. 22, 7824–7840.

Popper, H., Thung, S.N., Gerber, M.A., Hadler, S.C., Monzon, M., de Ponzetto, A., Anzola, E., Rivera, D., Mondolfi, A., Bracho, A., et al., 1983. Histologic studies of severe delta agent infection in Venezuelan Indians. Hepatology 3, 906–912.

Previsani, N., Lavanchy, D., 2001. World Health Organization. Hepatitis D. (WHO/CDS/CSR/NCS/2001.1).

Razavi, H., Waked, I., Sarrazin, C., Mters, R.P., et al., 2014. The present and future disease burden of hepatitis C virus (HCV) infection with today's treatment paradigm. J. Viral Hepat. 21 (Suppl. 1), 34–59.

Razavi, H., Elkhoury, A.C., Elbasha, E., Estes, C., Pasini, K., Poynard, T., Kumar, R., 2013. Chronic hepatitis C virus (HCV) disease burden and cost in the United States. Hepatology 57, 2164–2170.

Rein, D.B., Stevens, G.A., Theaker, J., Wittenborn, J.S., Wiersma, S.T., 2012. The global burden of hepatitis E genotypes 1 and 2 in 2005. Hepatology 55, 988–997.

Rizzetto, M., Verme, G., 1985. Delta hepatitis—present status. J. Hepatol. 1, 187–193.

Romeo, R., Del Ninno, E., Rumi, M., Russo, A., Sangiovanni, A., de Franchis, R., Ronchi, G., Colombo, M., 2009. A 28-year study of the course of hepatitis delta infection: a risk factor for cirrhosis and hepatocellular carcinoma. Gastroenterology 136, 1629–1638.

Samuel, D., Zignego, A.L., Reynes, M., Feray, C., Arulnaden, J.L., David, M.F., Gigou, M., Bismuth, A., Mathieu, D., Gentilini, P., et al., 1995. Long-term clinical and virological outcome after liver transplantation for cirrhosis caused by chronic delta hepatitis. Hepatology 21, 333–339.

Smedile, A., Casey, J.L., Cote, P.J., Durazzo, M., Lavezzo, B., Purcell, R.H., Rizzetto, M., Gerin, J.L., 1998. Hepatitis D viremia following orthotopic liver transplantation involves a typical HDV virion with a hepatitis B surface antigen envelope. Hepatology 27, 1723–1729.

Teshale, E.H., Hu, D.J., 2011. Hepatitis E: epidemiology and prevention. World J. Hepatol. 3 (12), 285–291.

Uchida, T., 1993. Genetic variations of the Hepatitis B Virus and their clinical relevance. Microbiol. Immunol. 37 (6), 425–439.

Wasley, A., Grytdal, S., Gallagher, K., 2008. Surveillance for acute viral hepatitis—United States, 2006. MMWR Surveill Summ. 57, 1–24.

Wedemeyer, H., Duberg, A.S., Buti, M., Rosenberg, W.M., Frankova, S., Esmat, G., Örmeci, N., et al., 2014. Strategies to manage hepatitis C virus (HCV) disease burden. J. Viral Hepat. 21 (Suppl. 1), 60–89.

Xiridou, M., Borkent-Raven, B., Hulshof, J., Wallinga, J., 2009. How hepatitis D virus can hinder the control of hepatitis B virus. PLoS One 4 (4), e5247.

Zoulim, F., Durantel, D., 2015. Antiviral therapies and prospects for a cure of chronic hepatitis B. Cold Spring Harb Perspect. Med 5. https://doi.org/10.1101/cshperspect.a021501.

CHAPTER 2

Hepatitis Viruses: Structure and Replication

1. INTRODUCTION

Let us first understand that viruses are not animals, plants, nor bacteria, but they constitute a typical class of parasites of the living beings. They are not living organisms and need a host cell to carry out their life-sustaining functions and to reproduce. Since they do not possess ribosomes, they cannot synthesize proteins. To translate viral messenger RNA into viral proteins, they use the ribosomes of their host cells. Similarly, they cannot store energy but depend for this and other metabolic functions on the host cell. Even for basic building materials, such as amino acids, nucleotides, and lipids, they depend on their hosts. All viruses contain either DNA or RNA, not both, called a genome, which is covered by a protein shell called a capsid (Fig. 2.1). This capsid is made of protein subunits, called capsomers. The capsid with a genome (nucleic acid) inside is called nucleocapsid, and this nucleocapsid is surrounded by an envelope made of a bilayer lipid membrane, and studding the envelope are protein "spikes." These

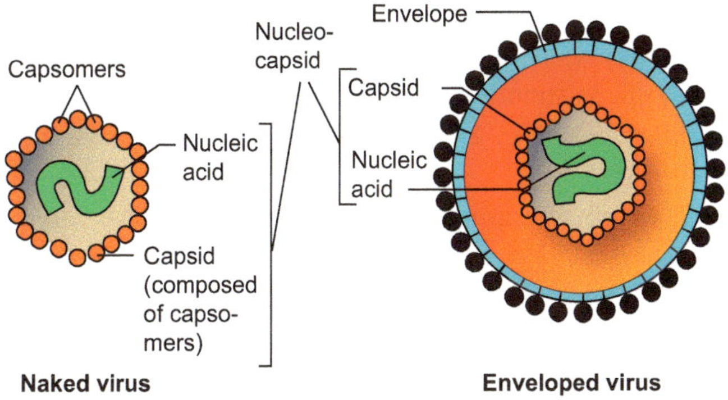

Figure 2.1 The general structure of a virus virion. *(www.pmbio.icbm.de/lehre/ws1011/vleuk/viruses1.pdf.)*

Studies on Hepatitis Viruses
ISBN 978-0-12-813330-9
https://doi.org/10.1016/B978-0-12-813330-9.00002-8

protrusions are essential for binding of the virus to certain receptors on the host cell as well as for host specificity and viral infectivity. The capsid is helical or has the shape of a 20-sided polygon called an icosahedral. This entire structure of the virus is called a "virion." Viruses multiply inside the host cells and utilize the cellular machineries for the synthesis of proteins, membranes, amino acids, carbohydrates, and lipids and for the generation of energy necessary for multiplication.

Viruses were not identified until the advances in live cell culture and microscopy in the 20th century were made. Advances in genetics dramatically improved the identification of the viruses. Virology is now a subject of great interest.

Viruses are classified based on four main characteristics: (1) nature of nucleic acid in virion, (2) symmetry of protein shell (capsid), (3) presence or absence of a lipid membrane, and (4) dimensions of virion and capsid (Fig. 2.1). Viruses that have an envelope are called "enveloped viruses," and those that do not have the envelope are called "nonenveloped viruses." In the enveloped viruses, the envelope is made of phospholipids, proteins, or glycoproteins. The glycoproteins identify the receptor sites on the host membrane and bind to them. Enveloped viruses have great adaptability and can change in a short time to evade the immune system. They can cause persistent infections. However, enveloped viruses are less virulent than nonenveloped viruses and rarely cause the host cell lysis, while nonenveloped viruses do cause the host cell lysis. Enveloped viruses are sensitive to heat, acid, and drying, while nonenveloped viruses are quite resistant to these factors. Enveloped viruses lose their infectivity on drying, but nonenveloped ones do not.

However, the most accepted and useful classification of viruses is the Baltimore classification, defined in 1971 by David Baltimore, a Nobel Prize-winning biologist. This classification is based on the way of synthesis of mRNA by single-stranded (ss) or double-stranded (ds) DNA or RNA present in the viruses (Fig. 2.2). According to this classification, there are seven classes of viruses:

Group I: viruses possessing dsDNA, e.g., Adenoviridae, Papovaviridae, Herpesviridae, and Poxviridae

Group II: viruses possessing ssDNA, e.g., Parvoviridae and Anelloviridae

Group III: viruses possessing ds RNA, e.g., Reoviridae

Group IV: viruses possessing positive-sense (+) sdRNA, e.g., Picornaviridae and Flaviviridae

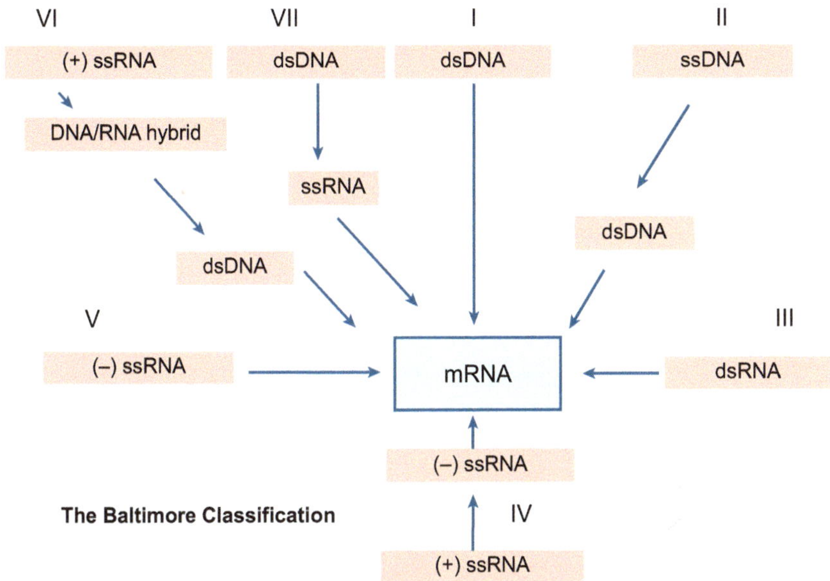

Figure 2.2 The Baltimore classification of viruses is based on the method of viral *mRNA* synthesis. *(https://en.wikipedia.org/wiki/Virus_classification. GrahamColm at English Wikipedia.)*

Group V: viruses possessing negative-sense (–) sdRNA, e.g., Arenaviridae and Filoviridae

Group VI: viruses possessing ssRNA but replicating through a DNA intermediate, e.g., retroviruses, such as HIV

Group VII: viruses possessing ds DNA but replicating through a reverse transcriptase (RT), e.g., Hepadnaviridae, such as hepatitis B virus (HBV)

Group I and II viruses come under DNA viruses, Group III to V viruses come under RNA viruses, and Group VI and VII viruses come under retroviruses.

On the basis of the preceding classifications the class and features of hepatitis viruses are described in Table 2.1.

After the virus has infected the host, it goes through several processes to multiply and survive, called replication. Replication generates abundant copies of the genome and packages them into viruses, which are able to infect new hosts. The mode of replication of viruses depends upon the genes present in them, e.g., most DNA viruses assemble in the nucleus, while most RNA viruses develop solely in cytoplasm (Dimmock et al., 2007). However, for the replication of any virus, it is essential that the host

Table 2.1 Class and features of hepatitis viruses

Hepatitis type	Size (nm)	Morphology	Genome	Classification	Antigen
HAV	27	Icosahedral, nonenveloped	linear ssRNA	Picornavirus	HAV-Ag
HBV	42	Double-shelled virion	partially circular dsDNA	Hepadnavirus	HBsAg, HBcAg, HBeAg
	27	Nucleocapsid core			HBcAg, HBeAg
	22	Virus coat material (spherical/filamentous)			HBsAg
HCV	40–60	Enveloped	linear ssRNA	Flavivirus	HCV C100–3, C33c, C22-3, NS5
HDV	35–37	Enveloped hybrid (HBsAg coat + HDV core)	circular ssRNA	Defective virus (□agent)	HBsAg, HDV-Ag
HEV	32–34	Icosahedral nonenveloped	linear RNA	Hepevirus	HEV-Ag

HAV, hepatitis A virus; *HBV*, hepatitis B virus; *HCV*, hepatitis C virus; *HDV*, hepatitis D virus; *HEV*, hepatitis E virus.

cell must provide the energy and the synthetic machinery. The replication of any virus involves seven steps: adsorption, entry, uncoating, transcription, synthesis of virus components, virion assembly, and release. Adsorption refers to the attachment of the virus to the cell membrane of the host cell, where it can inject its DNA or RNA. The cell membrane then houses it and protects its entry, after which cell enzymes uncoat the virus from its protein coat to release its genome. This step is followed by transcription, in which the virus protein is involved in the synthesis of mRNAs in the host cell nucleus or cytoplasm. The genomic template can be viral DNA or RNA, single- or double-stranded. The synthesized mRNA instructs the host cell to make the virus components, after which the virus takes the shape. Next, we describe the structure and replication of each hepatitis virus sequentially.

2. STRUCTURE AND REPLICATION OF HEPATITIS VIRUSES

2.1 Hepatitis A Virus

2.1.1 Structure

Hepatitis A virus (HAV) belongs to the Picornaviridae family of viruses. It has a single-stranded positive-sense linear RNA, written as ss(+)RNA, genome. A genome is the genetic material of an organism. It consists of DNA (or RNA in RNA viruses). HAV is an enterovirus. Enteroviruses are named by their transmission route through the intestine (enteric meaning intestinal). Humans appear to be the only reservoir for this virus. HAV can remain viable for many years.

The core structure of any virus, called a *nucleocapsid*, contains protein and nucleic acid, where the protein portion is called the *capsid*. This capsid is to coat and protect the virus genome. In some cases, the capsid also aids in viral entry. The nucleic acid contains the genes of the virus and can be RNA or DNA. The capsid of HAV is icosahedral and nonenveloped in structure and 28 nm in diameter (Fig. 2.3). An icosahedron is a three-dimensional structure with 20 faces, resembling a soccer ball. HAV's capsid is made of subunits called capsomeres, each made of five protomers, where every protomer is constituted of three proteins: vp1,vp2, and vp3.

As shown in Fig. 2.4, the genome of HAV is divided into three segments: P1(VP4 to VP1), P2 (2A to 2C), and P3 (Vpg to $3C_{pro}$). The first segment, P1, encodes genes to make capsid proteins. The other two segments make other genes needed for replication. At the 5′ end, Vpg refers to a viral protein, and VP4 + VP2 together are referred to as VP0.

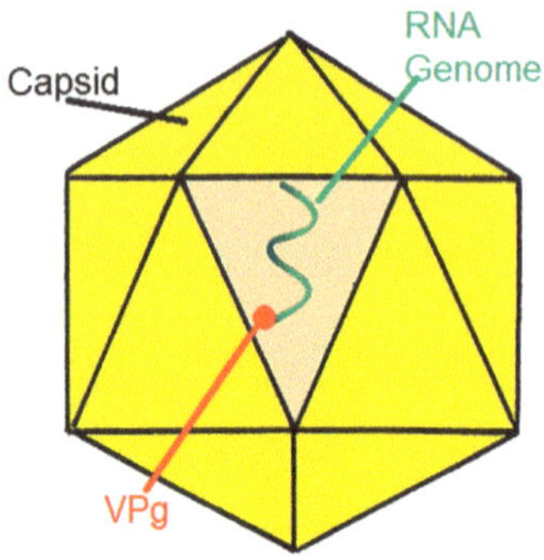

Figure 2.3 The icosahedral structure of the hepatitis A virus capsid with the genome inside. *(https://study.com/academy/lesson/hepatitisa-virus-structure-and-function.html.)*

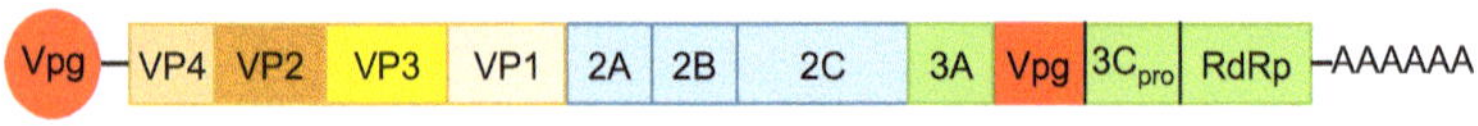

Figure 2.4 Genome of hepatitis A virus, which is divided into three segments: P1(VP4 to VP1), P2 (2A to 2C), and P3 (Vpg to $3C_{pro}$). Vpg refers to a viral protein. *(https://study.com/academy/lesson/hepatitis-a-virus-structure-and-function.html.)*

2.1.2 Replication

For HAV, the natural hosts are humans and vertebrates. Transmission routes are fecal-oral and blood. When transmitted in the body, HAV enters the bloodstream, which carries it to its target, the liver, where it multiplies within hepatocytes. Hepatocyte uptake involves a receptor, identified by Kaplan et al. (1996), on the plasma membrane of the cell. Viral replication is supposed to occur exclusively in hepatocytes, where the virus is attached to the receptor (Fig. 2.5). This step is followed by the release of the RNA genome from the protein coat. When the RNA genome is released from the protein coat, it is translated by the cell's own ribosomes, and proteolysis occurs. It leads to RNA replication that promotes the virion assembly (a late phase of the viral life cycle during which all the components necessary for the formation of a mature virion collect at a particular site in the cell, and the basic structure of the virus particle is formed). After the virions are matured, they are secreted into the bile and released in stool (feces). All these steps of HAV replication are shown in Fig. 2.5.

2.2 Hepatitis B Virus

2.2.1 Structure

HBV belongs to Hepadnavirus family of viruses (Zuckerman, 1996; Kaushik et al., 2010). The HBV virion (called the Dane particle) has an icosahedral nucleocapsid core enclosed by a lipid envelope (Fig. 2.6). This

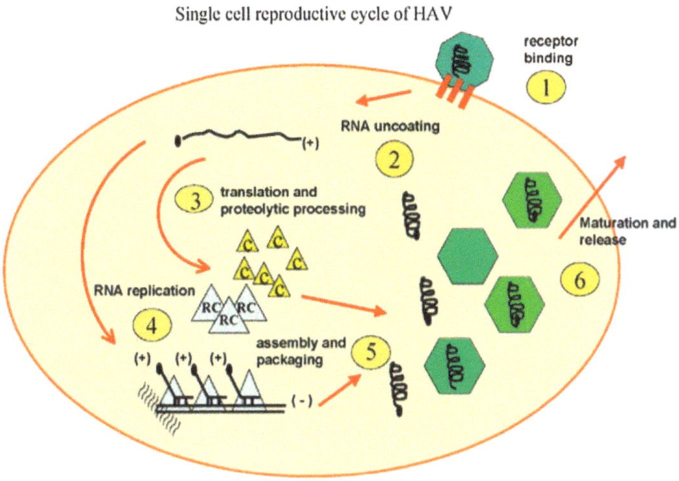

Figure 2.5 The life cycle of hepatitis A virus. *(http://resscientiae.wikia.com/wiki/ResScientiae_Wiki.)*

nucleocapsid is composed of protein. Within this capsid, there are the viral DNA and the DNA polymerase, equivalent to RT of retroviruses (Locarnini, 2004). HBV is one of the smallest enveloped animal viruses, whose virion has a spherical, double-shelled structure, 42 nm in diameter, consisting of a lipid envelope containing surface antigen (HBsAg). This lipid envelope surrounds an inner nucleocapsid composed of hepatitis B core antigen (HBcAg), within which are complexed virally encoded polymerase and viral DNA genome. The HBV genome is partially double-stranded circular DNA of about 3.2 kb pairs. The DNA polymerase is covalently attached to the 5′ end of the minus strand (Gerlich and Robinson, 1980).

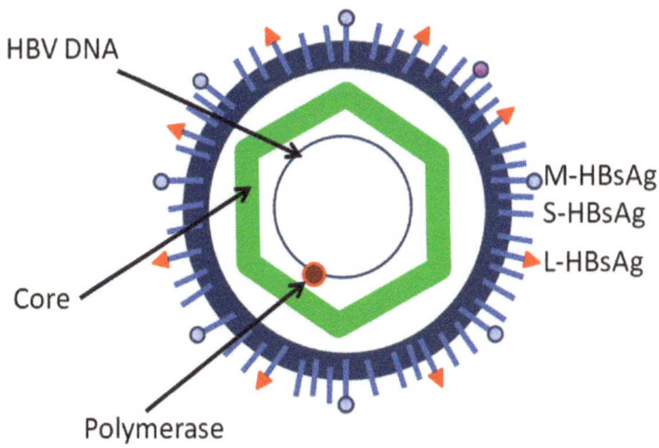

Figure 2.6 Structure of HBV (hepatitis B virus) virion. *(https://en.wikipedia.org/wiki/Hepatitis_B.)*

The surface antigen of the lipid envelope can be of large, medium, and small size, denoted as L-HBsAg, M-HBsAg, and S-HBsAg, respectively, as shown in Fig. 2.6. The former is translated from 2.4-kb subgenomic RNA, and the other two from the various forms of 2.1-kb RNAs. The smaller one is the major surface antigen, while L and M are in minority (Liang, 2009).

2.2.2 Replication

The life cycle of HBV where it replicates involves many steps, as shown in Fig. 2.7. Being a nonretroviral virus, it uses reverse transcription in its replication process. HBV enters the cells through the binding with a receptor on the surface of the cell and enters by clathrin-dependent endocytosis. Clathrin-dependent endocytosis is the uptake of the material into the cell from the surface using clathrin-coated vesicles. The virus membrane then fuses with the membrane of the host cell, releasing the DNA and the core proteins into the cytoplasm. This step is followed by the dissociation of core proteins from the partially double-stranded DNA. This partially double-stranded DNA is converted to fully double-stranded DNA by host DNA polymerase and transformed into covalently closed circular DNA (cccDNA). This cccDNA serves as a template for transcription of four viral mRNAs, of which the largest one is involved

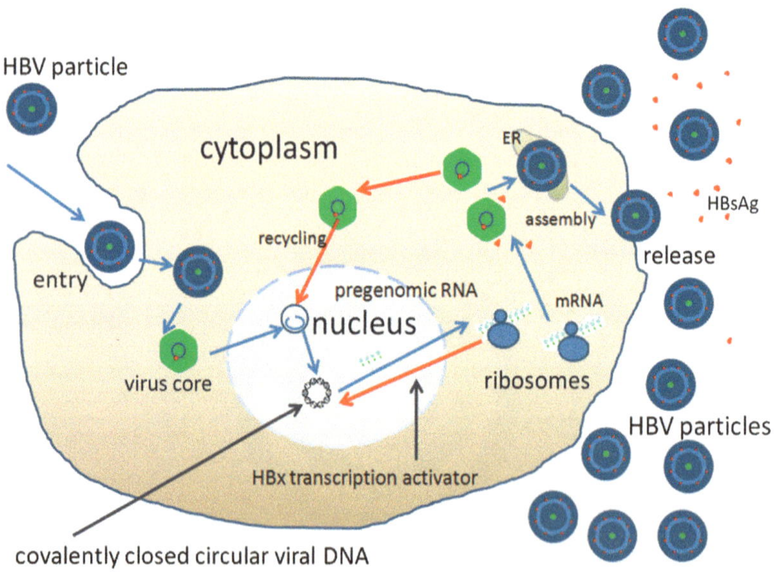

Figure 2.7 HBV (hepatitis B virus) replication cycle. *(https://en.wikipedia.org/wiki/Hepatitis_B.)*

to make the new copies of the genome, the capsid core protein, and the viral RNA-dependent-DNA-polymerase (RdDp). These four viral transcripts undergo additional processing to form progeny virions to be released from the cell. These virions can also be returned to the nucleus and recycled to produce even more copies (Beck and Nassal, 2007; Bruss, 2007).

2.3 Hepatitis C Virus

2.3.1 Structure

Hepatitis C virus (HCV) is a hepacivirus, which is a member of the Flaviviridae family. Before, it was supposed to be the only member of *Hepacivirus* genus, but a few more viruses of this genus have been discovered (Kapoor et al., 2011, 2013; Burbelo et al., 2012; Quan et al., 2013). As shown in Fig. 2.8, HCV has an icosahedral structure made of protein, in which there is a core of RNA. This core is further protected by a lipid envelope of cellular origin. In this lipid envelope are embedded two glycoproteins, E1 and E2 (Op De Beeck and Dubuisson, 2003). The HCV genome is a positive-sense ssRNA, consisting of a single open reading frame (ORF) of 96,000 nucleotide bases long (Kato, 2000). At the 5′ and 3′ ends of the RNA are the nontranslated regions (NTRs). These NTRs are not translated into proteins but play the role in the translation and replication of the viral RNA. The HCV makes structural proteins that

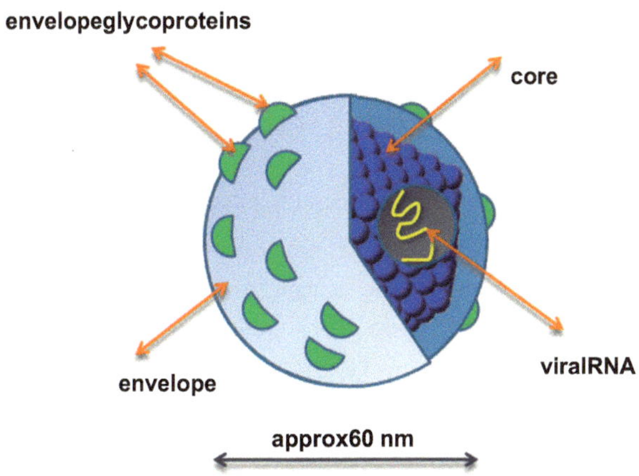

Figure 2.8 A diagrammatic representation of structure of hepatitis C virus. *(By GrahamColm at English Wikipedia, CC BY-SA 3.0, https://commons.wikimedia.org/w/index.php?curid=5624458.)*

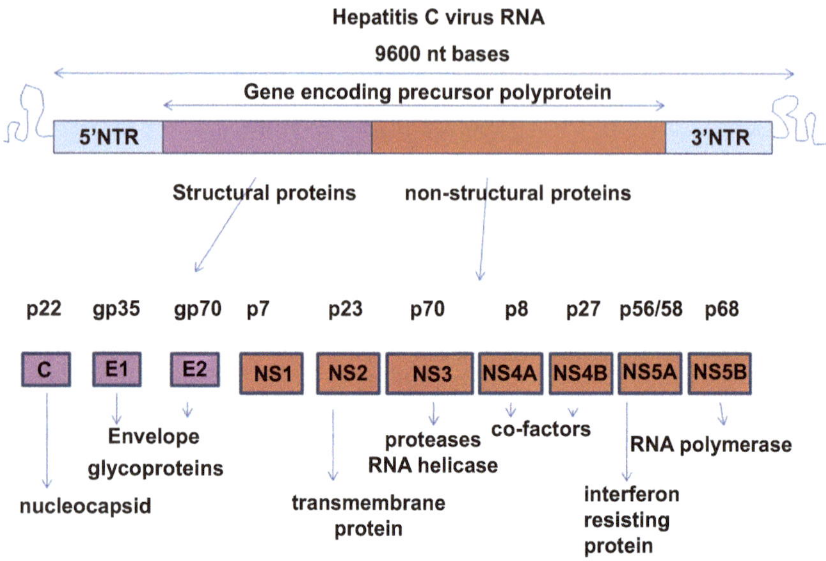

Figure 2.9 Structure of hepatitis C virus genome. *(By GrahamColm at English Wikipedia, CC BY-SA 3.0, https://commons.wikimedia.org/w/index.php?curid=5624458.)*

include core protein (C) and envelope glycoproteins E1 and E2, and the nonstructural (NS) proteins are designated as NS2, NS3, NS4A, NS4B, NS5A, and NS5B (Fig. 2.9). The core protein also includes a protein p7 (Fig. 2.9). Both the envelope proteins, E1 and E2, are highly glycosylated and important in the cell entry. All NS proteins, NS2 to NS5B, depend on the activity of viral proteases (De Francesco, 1999). The p7 protein has little role in genome replication but is important in virus morphogenesis.

2.3.2 Replication

Replication of HCV also involves several steps. It occurs mainly in the hepatocytes of the liver, though it may also occur in peripheral blood mononuclear cells. Entry of the virus into the host cells takes place involving several complex interactions between virions and cell surface molecules like CD81, low-density lipoprotein (LDL) receptors, SR-BI, DC-SIGN, Claudin-1, and Occludin (Zeisel et al., 2009). CD81 is a protein encoded by the *CD81* gene. It is also known as 26-kDa cell surface protein. LDL receptors are mosaic proteins of 839 amino acids. SR-BI, a scavenger receptor of class B type I, is an 82-kDa glycoprotein containing 509 amino acids. DC-SIGN stands for dendritic cell-specific intercellular adhesion molecule-3-grabbing nonintegrin. Also known as CD209 (cluster of differentiation 209), it is a C-type lectin receptor present on the surface of both macrophages and dendrite cells. It is a protein that in humans is encoded by the *CD209* gene. Claudin-1 is a protein that in humans is encoded by the *CLDN1* gene. It forms the paracellular tight junction seal in epithelial tissues (Halford et al., 2000; Morita et al., 1999). Occludin is a protein, encoded by *OCLN* gene, that was identified as the first tight junction integral membrane protein (Ploss et al., 2009).

As shown in Fig. 2.10, after the HCV is attached to the hepatocyte, it forces the intracellular machinery for its replication (Lindenbach and Rice, 2005). Its genome is translated to produce a single, very long protein, which is proteolytically processed by viral and cellular proteases to produce the three structural and seven NS proteins. The NS proteins then transfer the viral genome into an RNA replication complex associated with rearranged cytoplasmic membrane. RNA replication occurs with the help of viral RNA-dependent RNA polymerase NS5B, which produces a negative-strand RNA intermediate to serve as a template for the production of new positive-strand viral genomes, which then can be translated, further replicated, or packaged within new virus particles. These new virus particles bud into the secretary pathway to be released at the cell surface. Thus, virus replication occurs on intracellular lipid membranes.

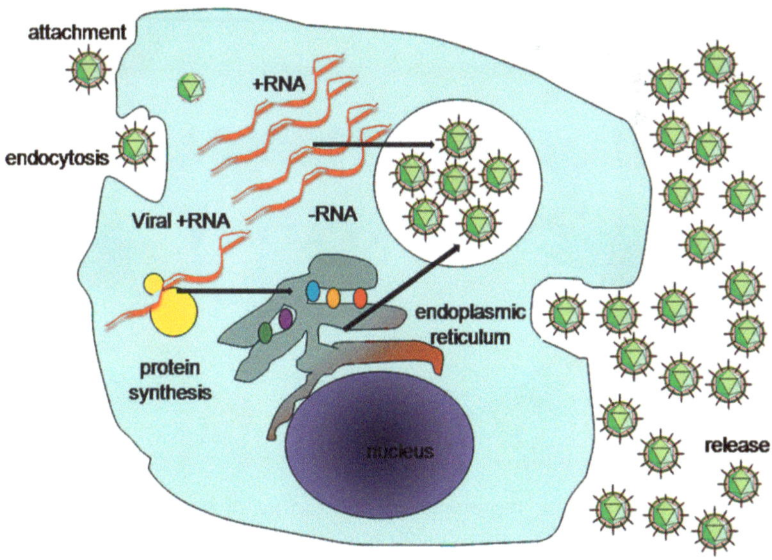

Figure 2.10 Replication of hepatitis C virus. *(By GrahamColm at English Wikipedia, CC BY-SA 3.0, https://commons.wikimedia.org/w/index.php?curid=5624458.)*

2.4 Hepatitis D Virus

2.4.1 Structure

Hepatitis D virus (HDV) is a virus that can affect only in the presence of HBV. For its replication and assembly of new virions, it requires HBV (Abbas and Afzal, 2013). Simultaneous introduction of HDV and HBV into a patient leads to the same effects as acute infection with HBV alone. Similarly, the risk of developing chronic HBV and HDV infection after acute exposure to both viruses is the same as the rate of developing chronic HBV infection after acute exposure to HBV. Introduction of HDV into an individual already infected with HBV may have dramatic consequences.

HDV is a small, spherical virus with a diameter of about 36 nm (Lin et al., 1990; Wang et al., 1991) that has an outer coat made of all three kinds of HBV envelope proteins, L-HBsAg, M-HBsAg, and S-HBsAg, and the host lipids surrounding the viral genome (Lai, 1995; Casaca et al., 2011). The HDV genome is a circular single-stranded negative-sense RNA (ssRNA) molecule (Fig. 2.11) with an internal core of hepatitis delta

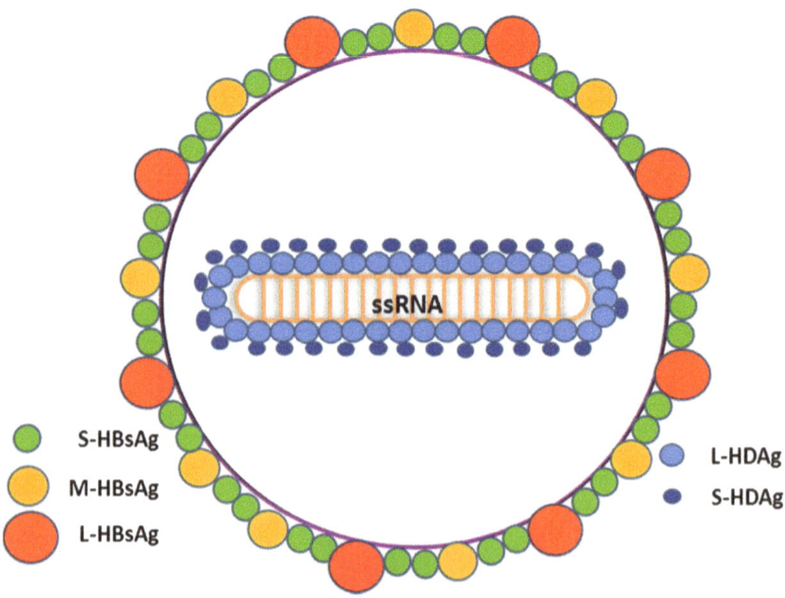

Figure 2.11 The schematic representation of structure of hepatitis D virus. *ssRNA*, single-stranded RNA. *(http://www.mdpi.com/pathogens/pathogens-04-00046/article_deploy/html/images/pathogens-04-00046-g001-1024.png.)*

antigen (HDAg). The genomic RNA is composed of about 1700 nucleotides and 200 molecules of HDAg. It is assumed that HDV may have originated from a class of plant pathogens called viroids, which are much smaller than viruses (Elena et al., 1991; Sureau, 2006). This is due to extensive base pairing within the RNA molecule, because of which the genome appears as a double-stranded rod-like structure to remarkably resemble viroids (plant pathogens) (Lin et al., 1990; Flores et al., 2012). However, viroids lack protein coding ability, whereas HDV codes for its own HDAg and requires HBV for its propagation (Flores et al., 2012). The virus is thus assigned the separate genus *Deltavirus*, though still studies are on to have complete taxonomic classification, including order and family (Abbas and Afzal, 2013).

2.4.2 Replication

The mechanism of entry of HDV into hepatocytes is not well understood, but it is thought that it gains entry into the liver cells as HBV does. HDV recognizes its receptor through the N-terminal domain of L-HBsAg (Engelke et al., 2006), in which amino acid residues 9—15 have been found to act as receptor binding site (Schulze et al., 2010). As soon as the HDV enters the hepatocytes, it is uncoated, and the nucleocapsid is translocated to the nucleus by a signal in HDAg (Xia et al., 1992). For the replication of the virus genome, HDV uses the cellular RNA polymerases (pol I—pol III), as its nucleocapsid does not contain any RNA polymerase (Lehmann et al., 2007; Filipovska and Konarska, 2000; Greco-Stewart et al., 2009). The RNA polymerases treat the RNA genome as dsDNA due to the folded rod-like structure.

Polymerase I (pol I) helps the transcription of antigenome from viral genome in the nucleus, whereas pol II catalyzes genome replication from antigenome and transcription of mRNA in the nucleoplasm (Abbas and Afzal, 2013; Huang et al., 2008). By intrinsic ribozyme activity the antigenomic RNA is self-cleaved and ligated to form a circular antigenome using cellular ligases. The antigenomic RNA then helps to produce genomic RNA in the nucleoplasm. This antigenomic RNA is mRNA containing the ORF for the HDAg. Once the HDAg is produced, it translocates the viral genome into the nucleus, where RNA polymerases I and II lead to the replication of the genome (Fig. 2.12) (Abbas and Afzal, 2013).

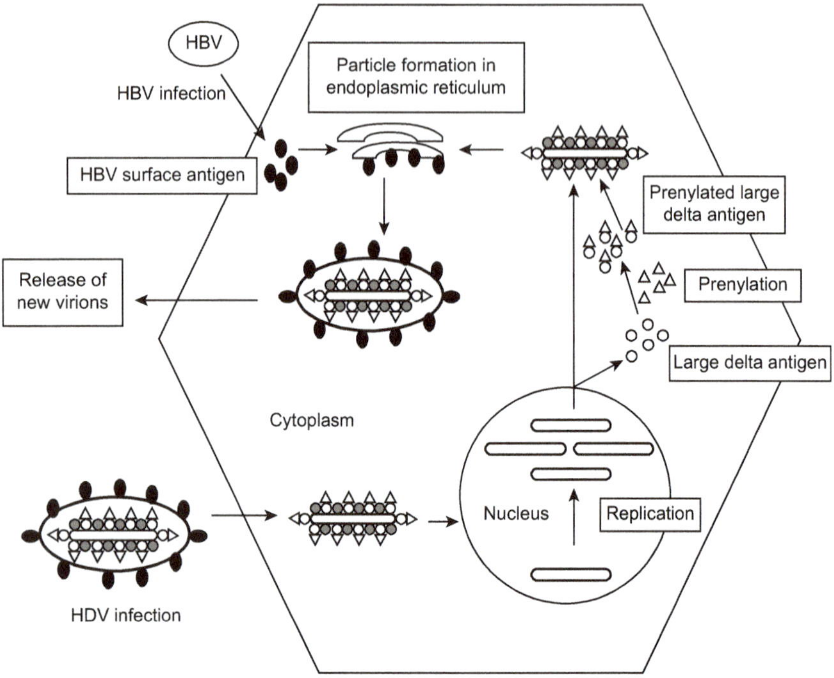

Figure 2.12 The replication cycle of HDV. Inside the large circle are HDV RNA genomes. *HBV*, hepatitis B virus; *HDV*, hepatitis D virus. *(Reprinted from Abbas, Z., Afzal, R., 2013. Life cycle and pathogenesis of hepatitis D virus: a review. World J. Hepatol. 5, 666–675. Online publication.)*

2.5 Hepatitis E Virus

2.5.1 Structure

Hepatitis E virus (HEV) is a positive-sense single-stranded nonenveloped RNA icosahedral virus. It has a spherical shape with a diameter of about 32 nm (Balayan et al., 1983; Bradley et al., 1988). Its virion is thought to be made of 180 copies of CP (capsid protein) with an approximate diameter of 27–34 nm (Guu et al., 2009; Xing et al., 1999). Like HAV, HEV does not have any viral envelope; rather, its genomic RNA is capped, polyadenylated containing three ORFs. ORF1, mapped at the 5′ end of the genome, has about 5124 bases comprising approximately 70% of the genome. It encodes several viral NS proteins such as a methyltransferase (Met), papain-like cysteine protease, macrodomain, helicase (Hel), and RNA-dependent RNA polymerase (RdRp) (Koonin et al., 1992). ORF2 at the 3′ end of the genome contains 1980 bases and encodes viral CP, while ORF3, partially overlapping with other two ORFs, encodes a

13.5-kDa regulatory phosphoprotein with multifunctions (Monga, 2011; Vasickova et al., 2007). However, due to lack of robust cell culture system, the exact structure of HEV could not be ascertained, but recent findings from HEV expression and purification, X-ray crystallography, and electron microscopy have given some hope to elucidate the structure and function of this virus.

2.5.2 Replication

The replication and life cycle of HEV is practically unknown. The CP obtains viral entry by binding to a cellular receptor, and ORF2 moderates viral entry by binding to heat shock cognate receptor 70 (HSC70).

Beyond this, our knowledge about the replication of HEV virion is limited. However, a hypothetical model of the life cycle of HEV based on some sources (Cao and Meng, 2012; Ahmad et al., 2011) has been given by Kenney and Meng (2015) (Fig. 2.13), which describes the following steps:

a. HEV particles attach to the cell via heparin sulfate proteoglycans, HSC70, and other potential receptors.
b. Particle undergoes clathrin-mediated endocytosis.
c. Entry complex moves into the cell potentially through interaction with heat shock protein 90 (HSP90) or glucose-regulated protein 78 (Grp78).
d. Positive-sense genomic RNA serves as the template for transcription of the nonstructural proteins.
e. Viral RdRp synthesizes an intermediate negative-sense RNA from the positive-sense genomic RNA.
f. Negative-stranded RNA serves as a template for the production of more positive-sense viral genomes.
g. ORF2 and ORF3 proteins are synthesized from the subgenomic, positive-sense RNA.
h. Particle assembly begins with ORF2 proteins binding to genomic RNA.
i. Nascent virions are trafficked to the plasma membrane for final assembly.
j. HEV interacts with the host endosomal sorting complexes required for transport machinery to transit, though the cellular membrane acquire and envelope in the process.
k. Enveloped virions lose their lipid association when passing through the digestive system and become ready to infect a new host.

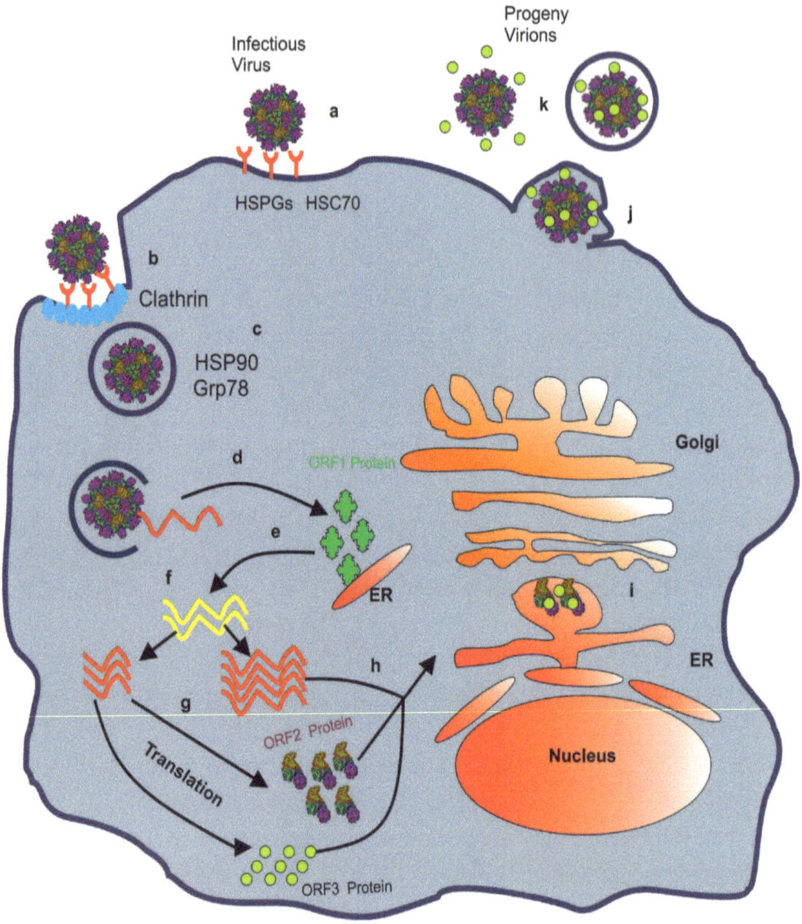

Figure 2.13 Hypothetical life cycle of the hepatitis E virus showing steps (a) to (k). *(Reprinted from Kenney, S.P., Meng, X.-J., 2015. Therapeutic targets for the treatment of hepatitis E virus infection. Expert Opin. Ther. Targets. 19, 1245–1260. Open access publication.)*

3. CONCLUSIONS

As far as the structure of the viruses is concerned, all viruses contain either DNA or RNA, not both, called a genome, which is covered by a protein shell called a capsid. The capsid containing a genome (nucleic acid) has the shape of a helix or a 20-sided polygon called an icosahedral. The entire structure of the virus is called a "virion." When viruses infect the host, they multiply there, utilizing the cellular machineries of the host to synthesize all necessary things for their replication, but the mode of their replication depends upon the genes present in them, e.g., most DNA viruses assemble in the nucleus, while most RNA viruses develop solely in cytoplasm.

REFERENCES

Abbas, Z., Afzal, R., 2013. Life cycle and pathogenesis of hepatitis D virus: a review. World J. Hepatol. 5, 666–675.

Ahmad, I., Holla, R.P., Jameel, S., 2011. Molecular virology of hepatitis E virus. Virus Res. 161, 47–58.

Balayan, M.S., Andjaparidze, A.G., Savinskaya, S.S., Ketiladze, E.S., Braginsky, D.M., Savinov, A.P., Poleschuk, V.F., 1983. Evidence for a virus in non-A, non-B hepatitis transmitted via the fecal-oral route. Intervirology 20, 23–31.

Beck, J., Nassal, M., 2007. Hepatitis B virus replication. World J. Gastroenterol. 13, 48–64.

Bradley, D., Andjaparidze, A., Cook, E.H., McCaustland, K., Balayan, M., Stetler, H., Velazquez, O., Robertson, B., Humphrey, C., Kane, M., Weisfuse, I., 1988. Aetiological agent of enterically transmitted non-A, non-B hepatitis. J. Gen. Virol. 69, 731–738.

Bruss, V., 2007. Hepatitis B virus morphogenesis. World J. Gastroenterol. 13, 65–73.

Burbelo, P.D., Dubovi, E.J., Simmonds, P., Medina, J.L., Henriquez, J.A., Mishra, N., Wagner, J., Tokarz, R., Cullen, J.M., Iadarola, M.J., Rice, C.M., Lipkin, W.I., Kapoor, A., 2012. Serology-enabled discovery of genetically diverse hepaciviruses in a new host. J. Virol. 86, 6171–6178.

Cao, D., Meng, X.J., 2012. Molecular biology and replication of hepatitis E virus. Emerg. Microbes Infect. e17. https://doi.org/10.1038/emi.2012.7.

Casaca, A., Fardilha, M., da Cruz e Silva, E., Cunha, C., 2011. The heterogeneous ribonuclear protein C interacts with the hepatitis delta virus small antigen. Virol. J. https://doi.org/10.1186/1743-422X-8-358.

De Francesco, R., 1999. Molecular virology of the hepatitis C virus. J. Hepatol. 31 (Suppl. 1), 47–53.

Dimmock, N.J., Easton, A.J., Leppard, K.N., 2007. Introduction to Modern Virology, seventh ed. Blackwell Publishing.

Elena, S.F., Dopazo, J., Flores, R., Diener, T.O., Moya, A., 1991. Phylogeny of viroids, viroidlike satellite RNAs, and the viroidlike domain of hepatitis delta virus RNA. Proc. Natl. Acad. Sci. U.S.A. 88, 5631–5634.

Engelke, M., Mills, K., Seitz, S., Simon, P., Gripon, P., Schnölzer, M., Urban, S., 2006. Characterization of a hepatitis B and hepatitis delta virus receptor binding site. Hepatology 43, 750–760.

Filipovska, J., Konarska, M.M., 2000. Specific HDV RNA-templated transcription by pol II in vitro. RNA 6, 41–54.

Flores, R., Ruiz-Ruiz, S., Serra, P., 2012. Viroids and hepatitis delta virus. Semin. Liver Dis. 32, 201–210.

Gerlich, W., Robinson, W.S., 1980. Hepatitis B virus contains protein attached to the 5′ end of its complete strand. Cell 21, 801–811.

Greco-Stewart, V.S., Schissel, E., Pelchat, M., 2009. The hepatitis delta virus RNA genome interacts with the human RNA polymerases I and III. Virology 386, 12–15.

Guu, T.S.Y., Liu, Z., Ye, Q.Z., Mata, D.A., Li, K.P., Yin, C.C., Zhang, J.Q., Tao, Y.J., 2009. Structure of the hepatitis E virus-like particle suggests mechanisms for virus assembly and receptor binding. Proc. Natl. Acad. Sci. U.S.A. 106 (31), 12992–12997.

Halford, S., Spencer, P., Greenwood, J., Winton, H., Hunt, D.M., Adamson, P., 2000. Assignment of claudin-1 (CLDN1) to human chromosome 3q28-q29 with somatic cell hybrids. Cytogenet. Cell Genet. 88 (217) https://doi.org/10.1159/000015553.

Huang, W.H., Chen, Y.S., Chen, P.J., 2008. Nucleolar targeting of hepatitis delta antigen abolishes its ability to initiate viral antigenomic RNA replication. J. Virol. 82, 692–699.
Kaplan, G., Totsuka, A., Thompson, P., Akatsuka, T., Moritsugu, Y., Feinstone, S.M., 1996. Identification of a surface glycoprotein on African green monkey kidney cells as a receptor for hepatitis A virus. EMBO J. 15, 4282–4296.
Kapoor, A., Simmonds, P., Gerold, G., Qaisar, N., Jain, K., Henriquez, J.A., Firth, C., Hirschberg, D.L., Rice, C.M., Shields, S., Lipkin, W.I., 2011. Characterization of a canine homolog of hepatitis C virus. Proc. Natl. Acad. Sci. U.S.A. 108, 11608–11613.
Kapoor, A., Simmonds, P., Scheel, T.K., Hjelle, B., Cullen, J.M., Burbelo, P.D., Chauhan, L.V., Duraisamy, R., Sanchez Leon, M., Jain, K., Vandegrift, K.J., Calisher, C.H., Rice, C.M., Lipkin, W.I., 2013. Identification of rodent homologs of hepatitis C virus and pegiviruses. mBio 4 (2) e00216–13.
Kato, N., 2000. Genome of human hepatitis C virus (HCV): gene organization, sequence diversity, and variation. Microb. Comp. Genomics 5, 129–151.
Kaushik, S., Gupta, S.P., Sharma, P.K., 2010. Design and development of anti-hepatitis B virus agents. Curr. Med. Chem. 17, 3377–3392.
Kenney, S.P., Meng, X.-J., 2015. Therapeutic targets for the treatment of hepatitis E virus infection. Expert Opin. Ther. Targets 19, 1245–1260.
Koonin, E.V., Gorbalenya, A.E., Purdy, M.A., Rozanov, M.N., Reyes, G.R., Bradley, D.W., 1992. Computer-assisted assignment of functional domains in the nonstructural polyprotein of hepatitis E virus: delineation of an additional group of positive-strand RNA plant and animal viruses. Proc. Natl. Acad. Sci. U.S.A. 89, 8259–8263.
Lai, M.M., 1995. Molecular biologic and pathogenetic analysis of hepatitis delta virus. J. Hepatol. 22, 127–131.
Lehmann, E., Brueckner, F., Cramer, P., 2007. Molecular basis of RNA-dependent RNA polymerase II activity. Nature. 450, 445–449.
Liang, T.J., 2009. Hepatitis B: the virus and disease. Hepatology 49, S13–S21.
Lin, J.H., Chang, M.F., Baker, S.C., Govindarajan, S., Lai, M.M., 1990. Characterization of hepatitis delta antigen: specific binding to hepatitis delta virus RNA. J. Virol. 64, 4051–4058.
Lindenbach, B., Rice, C., 2005. Unravelling hepatitis C virus replication from genome to function. Nature 436 (7053), 933–938.
Locarnini, S., 2004. Molecular virology of hepatitis B virus. Semin. Liver Dis. 24 (Suppl. 1), 3–10.
Monga, S.P.S, 2011. Molecular Pathology of Liver Diseases. Springer, New York, 9781441971067.
Morita, K., Furuse, M., Fujimoto, K., Tsukita, S., 1999. Claudin multigene family encoding four-transmembrane domain protein components of tight junction strands. Proc. Natl. Acad. Sci. U.S.A. 96, 511–516.
Op De Beeck, A., Dubuisson, J., 2003. Topology of hepatitis C virus envelope glycoproteins. Rev. Med. Virol. 13, 233–241.
Ploss, A., Evans, M.J., Gaysinskaya, V.A., Panis, M., You, H., deJong, Y.P., Rice, C.M., 2009. Human occludin is a hepatitis C virus entry factor required for infection of mouse cells. Nature. 457, 882–886.
Quan, P.L., Firth, C., Conte, J.M., Williams, S.H., Zambrana-Torrelio, C.M., Anthony, S.J., Ellison, J.A., et al., 2013. Bats are a major natural reservoir for hepaciviruses and pegiviruses. Proc. Natl. Acad. Sci. U.S.A. 110, 8194–8199.
Schulze, A., Schieck, A., Ni, Y., Mier, W., Urban, S., 2010. Fine mapping of pre-s sequence requirements for hepatitis B virus large envelope protein-mediated receptor interaction. J. Virol. 84 (4), 1989–2000.

Sureau, C., 2006. The role of the HBV envelope proteins in the HDV replication cycle. Curr. Top. Microbiol. Immunol. 307, 113–131.

Vasickova, P., Psikal, I., Kralik, P., Widen, F., Hubalek, Z., Pavlik, I., 2007. Hepatitis E virus: a review. Vet. Med. (Praha) 52, 365–384.

Wang, C.J., Chen, P.J., Wu, J.C., Patel, D., Chen, D.S., 1991. Small-form hepatitis B surface antigen is sufficient to help in the assembly of hepatitis delta virus-like particles. J. Virol. 65, 6630–6636.

Xia, Y.P., Yeh, C.T., Ou, J.H., Lai, M.M., 1992. Characterization of nuclear targeting signal of hepatitis delta antigen: nuclear transport as a protein complex. J. Virol. 66, 914–921.

Xing, L., Kato, K., Li, T.C., Takeda, N., Miyamura, T., Hammar, L., Cheng, R.H., 1999. Recombinant hepatitis e capsid protein self-assembles into a dual-domain T=1 particle presenting native virus epitopes. Virology 265, 35–45.

Zeisel, M., Barth, H., Schuster, C., Baumert, T., 2009. Hepatitis C virus entry: molecular mechanisms and targets for antiviral therapy. Front. Biosci. 14, 3274–3285.

Zuckerman, A.J., 1996. Hepatitis viruses. In: Baron's Medical Microbiology, fourth ed. Univ of Texas Medical Branch. ISBN:0-9631172-1-1.

FURTHER READING

Kohaar, I., Ploss, A., Korol, E., Mu, K., Schoggins, J., O'Brien, T., Rice, C., Prokunina-Olsson, L., 2010. Splicing diversity of the human OCLN gene and its biological significance for hepatitis C virus entry. J. Virol. 84, 6987–6994.

CHAPTER 3

Inhibition of Viruses: Promising Targets and Their Importance

1. INTRODUCTION

As already discussed in Chapter 2, the viral replication involves seven steps: adsorption, entry, uncoating, transcription, synthesis of virus components (nucleic acids and structural proteins), maturation of the virions, and release. Out of these seven steps, the two steps, namely the synthesis of virus components and maturation of viral proteins, have been well exploited to develop antiviral drugs, as the inhibitions of these two steps have been found to be quite handy. For these two steps, two important types of enzymes, polymerases and proteases, respectively, are involved. Polymerases catalyze the synthesis of the viral proteins, and once the viral proteins are synthesized, the identical copies of their genomes are synthesized and encapsulated. On the other hand, proteases catalyze the processing of the viral proteins or the maturation of the viral particles and the release from the cell. Now, we discuss here the structure and specific roles of these two kinds of enzymes.

2. POLYMERASES

The kind of a polymerase required for these actions of viruses depends upon the nature of their genome, whether it is DNA or RNA, double-stranded (ds) or single-stranded (ss), or whether it is plus- or minus-sense (Choi, 2012), and this can be found in the Baltimore classification of viruses, as discussed in Chapter 2.

A virus polymerase copies a template nucleic acid strand to produce a daughter strand, and then it transfers a nucleotidyl moiety of an incoming nucleoside triphosphate (NTP), complementary to the template strand, to the 3′-hydroxyl end of a growing daughter strand of RNA or DNA. However, for its catalytic activity, a virus polymerase requires two divalent metal ions, Mg^{2+} or Mn^{2+}, and it must have binding sites for a template

Studies on Hepatitis Viruses
ISBN 978-0-12-813330-9
https://doi.org/10.1016/B978-0-12-813330-9.00003-X

strand, the DNA or RNA primer terminus (the initiation "*i-site*") and the incoming NTP site ("*i+1*" site). There are four steps involved in the catalytic reactions of the polymerase: (1) binding of template-primer and NTP, (2) incorporation of nucleoside monophosphate into the growing daughter strand, (3) release of pyrophosphate, and (4) translocation along the template (Choi, 2012).

Viruses containing DNA (double-stranded or single-stranded) as their genetic material are called DNA viruses, and they use DNA polymerases for their replication. Similarly, viruses containing RNA (double-stranded or single-stranded) as their genetic material are called RNA viruses and they use RNA polymerases for their replication. DNA viruses replicate their genomes using a DNA polymerase called DNA-dependent DNA polymerase (DdDp) and transcribe their mRNA using an RNA polymerase called DNA-dependent RNA polymerase (DdRp). Viruses having RNA genomes (double-stranded or single-stranded) replicate and transcribe their genomes using an RNA polymerase called RNA-dependent RNA polymerase (RdRp). So far, we have discussed three types of polymerases: DdDp, DdRp, and RdRp. In addition to these three types of polymerases, there is one more type of polymerase known as RNA-dependent DNA polymerase (RdDp). This RdDp is also known as reverse transcriptase (RT), and this is used by retroviruses such as HIV. A retrovirus is a single-stranded positive-sense RNA virus, which once inside the host cell cytoplasm uses its own polymerase (RT) to produce DNA from its RNA genome, the reverse of the usual pattern (thus retro). This DNA produced in retrovirus from its RNA is then incorporated into the host cell genome by an enzyme called integrase. At this point, the retroviral DNA is referred to as a provirus. This DNA is treated by the host cell as the part of its own genome, which then undergoes usual transcription and translational processes to express the genes carried by the virus. Notwithstanding, in other viruses, DNA is transcribed into RNA, which is then translated into protein.

Almost all polymerases have a highly conserved structure with little difference in their catalytic sites. All four types of virus polymerases as described before share common structural features (Tsai et al., 2006; Cheetham and Steitz, 2000; Doublie et al., 1999; Ollis et al., 1985). Based on the homology of their primary sequences, DNA polymerases have been put into six families, A to D (Johansson and Dixon, 2013). The replicative DNA polymerases from Eukarya have been put in family B, those from Bacteria have been put in A and C families, and those from Archaea in families B and D. All known structures of polymerase resemble a right hand with thumb, fingers, and palm subdomains, as shown in Fig. 3.1 for a polymerase of family A, where the palm domain may be involved in catalyzing the transfer of phosphoryl groups in the phosphoryl transfer reactions, while the finger domain binds the NTP to the template base. The processivity (an enzyme's ability to catalyze consecutive reactions without releasing its substrate), translocation, and positioning of DNA is fixed by the thumb domain (Steitz, 1999). In the active state of the enzyme, DNA is bound to the palm domain. Studies have been mainly focused on the DNA polymerases of family A, which include *E. coli* DNA Pol I, *Taq* Pol I, and T7 DNA polymerase (Loh and Loeb, 2005).

As shown in Fig. 3.1, DNA polymerases possess six evolutionary conserved motifs—A, B, C, 1, 2, and 6—out of which A, B, and C are most highly conserved, and A and C exist in all known DNA and RNA polymerases (Loh and Loeb, 2005). Motif A and motif C contain the catalytic Asp residues that bind two metal ions in the active site. Motif C

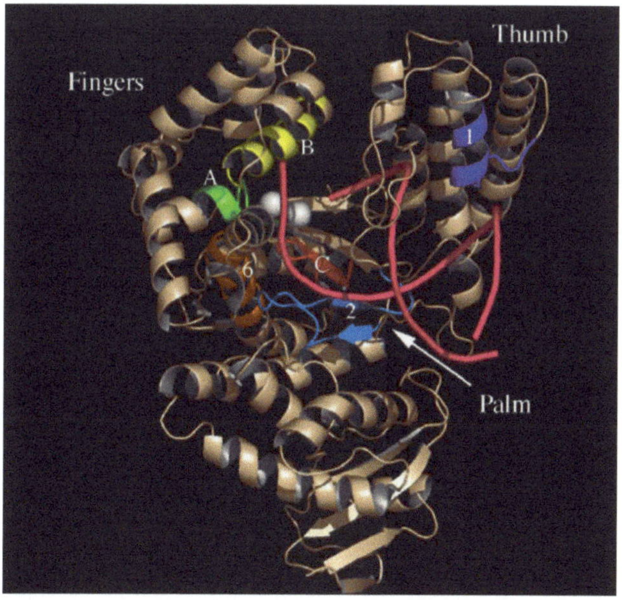

Figure 3.1 Structure of a polymerase showing its fingers, thumb, and palm domains. Its six motifs are also shown with *purple* (motif 1), *blue* (motif 2), *orange* (motif 6), *green* (motif A), *yellow* (motif B), and *red* (motif C). Two catalytic Mg^{2+} ions bound to the palm domain are shown with *white*. *(Reprinted from Loh, E., Loeb, L.A., 2005. Mutability of DNA polymerase I: implications for the creation of mutant DNA polymerases. DNA Repairs 4, 1390–1398. Open access publication.)*

contains the characteristic Gly-Asp-Asp (GDD) sequence embedded in hydrophobic residues (Choi, 2012). On the other hand, four residues—Arg659, Lys663, Phe667, and Tyr671—are evolutionary conserved in motif B. Motifs 1, 2, and 6 are also structurally conserved but have significant differences in their amino acid sequences and thus do not have many interactions with the substrates. Motif 1 consists of a helix and loop located at the tip of the thumb domain, motif 2 consists of two beta strands located in the palm domain, and motif 6 has a helix running parallel to the DNA template strand in the palm at the base of fingers (Doublie et al., 1998; Li et al., 1998; Kiefer et al., 1998; Loh and Loeb, 2005).

3. PROTEASES

Viral proteases play a crucial role in the replicative cycle of the viruses. As already discussed, they catalyze the processing of the viral proteins or the maturation of the viral particles and their release from the cell. Proteases have been found in almost all families of viruses, irrespective of the nature of their virions (Kräusslich and Wimmer, 1988). They can be found in both enveloped and nonenveloped viruses. In a recent review, Sharma and Gupta (2017) have vividly described the structure and function of proteases in different viruses. There are four families of proteases: serine proteases, cysteine proteases, aspartyl proteases, and metallo-proteases (belonging to any of the preceding three families but containing a metal ion). The main function of any protease is to perform proteolysis, i.e., protein catabolism by hydrolysis of peptide bonds. Different classes of proteases may adopt different mechanisms for this, but all possess an active site that consists of a binding site and a catalytic site, as shown in Fig. 3.2, both constituted of residues. Residues in the binding site interact with the substrate, and those in catalytic site catalyze the reaction. The active site in the form of a groove or pocket is usually located in a deep tunnel within the enzyme, where the catalytic site is very close to the binding site. The catalytic site can also help the binding of residues in the binding site with the substrate.

Usually, proteases possess a catalytic triad that refers to a group of three amino acid residues, such as Ser-His-Asp, Cys-His-Asp, Ser-His-His, Ser-Glu-Asp, and Ser-cisSer-Lys, functioning together at the center of the active site (Fig. 3.3). A catalytic triad may be constituted of an acidic residue, a basic residue, and a nucleophilic residue. Catalytic triads are used for one of two types of reaction: splitting a substrate (hydrolases) or transferring one portion of a substrate over to a second substrate (transferases). They perform covalent catalysis using a residue as a nucleophile.

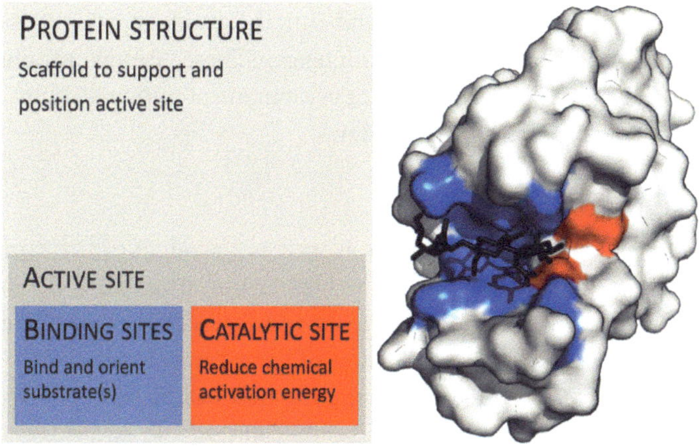

Figure 3.2 A schematic structure of a protease with its active site. The active site has a binding site (shown in *blue*) and a catalytic site (shown in *red*). A substrate of the protease (peptiglycon) crystallized with it is shown in *black* (PDB: 9LYZ). *(Reprinted from https://en.wikipedia.org/wiki/Active_site.)*

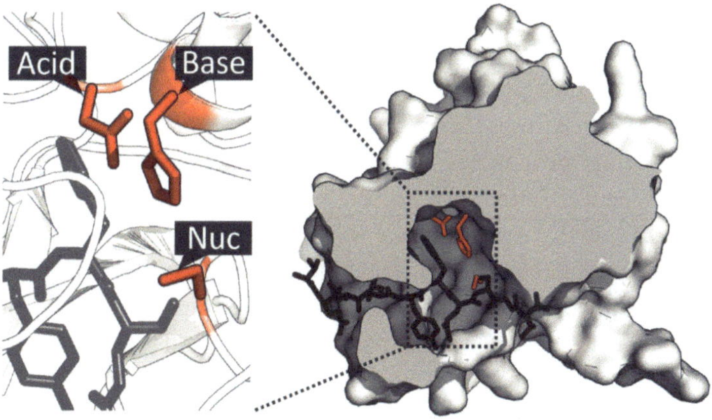

Figure 3.3 Example of a protease (tobacco etch virus protease) with a catalytic triad consisting of an aspartate (acid), histidine (base), and serine (nucleophile). The substrate (*black*) is bound by the binding site to orient it next to the triad. PDB: 1LVB. *(Reprinted from https://en.wikipedia.org/wiki/Catalytic_triad.)*

In addition to these polymerases and proteases, there are many other types of proteins that play crucial roles in the replication of the viruses and thus constitute attractive targets for the development of antiviral drugs. A few of them are described subsequently.

3.1 Capsid Proteins

Capsid proteins, designated as VP1, VP2, VP3, and VP4, are important components of infectious virions. They protect viral genomes during entry and exit from the host cells and can also modulate the activity and specificity of viral replication complexes. These capsid proteins can constitute the important targets for development of antiviral drugs. Capsid proteins are also known as coat proteins.

3.2 Viral Proteins

A viral protein can be a component as well as a product of the virus. Viral proteins include structural proteins as well as nonstructural (NS) proteins. Additionally, they also include regulatory and accessary proteins. These viral proteins are required for the replication of the viruses, but they may not be present in the viruses; rather, viruses use the host cell machinery to produce the proteins necessary for their replication (Slonczewski and Foster, 2013).

VPg is a viral protein that is linked to the genome of the virus (viral protein genome-linked). In a positive strand viral RNA, it is attached to the 5′ end of the RNA to act as primer during RNA synthesis in a variety of virus families, such as Picornaviridae, Potyviridae, and Caliciviridae. In some viruses, VPg also plays a role in translation initiation by acting like a 5′ mRNA cap. VPg was first observed in poliovirus RNA, covalently attached to the 5′ end of the genome (Flanegan et al., 1977; Lee et al., 1977) and later found in caliciviruses (Schaffer et al., 1980). VPg interacts directly with the cap-binding protein of the ribosome eIF4E for viral translation (Goodfellow et al., 2005).

Viral proteins that are components of mature assembled virus particles, such as nucleocapsid core proteins (gag proteins), enzymes packaged within the virus particles (pol proteins), and membrane components (env proteins), are called structural proteins. Gag, pol, and env proteins are the three major proteins encoded within the retroviral genome, such as the HIV genome. In retroviruses, gag encodes the matrix capsid and nucleocapsid proteins, which are structural core proteins, pol encodes the enzymes protease, RT, and integrase, and env is an envelope protein that resides in the lipid layer and determines viral tropism.

Proteins that are expressed in infected cells but not incorporated in the virus particles are called NS proteins. During the replication of the viruses, some NS proteins carry out important functions that affect the replication process (Uversky and Longhi, 2011). Some of these proteins also help the assembly process and play crucial roles in immunomodulation and transactivation of structural protein encoding genes (Uversky and Longhi, 2011). Replicon process refers to the formation of replication complex, interacting with host cell proteins, leading to generating replicons. In hepatitis C virus (HCV), the nonstructural 4B (NS4B) protein alters the host cell's membrane and starts the replication complex formation process (Uversky and Longhi, 2011; Gouttenoire et al., 2010a). Other viral NS proteins such as NS5A, NS5B, and NS3 are also recruited to the complex where NS4B interacts with them and binds to viral RNA (Uversky and Longhi, 2011; Gouttenoire et al., 2010b). Immunomodulation refers to the process where a protein can reduce the immune power of the infected host cell, e.g., the viral NS protein NS1 in the West Nile virus prevents complement activation through its binding to complement control protein, factor H, leading to a reduction in complement recognition of infected cells, so the infected cells remain unharmed by the host immune system (Chung et al., 2006).

3.3 Viroporins

Viroporins are the viral proteins that participate in several viral processes necessary for efficient production of virus progeny, in addition to interfering with a number of cellular processes (Nieva and Carrasco, 2015; Nieva et al., 2012; OuYang and Chou, 2014; Scott and Griffin, 2015). An example of a viroporin is p7 protein of HCV. Another example is M2 protein of influenza A virus. Viroporins are usually short proteins with at least one hydrophobic amphipathic helix. An example of a viroporin having three hydrophobic segments is rotavirus NS protein NSP4 (Hu et al., 2012). Viroporins have been found to be good targets for developing antiviral drugs. Attempts have been made to develop a number of specific inhibitors of HCV using its p7 viroporin (Gervais et al., 2011). Other viruses that have viroporins are poliovirus (2B, VP4), coxsackievirus (B2 and B3), EV71 (2B), and human rhinovirus (VP4). All these viruses belong to Picornaviridae family and have ssRNA (+) genome. HIV-1 virus, which is a retrovirus, has a viroporin Vpu.

Viroporins were first identified in RNA viruses but later were observed in several other families of viruses such as Flaviviridae, Picornaviridae, Togaviridae, Coronaviridae, Paramyxoviridae, Orthomyxoviridae, Reoviridae, and Retroviridae (Scott and Griffin, 2015). Attentions have been paid to the recognition of HIV-1 Vpu, HCV p7, and picornavirus 2B viroporins. However, IAV (influenza A virus) M2 remains so far the best characterized viroporin. The channel-forming hydrophobic integral membrane proteins, viroporins, have been found to be essential for the life cycle of several families of RNA and DNA viruses and thus have drawn great attention to be exploited as antiviral drug targets. In context to hepatitis therapeutics, p7 inhibitors have been well studied to treat HCV infection, some of which already have been licensed for clinical use (Foster et al., 2011a,b, 2014).

3.4 Other Proteins

Certain families of viruses, such as Picornaviridae, have some other proteins that have been exploited for development of antiviral drugs. Take, for example, one member of its family, hepatitis A virus (HAV). As shown in Fig. 2.4 (Chapter 2), its genome is divided into three segments—P1, P2, and P3—where P1 covers the capsid proteins VP1, VP2, VP3, and VP4, in which VP4 is located on the inner surface of the capsid formed by VP1, VP2, and VP3. P2 is a regulatory segment that covers 2A, 2B, and 2C proteins, and P3 covers growth functional proteins, Vpg and 3Cpro (3C protease). At the 5′ end, Vpg refers to a viral protein. Many of these proteins have been useful targets for developing antiviral drugs, e.g., 2B and 3Cpro have been studied for investigating anti-HAV drugs (Debing et al., 2014; Lall et al., 2004).

4. IMPORTANT DRUG TARGETS IN HEPATITIS VIRUSES

4.1 Hepatitis A Virus

HAV is the only species and the only serotype in the *Hepatovirus* genus that has three genotypes circulating in humans and whose outbreaks occur in developed countries in both adults and children (Gervelmeyer et al., 2006; Hauri et al., 2006; Pontrelli et al., 2008). As shown in Fig. 2.4 (Chapter 2), the genome of HAV has several polyproteins, all of which play very important roles in the replication of the virus. All polyproteins are processed to play different roles. The primary processing takes place between 2A and 2B, carried out by a single proteinase 3Cpro. HAV 3Cpro cleaves the 2A-2B junction between residues Gln836 and Ala837, resulting in smaller 2A protein and a larger, 251-amino acid–long 2B protein (Martin et al., 1995; Jia et al., 1993; Gosert et al., 1996; Graff et al., 1999). This 2B protein has been found to play a very crucial role in the replication of HAV virus, after 3Cpro has played its role. The 3Cpro and 2B have been well studied as important drug targets (Vives-Adrián et al., 2015; Paulmann et al., 2008). We present here the structural details of these two targets.

4.1.1 3C Protease

Structural studies on 3Cpro of HAV and some other picornaviruses have shown that this protease belongs to family of cysteine proteases that have an overall fold similar to chymotrypsin (Allaire et al., 1994; Matthews et al., 1994; Mosimann et al., 1997; Birtley et al., 2005). 3C proteases possess a conserved Cys172-His44-Asp84 (C172-H44-D84) catalytic triad at the active site, which is similar to the Ser-His-Asp triad found in the vast majority of serine proteases, where His residue serves to prime the Ser side chain for nucleophilic attack on the scissile bond by abstracting a proton. The role of the Asp is to provide electrostatic stabilization of the resulting positive charge on His (Hedstrom, 2002). However, this Asp84 (D84) in HAV 3Cpro is found to be directed away from the active site and to play little role in catalysis. The crystal structure of HAV 3Cpro (Fig. 3.4) bound with an irreversible inhibitor N-Cbz-L-serine β-lactone **(1a)** shows that the side-chain of Asp84 forms a hydrogen bond with the imidazole ring of His44 and the inhibitor **1a** alkylates the active site thiol of Cys172 leading to complete loss of the enzymatic activity (Yin et al., 2005).

HAV 3Cpro has cleavage specificity. Though it has been suggested that 3Cpro cleavage occurs at 2A/2B, 2B/2C, 2C/3A, or 3A/3B (Ryan and Flint, 1997; Seipelt et al., 1999; Stanway et al., 1984; Lawson and Semler, 1992), the most preferred sites have been only 2A/2B and 2C/3A. Some authors have suggested that 2B/2C site is also a preferred cleavage site (Jewell et al., 1992). The cleavage site was, in fact, found to depend upon the 3Cpro concentration (Schultheiss et al., 1995). However, Schulthiess et al. (2005) have proposed that polyprotein processing of HAV relies totally on 3C as a single proteinase presumably controlled by concentration.

4.1.2 2B Protease

As already discussed, HAV 3Cpro cleaves 2A/2B site to produce 2A and 2B peptides, in which 2B has been extensively studied as a target for drug design. The N-terminal amino acid residues of this 2B peptide are Ala-Lys-Ile-Ser-Leu-Phe (Gosert et al., 1996). HAV 2B is supposed to have at least one hydrophobic transmembrane α-helix (Garriga et al., 2011) to bind to ER (endoplasmic reticulum) membranes (De Jong et al., 2008). It prevents the synthesis of beta interferon (IFN-β) and thus inhibits the host innate immune response (Paulmann et al., 2008) and thus helps the HAV infection to occur. It has also been observed that 2B weakens the cytoplasmic membrane to help the release of the virus particles (Jecht et al., 1998). However, the three-dimensional structure of HAV 2B is not so well studied as compared to that of 3Cpro (Bergmann et al., 1997). In a recent communication, however, Vives-Adrián et al. (2015) attempted to find the three-dimensional structure of the soluble domain of HAV 2B ($2B_{sol}$). These authors found that $2B_{sol}$ is folded into 5 α-helices and 8 β-strands in subdomains, where 5 β-strands (β1, β4, β5, β6, and β7) constitute a pseudo-β-barrel at N-terminal. These β-strands are in a curved

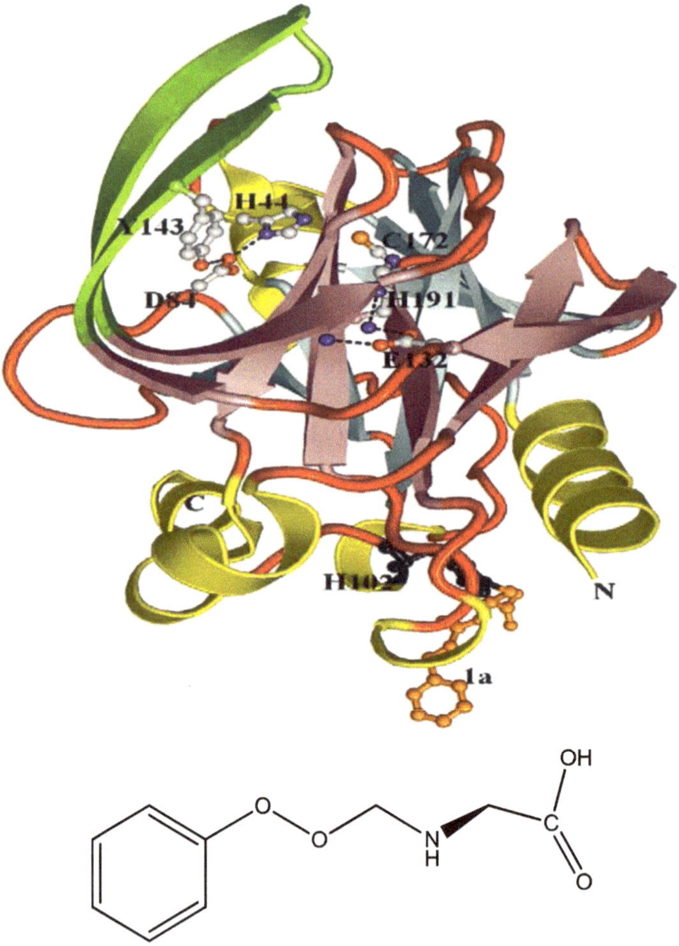

(1a)

Figure 3.4 Structural overview of the HAV 3C complexed with compound with an irreversible inhibitor N-Cbz-L-serine β-lactone (1a). The catalytic triad, Cys172 (C172), His44 (H44) and Asp84 (D84) are represented in ball and stick models with C atoms in grey, O in red, N in blue, and S in orange. *(Reprinted with permission from Yin, J., Bergmann, E.M., Cherney, M.M., Lall, M.S., Jain, R.P., Vederas, J.C., James, M.N.G., 2005. Dual modes of modification of Hepatitis A virus 3C protease by a serine-derived β-lactone: selective crystallization and formation of a functional catalytic triad in the active site. J. Mol. Biol, 354 (4), 854–871. Copyright 2005 Elsevier.)*

antiparallel β-sheet. The two short β-strands, β2 and β3, form a β-hairpin to prolong β1- and β4-strands. Four α-helices (α1 to α4) constitute a second subdomain. The C-terminal tail has a short β-strand (β8) followed by short α5-helix projecting perpendicular to the main structure of 2B protein (Vives-Adrián et al., 2015). Shukla et al. (2015) have identified and functionally characterized a putative viroporin motif in the C-terminus of the NS protein 2B from HAV. This motif encompasses amino acids 174–233 of 2B and can be expected to fulfill all structural and functional requirements of a viroporin. It can fold into a hydrophobic α-helical hairpin with a marked tendency to dimerize and form small pores in artificial membranes. Shukla et al. (2015) also tried to show that the membrane interacting activity of HAV 2B is localized in its C-terminal region that contains an α-helical hairpin motif. This C-terminal region has lipid specific activity, which partially rationalizes the intracellular localization of full length 2B. Through molecular dynamics studies, Shukla et al. established that the preferred mode of membrane interaction by the 2B peptide is in the form of a dimer, in which both termini of the peptides are facing the membrane. In this orientation, both peptides sink deep into the membrane (Fig. 3.5A), and form numerous H-bonding associations with each other and with the membrane. The N- and C-termini of both the peptides may be oriented in the same (Fig. 3.5B) or opposite (Fig. 3.5C) directions to each other.

4.2 Hepatitis B Virus

For design and development of potent antihepatitis B virus (HBV) drugs, the most important targets have been the HBV antigens and its polymerase. The importance of these targets in HBV replication is described next.

4.2.1 Antigens

As shown in Fig. 3.6, there are three important antigens that play crucial roles in HBV infection. These are surface antigen (HBsAg), core antigen (HBcAg), and e-antigen (HBeAg). HBsAg, the surface antigen, is also called "Australia antigen." HBsAg is required to assemble viral particles. It is also suggested that HBsAg may be involved in the regulation of apoptosis (a process of programmed cell death that occurs in multicellular organisms), as both up- and down-regulation of specific proteins is involved in the process of apoptosis. Further, HBsAg protein is also supposed to be involved in HCC (hepatocellular carcinoma) development. Thus, HBsAg is taken as a potential target for HCC gene therapy. HBsAg is

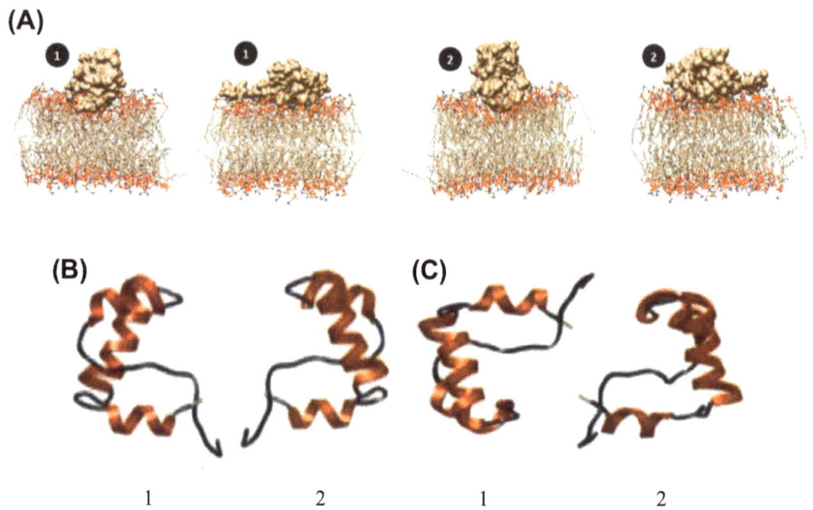

Figure 3.5 A diagram showing the preferred mode of membrane interaction by the 2B peptide in the form of a dimer (A), where the N- and C-termini of both the peptides may be oriented in the same (Fig. 3.5B) or opposite (Fig. 3.5C) directions to each other. *(Reprinted with permission from Shukla, A., Dey, D., Banerjee, K., Nain, A., Banerjee, M., 2015. The C-terminal region of the non-structural protein 2B from Hepatitis A Virus demonstrates lipid-specific viroporin-like activity. Sci. Rep. 5, 15884. Open Access publication.)*

overwhelmingly produced and stably maintained in chronically HBV-infected patients (Kang et al., 2015).

HBeAg is an indicator of active viral replication, meaning that the person infected with HBV can likely transmit the virus to another person. HBeAg may be present between the nucleocapsid core and the lipid

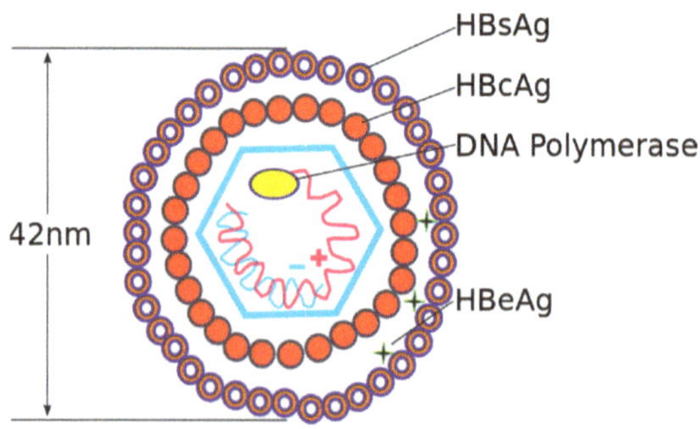

Figure 3.6 Structure of HBV (hepatitis B virus) virion showing the location of DNA polymerase and different antigens important in the replication and transmission of HBV. Structure in *purple* refers to lipid envelope, and the one in *red* refers to the nucleocapsid. *(Public domain, https://commons.wikimedia.org/w/index.php?curid=5670720.)*

envelope (Fig. 3.6). It is secreted and accumulates in serum as an immunologically distinct soluble antigen and serves as a marker of active replication in chronic hepatitis. The core antigen, HBcAg, is also an indicator of viral replication and can make the HBV-infected person infectious (Kimura et al., 2003; Cao et al., 2001). While both HBcAg and HBeAg are made from the same open reading frame (ORF), HBcAg is not secreted. It remains in the form of minute separate particles and does not circulate in the blood. However, the presence of both HBcAg and HBeAg proteins act as a marker of viral replication.

4.2.2 Hepatitis B Virus Polymerase

HBV polymerase is an RT and contains several important domains and motifs that define its functions and reveal ways to further target it. Several functions required for HBV replications, such as viral RNA binding, RNA packaging, protein priming, template switching, DNA synthesis, and RNA degradation, are associated with it (Clark and Hu, 2015). However, HBV polymerase interacts with host proteins for its functions. Currently available anti-HBV drugs inhibit the steps of DNA synthesis carried out by RT, but attempts are being made to find the drugs that not only inhibit the DNA synthesis but also any of other domains, functions, and interactions that are essential for the HBV replication cycle. HBV polymerase is the only enzymatically active protein and participates in many steps of the viral replication cycle. It has four domains: (1) terminal protein (TP) at the N-terminal, (2) a spacer domain, (3) RT domain, and (4) an RNase H domain at C-terminal (Fig. 3.7). TP is essential in initial DNA synthesis, RT domain is responsible for polymerase activity, and RNase H domain is required to remove template RNA. These three domains are critical for polymerase function and can be exploited as potential targets for the design of antivirals (Clark and Hu, 2015). As shown in Fig. 3.7, RT domain has three subdomains: finger, palm, and thumb domains. RT is required to reverse transcribe pregenomic RNA (pgRNA) to viral (−) strand DNA.

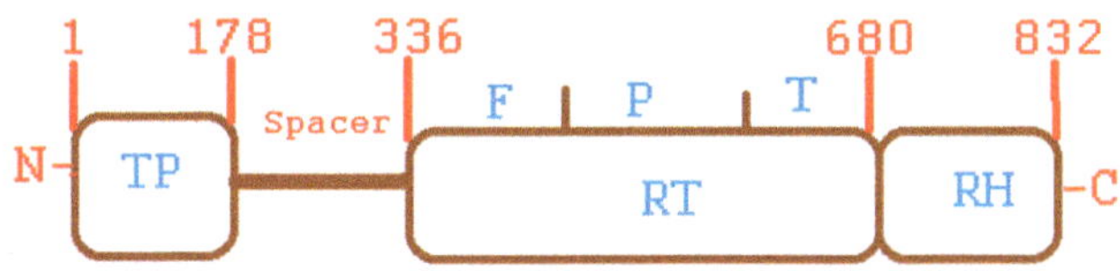

Figure 3.7 Schematic representation of hepatitis B virus polymerase domains, where reverse transcriptase (RT) domain is shown to have three subdomains: fingers (F); palm (P), and thumb (T).

Once the virus enters the host cell, it is decoded, and the released capsid is delivered to the nucleus to release a viral genome (Kang et al., 2015). The partially double-stranded DNA is converted to covalently closed circular DNA (cccDNA), which serves as a transcription template. It is transcribed by host factors into all viral mRNAs, including the pgRNA. The functions of different domains of HBV protease are as follows.

In HBV polymerases, TP domain is essential for ε-RNA binding, RNA packaging, and protein priming. ε-RNA is a structural motif in pgRNA at which the polymerase binds (Jones et al., 2012; Hu and Anselmo, 2000; Hu et al., 1997, 2004). The interaction of both polymerase and ε-RNA is necessary for their packaging into assembling nucleocapsids. Further, ε-RNA contains the initial template for reverse transcription. The spacer domain is not very critical in the role of HBV polymerase; mutations in the spacer may affect the function of HBsAg in the context of an infection. The main catalytic center in HBV polymerase is RT domain. The three subdomains of RT in HBV have a high degree of homology with HIV's RT enzyme. The RT domain remains an excellent target for antiviral drug development, even though the drug resistance remains a problem. Next to it, the attractive target for the next generation of anti-HBv drugs is the fourth domain of the HBV polymerase, i.e., RNase H domain. It plays the critical role in the replication cycle of HBV. It removes the pgRNA template but leaves its 5′ end intact, so it clears the way for the plus-stranded DNA synthesis, as well as creating the plus strand primer from the 5′ end of pgRNA. HBV RNase is expected to act as a good target for antiviral therapy; it has been found to be sensitive to compounds first designed as anti-HIV agents (Tavis et al., 2013).

The crystal structure of HBV polymerase is not well studied. However, its structure has been deduced from HIV-1 RT based on the homology (van Hemert et al., 2014). Irrespective of their amino acid sequences and difference in domain structures, all polymerases are found to have a common right-handed structure having a thumb, a palm, and a fingers subdomain (Steitz, 1999; Doublie et al., 1999), as shown in Fig. 3.7 as well as Fig. 3.8. In these subdomains, the active site is represented by the palm domain that catalyzes the phosphoryl transfer reaction. The other two subdomains facilitate the catalytic reactions. The polymerase inhibitors (nucleoside analogs, NAs) or substrates (deoxynucleotide triphosphates) bind in the palm domain at a site that is located adjacent to the 3′ terminus of the primer strand (Tantillo et al., 1994). HBV polymerase differs from other polymerases in the sense that, while the other polymerases prefer the

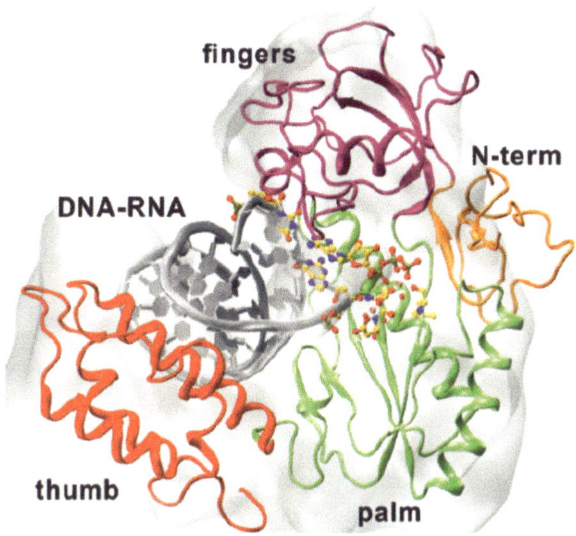

Figure 3.8 The homology model of HBV-RT interacting with the substrate dATP with DNA-RNA in place. The subdomains are shown in different colors. The substrate dATP and the residues involved in direct interaction with RT are shown in stick and ball representation. The corresponding part from the template HIV-RT/DNA-RNA/dATP (PDB code: 4PQU) is shown in *gray* surface representation, structurally aligned to the HBV-RT/DNA-RNA/dATP model. *dATP*; deoxyadenosine triphosphate; *HBV*, hepatitis B virus; *RT*, reverse transcriptase. *(Reprinted from Xu, X., Thai, H., Kitrinos, K.M., Xia, G., Gaggar, A., Paulson, M., Ganova-Raeva, L., Khudyakov, Y., Lara, J., 2016. Modeling the functional state of the reverse transcriptase of hepatitis B virusand its application to probing drug-protein interaction. BMC Bioinform. 17 (Suppl. 8), 280. Open access publication.)*

nucleotides in D-configuration, it prefers L-configuration (Davis et al., 1996).

While many studies have been made to find crystal structures of HIV-RT with or without ligands, no successful experimental efforts could be made to illustrate the crystal structure of HBV-RT. However, since HBV-RT is homologous to HIV-RT with similar multiple catalytic activities (van Hemert et al., 2014), some computational efforts have been made to define the HBV-RT structure using HIV-RT as a template (Das et al., 2001; Xu et al., 2016). In a recent study, Xu et al. (2016) used a hybrid computational approach to find the structures of HBV-RT complexed with deoxyadenosine triphosphate (dATP) and tenofovir diphosphate with the native hybrid DNA-RNA substrate in place. The model obtained showed the strong coupling of the natural nucleotide substrate (dATP) to the active site of RT (Fig. 3.8). These authors also found the differential involvement of two putative Mg^{2+} ions at the active site, one directly bridging the interaction between dATP and HBV-RT and the other serving as a coordinator to maintain an optimal configuration of the active site.

4.3 Hepatitis C Virus

As discussed in Chapter 2, HCV is a single-stranded RNA virus and encodes structural proteins (core, E1, E2, p7) and NS proteins (NS2, NS3, NS4A, NS4B, NS5A, and NS5B) (Fig. 2.9). In the NS proteins, NS2 and NS3 constitute an NS2/3 protein that undergoes autocleavage to separate NS2 and NS3 (Fig. 3.9) (Welbourn and Pause, 2007). This NS2/3 protein is an autoprotease, whose activity is different from that of NS3. NS2/3 protease has been an attractive target for antiviral drug discovery, as its cleavage is required for viral replication in vivo. Also, efficient NS2/3 cleavage is required for the replication of the viral gene in an NS2-3 replicon system (Welbourn et al., 2005). While uncleaved NS2/3 is required for viral infectivity (Agapov et al., 2004), the amount of cleaved NS3 protein is a major of RNA replication levels (Lackner et al., 2004). Though not required for RNA replication, the presence of NS2 in the system as an accessary protein is plausible, but it requires further investigation. Roles of mature cleaved NS2 are not well explored.

4.3.1 NS3/4A Protease

The NS protease NS3 is a bifunctional protein with a serine protease domain. It cleaves at the NS3/4A, 4 A/4B, 4 B/5A, and 5 A/5B junctions

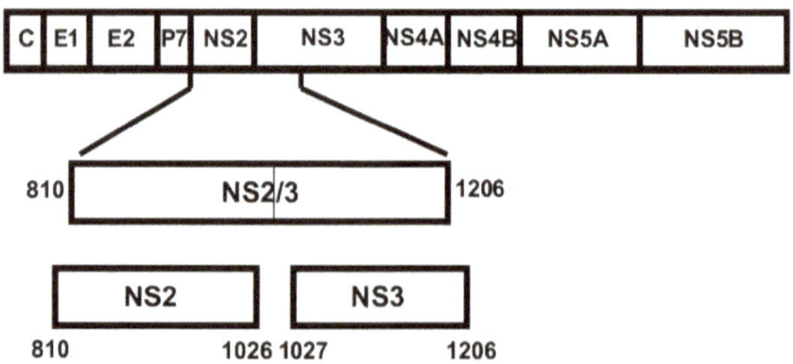

Figure 3.9 The hepatitis C virus (HCV) NS2/3 protease. The NS2/3 protease is shown in the context of the HCV polyprotein. NS2 and the protease domain of NS3 (from aa 810 to 1206) constitute NS2/3, which undergoes autocatalytic cleavage between aa 1026 and 1027. *(Reprinted from Welbourn, S., Pause, A., 2007. The hepatitis C virus NS2/3 protease. Curr. Issues Mol. Biol. 9, 63–70. Open access publication.)*

and requires the NS4A protein as an essential cofactor for an efficient enzymatic function (Raney et al., 2010). NS3 exerts multiple enzymatic functions via serine protease and NTPase/helicase (NS3 helicase) domains at the N- and C-terminus, respectively (Gallinari et al., 1998). The NS3 helicase is responsible for HCV replication both in vitro and in vivo (Lam and Frick, 2006; Kolykhalov et al., 2000), so it is assumed to be a potential target for anti-HVC drug design, but so far, no inhibitors of it have entered the clinical trials, as they exhibited low efficacy and high toxicity. It has been established that NS3 requires NS4A to be functional and for proteolytic processing of HCV NS proteins (Failla et al., 1994).

In a recent review, Sharma et al. (2017) have discussed in detail the structural aspects of NS3/4A protease and its importance in drug design. The crystal structure and specificity of HCV protease have been characterized in detail (Lin, 2006; Di Marco et al., 2000; Kim et al., 1996; Raney et al., 2010). However, the design and development of NS3/4A inhibitors is more challenging than design of inhibitors of other proteases, because its binding pocket is more solvent exposable, shallow, and hydrophobic than that of other proteases. It has been found that HCV NS3 protease is a chymotrypsin-like serine protease (Lin, 2006; Kramer et al., 2014; Barbato et al., 1999), where Ser139, His57, and Asp81 form the catalytic triad (Fig. 3.10). The protease contains two β barrels and 30-residue extension at the N-terminus. In the stabilization of the protease structure, a zinc ion present in it plays a vital role. The zinc ion is coordinated by three cysteine residues (Cys97, Cys99, and Cys194) and a histidine residue (His149) (Fig. 3.11). NS4A acts as cofactor of NS3, which

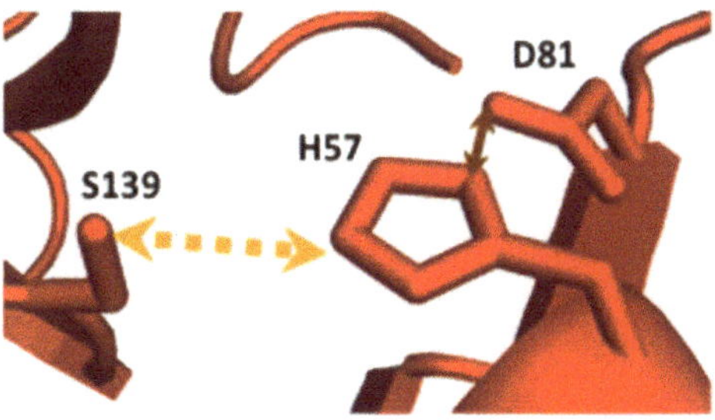

Figure 3.10 Catalytic triad of NS3 (S139, H57, and D81) in ball and stick model. The arrow in orange indicates the distance between two nearest residues. *(Reprinted from Kramer, M., Halleran, D., Rahman, M., Iqbal, M., Anwar, M.I., Sabet, S., Ackad, E., Yousef, M., 2014. Comparative molecular dynamics simulation of hepatitis C virus NS3/4A protease (genotypes 1b, 3a and 4a) predicts conformational instability of the catalytic triad in drug resistant strains. PLOS ONE 9, e104425. Open access publication (Plos 1).*

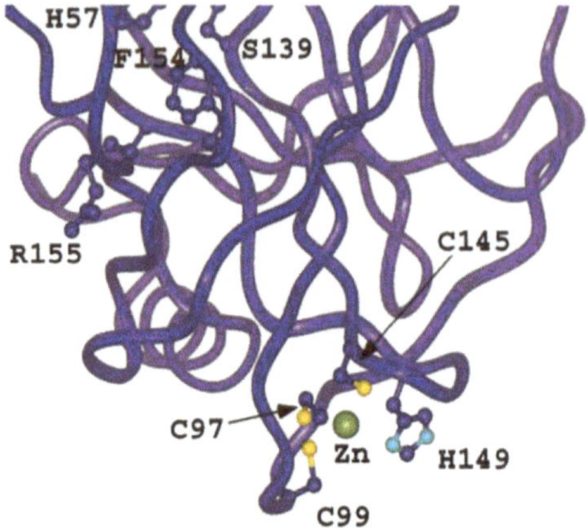

Figure 3.11 Zinc-binding site in NS3/4A coordinated with C99, C145, C97, and H149, where "C" stands for Cys and "H" for His residues. *(Reprinted from Barbato, G., Cicero, D.O., M. Nardi, M.C., Steinkühler, C., Cortese, R., Francesco, R.D., Bazzo, R., 1999. The solution structure of the N-terminal proteinase domain of the hepatitis C virus (HCV) NS3 protein provides new insights into its activation and catalytic mechanism. J. Mol. Biol. 289 (2) 371-384. Copyright 1999 Elsevier.)*

allows the proper folding of NS3. It is a peptide of both hydrophobic N-terminus and hydrophilic C-terminus that leads to the activation of NS3 (Bartenschlager et al., 1994).

4.3.2 NS4B Protease

NS4B, i.e., nonstructural 4B protein of HCV, has recently drawn the attention for design of anti-HCV inhibitors. It is a 27-kDa integral membrane protein with an N-terminal portion (aa 1—69), a central transmembrane part (aa 70—190), and a C-terminal position (aa 191—261) (Fig. 3.12) (Dvory-Sobol et al., 2010; Li et al., 2012). The terminal domain

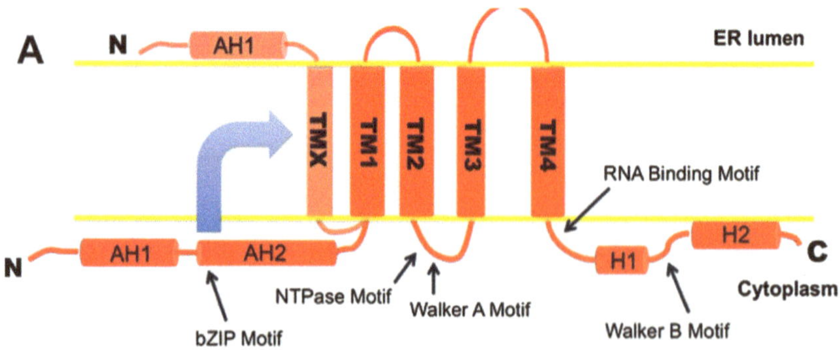

Figure 3.12 A schematic representation of the hepatitis C virus NS4B protease, showing its three terminals and helices and motifs in them. *(Reprinted from Dvory-Sobol, H., Pang, P.S., Glenn, J.S., 2010. The future of HCV therapy: NS4B as an antiviral target. Viruses 2, 2481—2492. Open access publication.)*

possesses two amphipathic α-helices, AH1 and AH2, where AH2 has been found to be conserved in all HCV genotypes. It undergoes oligomerization and stimulates aggregation of lipid vesicles, critical for HCV replication (Gouttenoire et al., 2009, 2010a, 2010b; Cho et al., 2010).

The NS4B N-termini is predicted to lie in the cytosolic face and is implicated in membrane association and retention, which are required for NS4B-mediated formation of HCV replication (Elazar et al., 2004). The C-terminal domain has been found to be composed of two transmembrane helices, H1 and H2, where H2 is involved in the formation of functional HCV replication complexes (Dvory-Sobol et al., 2010). An arginine-rich motif is also contained by C-terminal domain to bind viral RNA and a nucleotide-binding Walker B motif (228DAAA231), both important for HCV replication assay (Einav et al., 2004). A nucleotide-binding A motif (129GSIGLK135) is associated with the central domain of NS4B. Its function is not well defined.

NS4B plays a lead role in the formation of membranous web (MW) to facilitate RNA replication (Ashfaq et al., 2011; Egger et al., 2002; Gosert et al., 2003). The MW is formed through aggregation of membranous vesicles derived from ER. MW is the platform where the NS viral proteins and host factors associate to the HCV RNA (Gouttenoire et al., 2010a; Cannalire et al., 2016; Moradpour et al., 2003).

4.3.3 NS5A Protease

Another protease important in the replication of HCV and constituting an important target for HCV therapy is NS5A (Sheridan, 2011). This protease, however, encodes no known enzymatic function, but it has been found to be important in both viral replication and viral assembly (Sheridan, 2011; Yu et al., 2016). Over the past decades, it has drawn great attention for antiviral drug design. In particular, its inhibitors have become very important for fixed-dose combination regimens (Schinazi et al., 2010). To date, four NS5A inhibitors are available for use in clinical practice for HCV treatment: daclatasvir (Gao et al., 2010; Belema et al., 2014), ledipasvir (Link et al., 2014), and elbasvir (Coburn et al., 2013).

As far as the structure of NS5A is concerned, it has been found to be comprised of approximately 447 amino acids and localized to ER-derived membranes (Pawlotsky, 2013). It basically exists in phosphorylated (p56) and hyperphosphorylated (p58) forms, which are implicated in different forms (Evans et al., 2004; Neddermann et al., 2004; Fridell et al., 2011). Both in vivo and in vitro, NS5A has been found to directly interact with RdRp (Shirota et al., 2002), and in vitro this interaction stimulates

RdRp-catalyzed synthesis of the negative RNA strand (Quezada and Kane, 2009). The cytoplasmic moiety of NS5A has been found to contain three domains, all of which have been shown to bind to RNA (Kolykhalov et al., 1996). The domain 1 has been found to be most conserved (Tellinghuisen et al., 2008), and its interactions with polypyrimidine tract of 30 UTR suggested that it could affect the efficiency of RNA replication by RdRp. On the other hand, the interaction of domain 2 with cyclophilin A, a host cell protein required for replication, is vital for RNA binding (Foster et al., 2011). Domain 3 has been found to be essential for NS5A for its role in viral packaging and assembly (Cun et al., 2010; Appel et al., 2008).

4.3.4 NS5B Polymerase

NS5B, another nonstructural protein of HCV, has been most widely exploited for designing anti-HCV agents because it has no counterpart in mammalian cells, so its inhibition has little chance to cause any side effects (Lindenbach and Rice, 2005). It is an RdRp that plays a pivotal role in HCV RNA genome replication, and the host lacks its functional equivalent (Behrens et al., 1996; Moradpour et al., 2007). It contains no nucleases or proofreading domain. In its active site, it has three aspartate residues (Asp 220, Asp318, and Asp319) that coordinate divalent cations, Mg^{2+} or Mn^{2+}, present in it. NS5B contains a fully encircled active site and thus is similar to bacteriophage polymerase, though HCV belongs to Flaviviridae family. NS5B contains several domains, finger, palm, thumb, and a C-terminal (Fig. 3.13). The palm domain contains the active site, and there are several contacts between the finger and thumb domain that give the

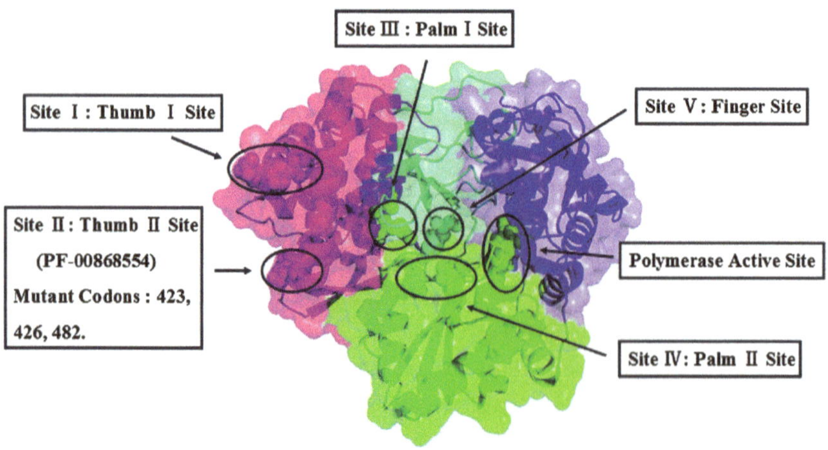

Figure 3.13 Hepatitis C virus NS5B structure and its different domains. *(Taken from Wikipedia.)*

active site an encircled structure. The completely encircled active site is responsible for extensive interactions between the finger and thumb subdomains (Delang et al., 2010). It has been found that NS5B can oligomerize and is thus important for modulating the polymerase (Qin et al., 2002). Studies have identified five distinct binding pockets for NS5B inhibitors, three of which are allosteric binding sites (Huang et al., 2006; Soriano et al., 2009; Beaulieu, 2007; Koch and Narjes, 2007): AP-1 (located on the surface of the thumb domain) (Bressanelli et al., 2002; Di Marco et al., 2005), AP-2 (located in the thumb domain, next to AP-1) (Biswal et al., 2005; Love et al., 2003; Wang et al., 2003), and AP-3 (located adjacent to the active site) (Tedesco et al., 2006). Since NS5B has been most attractive target for anti-HCV therapy, its crystal structure has been well studied, and several reviews have been written on its structure and prominent inhibitors (e.g., Patil et al., 2011; Ishii et al., 1999; Leveque et al., 2003; Mayhoub, 2012; Bressanelli et al., 2002; Das, 2017).

4.4 Hepatitis Delta Virus

Hepatitis D virus (HDV), being a defective RNA virus, is not able to replicate and assemble its new virions on its own and thus requires the help of HBV (Abbas and Afzal, 2013). HDV genome contains only one actively transcribed ORF, which encodes for two isoforms of hepatitis delta antigen (HDAg), small delta antigen (S-HDAg), and large delta antigen (L-HDAg), where the former is required for the initiation of the viral genome replication and the latter serves as a principal inhibitor of replication and is essential for the assembly of new virion particles. Methylation of the arginine residue of S-HDAg has been observed to influence HDV replication (Wang et al., 2009; Li et al., 2004), so inhibition of methylation has been attempted to design anti-HIV drugs. Prenylation and farnesylation are the two chemical transformations that take place with HDAgs and modulate their actions. Prenylation, also called isoprenylation or lipidation, is the addition of hydrophobic molecules to a protein or chemical compound. It is usually assumed that prenyl groups (3-methyl-but-2-en-1-yl) facilitate attachment to cell membranes. Farnesylation is a type of prenylation, a posttranslational modification of proteins by which an isoprenyl group is added to a cysteine residue (Maltese, 1990). It is an important process to mediate protein—protein interactions and protein—membrane interactions (Novelli and D'Apice, 2012). Farnesylated L-HDAg has been found to inhibit viral RNA replication and to facilitate virion assembly by mediating the direct binding of L-HDAg with HBV envelope proteins (Greco-Stewart and Pelchat, 2010), so inhibition of farnesylation has attracted attention for anti-HDV drug investigation. Thus there have been three important targets to inhibit the replication of HDV, S-HDAg, prenylation, and farnesylation.

4.5 Hepatitis E Virus

For antihepatitis E virus (HEV) drug development, any step in its life cycle (Fig. 2.11) can be attempted, but only its protease has been fruitfully attempted. The genome of HEV codes three proteins: a multifunctional protein (pORF1), a capsid protein (pORF2), and a multiregulatory protein (pORF3). pORF1 encodes methyltransferase, putative papain-like cysteine protease, helicase, and replicase enzymes, but its domain is not characterized. Recently, Paliwal et al. (2014) made some attempts in this direction. On the basis of sequence analysis, these authors found that the folded form of pORF1 had protease activity in gelatin zymography. Cleavage sites were observed in pORF1 as well as pORF2, which were alanine15/isoleucine16 and alanine1364/valine1365 in pORF1 and leucine197/valine198 in pORF2. Specificity of these cleavage sites was validated by site-directed mutagenesis. Characterization of the HEV protease, carried out using 12 inhibitors, indicated this protein as a chymotrypsin-like protease, because out of these 12 inhibitors, the two inhibitors-chymostatin and phenylmethylsulfonyl fluoride, which are potent chymotrypsin inhibitors, were found to be the most efficient inhibitors. The specificity was further confirmed by cleavage of the chymotrypsin-specific fluorogenic peptide N-succinyl-Leu-Leu-Val-Tyr-7-amido-4-methylcoumarin. Mutational analysis of the conserved serine/cysteine/histidine residues by Paliwal et al. (2014) suggested that H443 and C472/C481/C483 were possibly the active site residues. However, the nature and activity of a functional HEV protease, which may be used as potential target for drug design, has not been so far well characterized experimentally. By homology analysis of 36 HEV strains, Paliwal et al. (2014) showed that N-terminus of the protease domain (aa 440–520) across the strains are conserved. The study of crystal structure of HEV protease and the development of its specific inhibitors may be of great value for anti-HEV therapy.

5. CONCLUSIONS

The chapter has discussed the structures and functions of different polymerases and proteases that are important for the replication of different hepatitis viruses and that constitute the important targets for the development of potent antihepatitis drugs. Conclusively, the most important drug targets in HAV are 3C protease and 2B protease, in HBV the three antigens, namely surface antigen (HBsAg), core antigen (HBcAg), and e-antigen (HBeAg), and HBV polymerase, in HCV the NS3/4A protease, NS4B and NS5A proteases, and NS5B polymerase, in HDV the S-HDAg, prenylation, and farnesylation, and in HEV, the HEV protease. The polymerase and proteases have been well exploited to develop the potent antihepatitis agents, most of which are now FDA approved.

REFERENCES

Abbas, Z., Afzal, R., 2013. Life cycle and pathogenesis of hepatitis D virus: a review. World J. Hepatol. 5, 666–675.

Agapov, E.V., Murray, C.L., Frolov, I., Qu, L., Myers, T.M., Rice, C.M., 2004. Uncleaved NS2–3 is required for production of infectious bovine viral diarrhea virus. J. Virol. 78, 2414–2425.

Allaire, M., Chernaia, M.M., Malcolm, B.A., James, M.N., 1994. Picornaviral 3C cysteine proteinases have a fold similar to chymotrypsin-like serine proteinases. Nature 369, 72–76.

Appel, N., Zayas, M., Miller, S., Krijnse-Locker, J., Schaller, T., Friebe, P., Kallis, S., Engel, U., Bartenschlager, R., 2008. Essential role of domain III of nonstructural protein 5A for hepatitis C virus infectious particle assembly. PLoS Pathog. 4, e1000035.

Ashfaq, U.A., Jaued, T., Rehman, S., Nawaz, Z., Riazuddin, S., 2011. An overview of HCV molecular biology, replication and immune responses. Virol. J. 8, 161–171.

Barbato, G., Cicero, D.O., M. Nardi, M.C., Steinkühler, C., Cortese, R., Francesco, R.D., Bazzo, R., 1999. The solution structure of the N-terminal proteinase domain of the hepatitis C virus (HCV) NS3 protein provides new insights into its activation and catalytic mechanism. J. Mol. Biol. 289 (2), 371–384.

Bartenschlager, R., Ahlborn-Laake, L., Mous, J., Jacobsen, H., 1994. Kinetic and structural analyses of hepatitis C virus polyprotein processing. J. Virol. 68, 5045–5055.

Beaulieu, P.L., 2007. Non-nucleoside inhibitors of the HCV NS5B polymerase: progress in the discovery and development of novel agents for the treatment of HCV infections. Curr. Opin. Invest. Drugs 8, 614–634.

Behrens, S.E., Tomei, L., De Francesco, R., 1996. Identification and properties of the RNA-dependent RNA polymerase of hepatitis C virus. EMBO J. 15, 12–22.

Belema, M., Nguyen, V.N., Bachand, C., Deon, D.H., Goodrich, J.T., James, C.A., Lavoie, R., Lopez, O.D., Martel, A., Romine, J.L., Ruediger, E.H., Snyder, L.B., St. Laurent, D.R., Yang, F., Zhu, J., Wong, H.S., Langley, D.R., Adams, S.P., Cantor, G.H., Chimalakonda, A., Fura, A., Johnson, B.M., Knipe, J.O., Parker, D.D., Santone, K.S., Fridel, R.A., Lemm, J.A., O'Boyle, D.R., Colonno, R.J., Gao, M., Meanwell, N.A., Hamann, L.G., 2014. Hepatitis C virus NS5A replication complex inhibitors: the discovery of Daclatasvir. J. Med. Chem. 57, 2013–2032.

Bergmann, E.M., Mosimann, S.C., Chernaia, M.M., Malcolm, B.A., James, M.N., 1997. The refined crystal structure of the 3C gene product from hepatitis A virus: specific proteinase activity and RNA recognition. J. Virol. 71, 2436–2448.

Biswal, B.K., Cherney, M.M., Wang, M., Chan, L., Yannopoulos, C.G., Bilimoria, D., Nicolas, O., Bedard, J., James, M.N.G., 2005. Crystal structures of the RNA-dependent RNA polymerase genotype 2a of hepatitis C virus reveal two conformations and suggest mechanisms of inhibition by non-nucleoside inhibitors. J. Biol. Chem. 280, 18202–18210.

Birtley, J.R., Knox, S.R., Jaulent, A.M., Brick, P., Leatherbarrow, R.J., Curry, S., 2005. Crystal structure of foot-and-mouth disease virus 3C protease: new insights into catalytic mechanism and cleavage specificity. J. Biol. Chem. 280, 11520–11527.

Bressanelli, S., Tomei, L., Rey, F.A., De Francesco, R., 2002. Structural analysis of the hepatitis C virus RNA polymerase in complex with ribonucleotides. J. Virol. 76, 3482–3492.

Cannalire, R., Barreca, M.L., Manfroni, G., Cecchetti, V., 2016. A journey around the medicinal chemistry of hepatitis C virus inhibitors targeting NS4B: from target topreclinical drug candidates. J. Med. Chem. 59, 16–41.

Cao, T., Meuleman, P., Desombere, I., Sällberg, M., Leroux-Roels, G., 2001. In vivo inhibition of anti-hepatitis B virus core antigen (HBcAg) immunoglobulin G production by HBcAg-specific CD4(+) Th1-type T-cell clones in a hu-PBL-NOD/SCID mouse model. J. Virol. 75, 11449–11456.

Cheetham, G.M., Steitz, T.A., 2000. Insights into transcription: structure and function of single-subunit DNA-dependent RNA polymerases. Curr. Opin. Struct. Biol. 10, 117–123.

Cho, N.J., Dvory-Sobol, H., Lee, C., Cho, S.J., Bryson, P., Masek, M., et al., 2010. Identification of a class of HCV inhibitors directed against the nonstructural protein NS4B. Sci. Translat. Med. 2, 15–16.

Choi, C.H., 2012. Viral polymerases. Adv. Exp. Med. Biol. 726, 267–304.

Chung, K.M., Liszewski, M., Kathryn, L.M., Grant, N., Davis, A.E., Townsend, R., Reid, T.R., Fremont, D.H., Atkinson, J.P., Diamond, M.S., 2006. West Nile virus nonstructural protein NS1 inhibits complement activation by binding the regulatory protein factor H. Proc. Natl. Acad. Sci. U.S.A. 103, 19111–19116.

Clark, D.N., Hu, J., 2015. Unveiling the roles of HBV polymerase for new antiviral strategies. Future Virol. 10, 283–295.

Coburn, C.A., Meinke, P.T., Chang, W., Fandozzi, C.A., Graham, D.J., Hu, B., Huang, Q., Kargman, S., Kozlowski, J., Liu, R., McCauley, J.A., Nomeir, A.A., Soll, R.M., Vacca, J.P., Wang, D., Wu, H., Zhong, B., Olsen, D.B., Ludmerer, S.W., 2013. Discovery of MK-8742: an HCV NS5A inhibitor with broad genotype activity. ChemMedChem 8, 1930–1940.

Cun, W., Jiang, J., Luo, G., 2010. The C-terminal alpha-helix domain of apolipoprotein E is required for interaction with nonstructural protein 5A and assembly of hepatitis C virus. J. Virol. 84, 11532–11541.

Das, D., 2017. Design and development of HCV NS5B polymerase inhibitors. In: Gupta, S.P. (Ed.), Viral Proteases and Their Inhibitors. Elsevier Academic, pp. 189–219.

Das, K., Xiong, X., Yang, H., Westland, C.E., Gibbs, C.S., Sarafianos, S.G., Arnold, E., 2001. Molecular modeling and biochemical characterization reveal the mechanism of hepatitis B virus polymerase resistance to lamivudine (3TC) and emtricitabine (FTC). J. Virol. 75, 4771–4779.

Davis, M.G., Wilson, J.E., VanDraanen, N.A., Miller, W.H., Freeman, G.A., Daluge, S.M., Boyd, F.L., Aulabaugh, A.E., Painter, G.R., Boone, L.R., 1996. DNA polymerase activity of hepatitis B virus particles: differential inhibition by L-enantiomers of nucleotide analogs. Antiviral Res. 30, 133–145.

De Jong, A.S., de Mattia, F., Van Dommelen, M.M., Lanke, K., Melchers, W.J., Willems, P.H., van Kuppeveld, F.J., 2008. Functional analysis of picornavirus 2B proteins: effects on calcium homeostasis and intracellular protein trafficking. J. Virol. 82, 3782–3790.

Debing, Y., Neyts, J., Thibaut, H.J., 2014. Molecular biology and inhibitors of hepatitis A virus. Med. Res. Rev. 34, 895–917.

Delang, L., Coelmont, L., Neyts, J., 2010. Antiviral therapy for hepatitis C virus: beyond the standard of care. Viruses 2, 826–866.

Di Marco, S., Rizzi, M., Volpari, C., Walsh, M.A., Narjes, F., Colarusso, S., De Francesco, R., Matassa, V.G., Sollazzo, M., 2000. Inhibition of the hepatitis C virus NS3/4A protease. The crystal structures of two protease-inhibitor complexes. J. Biol. Chem. 275, 7152–7157.
Di Marco, S., Volpari, C., Tomei, L., Altamura, S., Harper, S., Narjes, F., Koch, U., Rowley, M., De Francesco, R., Migliaccio, G., Carfi, A., 2005. Interdomain communication in hepatitis C virus polymerase abolished by small-molecule inhibitors bound to a novel allosteric site. J. Biol. Chem. 280, 29765–29770.
Doublie, S., Sawaya, M.R., Ellenberger, T., 1999. An open and closed case for all polymerases. Structure 7, R31–R35.
Doublie, S., Tabor, S., Long, A.M., Richardson, C.C., Ellenberger, T., 1998. Crystal structure of a bacteriophage T7 DNA replication complex at 2.2 Å resolution. Nature 391, 251–258.
Dvory-Sobol, H., Pang, P.S., Glenn, J.S., 2010. The future of HCV therapy: NS4B as an antiviral target. Viruses 2, 2481–2492.
Egger, D., Wolk, B., Gosert, R., Bianchi, L., Blum, H.E., Moradpour, D., Bienz, K., 2002. Expression of hepatitis C virus proteins induces distinct membrane alterations including a candidate viral replication complex. J. Virol. 76, 5974–5984.
Einav, S., Elazar, M., Danieli, T., Glenn, J.S., 2004. A nucleotide binding motif in hepatitis C virus (HCV) NS4B mediates HCV RNA replication. J. Virol. 78, 11288–11295.
Elazar, M., Liu, P., Rice, C.M., Glenn, J.S., 2004. An N-terminal amphipathic helix in hepatitis C virus (HCV) NS4B mediates membrane association, correct localization of replication complex proteins, and HCV RNA replication. J. Virol. 78, 11393–11400.
Evans, M.J., Rice, C.M., Goff, S.P., 2004. Phosphorylation of hepatitis C virus nonstructural protein 5A modulates its protein interactions and viral RNA replication. Proc. Natl. Acad. Sci. U.S.A. 101, 13038–13043.
Failla, C., Tomei, L., De Francesco, R., 1994. Both NS3 and NS4A are required for proteolytic processing of hepatitis C virus nonstructural proteins. J. Virol. 68, 3753–3760.
Flanegan, J.B., Petterson, R.F., Ambros, V., Hewlett, M.J., Baltimore, D., 1977. Covalent linkage of a protein to a defined nucleotide sequence at the 5′-terminus of virion and replicative intermediate RNAs of poliovirus. Proc. Natl. Acad. Sci. U.S.A. 74, 961–965.
Foster, T.L., Gallay, P., Stonehouse, N.J., Harris, M., 2011a. Cyclophilin A interacts with domain II of hepatitis C virus NS5A and stimulates RNA binding in an isomerase-dependent manner. J. Virol. 85, 7460–7464.
Foster, T.L., Thompson, G.S., Kalverda, A.P., Kankanala, J., Bentham, M., Wetherill, L.F., Thompson, J., Barker, A.M., Clarke, D., Noerenberg, M., Pearson, A.R., Rowlands, D.J., Homans, S.W., Harris, M., Foster, R., Griffin, S., 2014. Structure-guided design affirms inhibitors of hepatitis C virus p7 as a viable class of antivirals targeting virion release. Hepatology 59, 408–422.
Foster, T.L., Verow, M., Wozniak, A.L., Bentham, M.J., Thompson, J., Atkins, E., Weinman, S.A., Fishwick, C., Foster, R., Harris, R., Griffin, S., 2011b. Resistance mutations define specific antiviral effects for inhibitors of the hepatitis C virus p7 ion channel. Hepatology 54, 79–90.
Fridell, R.A., Qiu, D., Valera, L., Wang, C., Rose, R.E., Gao, M., 2011. Distinct functions of NS5A in hepatitis C virus RNA replication uncovered by studies with the NS5A inhibitor BMS-790052. J. Virol. 85, 7312–7320.
Gallinari, P., Brennan, D., Nardi, C., Brunetti, M., Tomei, L., Steinkühler, C., de Francesco, R., 1998. Multiple enzymatic activities associated with recombinant NS3 protein of hepatitis C virus. J. Virol. 72, 6758–6769.
Gao, M., Nettles, R.E., Belema, M., Snyder, L.B., Nguyen, V.N., Fridell, R.A., Serrano-Wu, M.H., Langley, D.R., Sun, J.-H., O'Boyle II, D.R., Lemm, J.A.,

Wang, C., Knipe, J.O., Chien, C., Colonno, R.J., Grasela, D.M., Meanwell, N.A., Hamman, L.G., 2010. Chemical genetics strategy identifies an HCV NS5A inhibitor with a potent clinical effect. Nature 465, 96–100.
Garriga, D., Vives-Adrián, L., Buxaderas, M., Ferreira-da-Silva, F., Almeida, B., Macedo-Ribeiro, S., Pereira, P.J., Verdaguer, N., 2011. Cloning, purification and preliminary crystallographic studies of the 2AB protein from hepatitis A virus. Acta Crystallogr. Sec. F Struct. Biol. Crystallogr. Comm. 67, 1224–1227.
Gervais, C., Do, F., Cantin, A., Kukolj, G., White, P.W., Gauthier, A., Vaillancourt, F.H., 2011. Development and validation of a high-throughput screening assay for the hepatitis C virus p7 viroporin. J. Biomol. Screen 16, 363–369.
Gervelmeyer, A., Nielsen, M.S., Chemnitz-Frey, L., Sckerl, H., Damberg, E., Molbak, K., 2006. An outbreak of hepatitis A among children and adults in Denmark, August 2002 to February 2003. Epidemiol. Infect. 134, 485–491.
Goodfellow, I., Chaudhry, Y., Gioldasi, I., Gerondopoulos, A., Natoni, A., Labrie, L., Laliberte, J.F., Roberts, L., 2005. Calicivirus translation initiation requires an interaction between vpg and eif 4 e. EMBO Rep. 6 (10), 968–972.
Gosert, R., Cassinotti, P., Seigl, G., Weitz, M., 1996. Identification of hepatitis non-structural P protein 2B and its release by the major viral protease 3C. J. Gen. Virol. 77, 247–255.
Gosert, R., Egger, D., Lohmann, D., Bartenschlager, R., Blum, H.E., Bienz, K., Moradpour, D., 2003. Identification of the hepatitis C virus RNA replication complex in Huh-7 cells harbouring subgenomic replicons. J. Virol. 77, 5487–5492.
Gouttenoire, J., Castet, V., Montserret, R., Arora, N., Raussens, V., Ruysschaert, J.M., et al., 2009. Identification of a novel determinant for membrane association in hepatitis C virus nonstructural protein 4B. J. Virol. 83, 6257–6268.
Gouttenoire, J., Penin, F., Moradpour, D., 2010a. Hepatitis C virus nonstructural protein 4B: a journey into unexplored territory. Rev. Med. Virol. 20, 117–129.
Gouttenoire, J., Roingeard, P., Penin, F., Moradpour, D., 2010b. Amphipathic alpha-helix AH2 is a major determinant for the oligomerization of hepatitis C virus nonstructural protein 4B. J. Virol. 84, 12529–12537.
Graff, J., Richards, O.C., Swiderek, K.M., Davis, M.T., Rusnak, F., Harmon, S.A., Jia, X.Y., Summers, D.F., Ehrenfeld, E., 1999. Hepatitis A virus capsid protein VP1 has a heterogeneous C terminus. J. Virol. 73, 6015–6023.
Greco-Stewart, V., Pelchat, M., 2010. Interaction of host cellular proteins with components of the hepatitis delta virus. Viruses 2, 189–212.
Hauri, A.M., Fischer, E., Fitzenberger, J., Uphoff, H., Koenig, C., 2006. Active immunisation during an outbreak of hepatitis A in a German day-care centre. Vaccine 24, 5684–5689.
Hedstrom, L., 2002. Serine protease mechanism and specificity. Chem. Rev. 102, 4501–4524.
Hu, J., Anselmo, D., 2000. In vitro reconstitution of a functional duck hepatitis B virus reverse transcriptase: posttranslational activation by Hsp90. J. Virol. 74, 11447–11455.
Hu, J., Flores, D., Toft, D., Wang, X., Nguyen, D., 2004. Requirement of heat shock protein 90 for human hepatitis B virus reverse transcriptase function. J. Virol. 78, 13122–13131.
Hu, J., Toft, D.O., Seeger, C., 1997. Hepadnavirus assembly and reverse transcription require a multi-component chaperone complex which is incorporated into nucleocapsids. EMBO J. 16, 59–68.
Hu, L., Crawford, S.E., Hyser, J.M., Estes, M.K., Venkataram Prasad, B.V., 2012. Rotavirus non-structural proteins: structure and function. Curr. Opin. Virol. 2, 380–388.
Huang, Z., Murray, M.G., Secrist, J.A., 2006. Development of therapeutics for chronic HCV infection. Antiviral Res. 71, 351–362.

Ishii, K., Tanaka, Y., Yap, C.C., Aizaki, H., Matsuura, Y., Miyamura, T., 1999. Expression of hepatitis C virus NS5B protein: characterization of its RNA polymerase activity and RNA binding. Hepatology 29, 1227–1235.

Jecht, M., Probst, C., Gauss-Muller, V., 1998. Membrane permeability induced by hepatitis A virus proteins 2B and 2BC and proteolytic processing of HAV 2BC. Virology 252, 218–227.

Jewell, D.A., Swietnicki, W., Dunn, B.M., Malcolm's, B.A., 1992. Hepatitis A virus 3C proteinase substrate specificity. Biochemistry 31, 7862–7869.

Jia, X.Y., Summers, D.F., Ehrenfeld, E., 1993. Primary cleavage of the HAV capsid protein precursor in the middle of the proposed 2Å coding region. Virology 193, 515–519.

Johansson, E., Dixon, N., 2013. Replicative DNA polymerases. Cold Spring Harb. Perspect. Biol. 5, a012799.

Jones, S.A., Boregowda, R., Spratt, T.E., Hu, J., 2012. In vitro epsilon RNA-dependent protein priming activity of human hepatitis B virus polymerase. J. Virol. 86, 5134–5150.

Kang, L., Pan, J., Wu, J., Hu, J., Sun, Q., Tang, J., 2015. Anti-HBV drugs: progress, unmet needs, and new hope. Viruses 7, 4960–4977.

Kiefer, J.R., Mao, C., Braman, J.C., Beese, L.S., 1998. Visualizing DNA replication in a catalytically active Bacillus DNA polymerase crystal. Nature 391, 304–307.

Kim, J.L., Morgenstern, K.A., Lin, C., Fox, T., Dwyer, M.D., Landro, J.A., Chambers, S.P., Markland, W., Lepre, C.A., O'Malley, E.T., Harbeson, S.L., Rice, C.M., Murcko, M.A., Caron, P.R., Thomson, J.A., 1996. Crystal structure of the hepatitis C virus NS3 Protease domain complexed with a synthetic NS4A cofactor peptide. Cell 87 (2), 343–355.

Kimura, T., Rokuhara, A., Matsumoto, A., Yaqi, S., Tanaka, E., Kiyosawa, K., Maki, N., 2003. New enzyme immunoassay for detection of hepatitis B virus core antigen (HBcAg) and relation between levels of HBcAg and HBV DNA. J. Clin. Microbiol. 41, 1901–1906.

Koch, U., Narjes, F., 2007. Recent progress in the development of inhibitors of the hepatitis C virus RNA-dependent RNA polymerase. Curr. Top. Med. Chem. 7, 1302–1329.

Kolykhalov, A.A., Feinstone, S.M., Rice, C.M., 1996. Identification of a highly conserved sequence element at the 30 terminus of hepatitis C virus genome RNA. J. Virol. 70, 3363–3371.

Kolykhalov, A.A., Mihalik, K., Feinstone, S.M., Rice, C.M., 2000. Hepatitis C virus-encoded enzymatic activities and conserved RNA elements in the 3′ nontranslated region are essential for virus replication in vivo. J. Virol. 74, 2046–2051.

Kramer, M., Halleran, D., Rahman, M., Iqbal, M., Anwar, M.I., Sabet, S., Ackad, E., Yousef, M., 2014. Comparative molecular dynamics simulation of hepatitis C virus NS3/4A protease (genotypes 1b, 3a and 4a) predicts conformational instability of the catalytic triad in drug resistant strains. PLOS ONE 9, e104425.

Kräusslich, H.G., Wimmer, E., 1988. Viral proteinases. Annu. Rev. Biochem. 57, 701–754.

Lackner, T., Muller, A., Pankraz, A., Becher, P., Thiel, H.J., Gorbalenya, A.E., Tautz, N., 2004. Temporal modulation of an autoprotease is crucial for replication and pathogenicity of an RNA virus. J. Virol. 78, 10765–10775.

Lall, M.S., Jain, R.P., Vederas, J.C., 2004. Inhibitors of 3C cysteine proteinases from Picornaviridae. Curr. Topics Med. Chem. 4, 1239–1253.

Lam, A.M.I., Frick, D.N., 2006. Hepatitis C virus subgenomic replicon requires an active NS3 RNA helicase. J. Virol. 80, 404–411.

Lawson, M.A., Semler, B.L., 1992. Alternate poliovirus nonstructural protein processing cascades generated by primary sites of 3C proteinase cleavage. Virology 191, 309–320.

Lee, Y.F., Nomoto, A., Detjen, B.M., Wimmer, E., 1977. The genome-linked protein of picornaviruses: a protein covalently linked to poliovirus genome RNA. Proc. Natl. Acad. Sci. U.S.A. 74, 59–63.
Leveque, V.J., Johnson, R.B., Parsons, S., Ren, J., Xie, C., Zhang, F., Wang, Q.M., 2003. Identification of a C-terminal regulatory motif in hepatitis C virus RNA-dependent RNA polymerase: structural and biochemical analysis. J. Virol. 77, 9020–9028.
Li, S., Yu, X., Guo, Y., Kong, L., 2012. Interaction networks of hepatitis C virus NS4B: implications for antiviral therapy. Cell Microbiol. 14, 994–1002.
Li, Y., Korolev, S., Waksman, G., 1998. Crystal structures of open and closed forms of binary and ternary complexes of the large fragment of Thermus aquaticus DNA polymerase I: structural basis for nucleotide incorporation. EMBO J. 17, 7514–7525.
Li, Y.J., Stallcup, M.R., Lai, M.M., 2004. Hepatitis delta virus antigen is methylated at arginine residues, and methylation regulates subcellular localization and RNA replication. J. Virol. 78, 13325–13334.
Lin, C., 2006. HCV NS3-4A serine protease. In: Tan, S.L. (Ed.), Hepatitis C Viruses: Genomes and Molecular Biology Horizon Bioscience, pp. 163–206.
Lindenbach, B.D., Rice, C.M., 2005. Unravelling hepatitis C virus replication from genome to function. Nature 436, 933–938.
Link, J.O., Taylor, J.G., Xu, L., Mitchell, M., Guo, H., Liu, H., Kato, D., Kirschberg, T., Sun, J., Squires, N., Parrish, J., Keller, T., Yang, Z., Yang, C., Matles, M., Wang, Y., Wang, K., Cheng, G., Tian, Y., Mogalian, E., Mondou, E., Cornpropst, M., Perry, J., Desai, M.C., 2014. Discovery of ledipasvir (GS-5885): a potent, once daily oral NS5A inhibitor for the treatment of hepatitis C virus infection. J. Med. Chem. 57, 2033–2046.
Loh, E., Loeb, L.A., 2005. Mutability of DNA polymerase I: implications for the creation of mutant DNA polymerases. DNA Repairs 4, 1390–1398.
Love, R.A., Parge, H.E., Yu, X., Hickey, M.J., Diehl, W., Gao, J., Wriggers, H., Ekker, A., Wang, L., Thomson, J.A., Dragovich, P.S., Fuhrman, S.A., 2003. Crystallographic identification of a noncompetitive inhibitor binding site on the hepatitis C virus NS5B RNA polymerase enzyme. J. Virol. 77, 7575–7581.
Maltese, W.A., 1990. Posttranslational modification of proteins by isoprenoids in mammalian cells. FASEB (Fed. Am. Soc. Exp. Biol.) J. 4, 3319–3328.
Martin, A., Escriou, N., Chao, S.-F., Girard, M., Lemon, S.M., Wychowski, C., 1995. Identification and site-directed mutagenesis of the primary (2A/2B) cleavage site of the hepatitis A virus polyprotein: functional impact on the infectivity of HAV RNA transcripts. Virology 213, 213–222.
Matthews, D.A., Smith, W.W., Ferre, R.A., Condon, B., Budahazi, G., Sisson, W., Villafranca, J.E., Janson, C.A., McElroy, H.E., Gribskov, C.L., Worland, S., 1994. Structure of human rhinovirus 3C protease reveals a trypsin-like polypeptide fold, RNA-binding site, and means for cleaving precursor polyprotein. Cell 77, 761–777.
Mayhoub, A.S., 2012. Hepatitis C RNA-dependent RNA polymerase inhibitors: a review of structure–activity and resistance relationships; different scaffolds and mutations. Bioorg. Med. Chem. 20, 3150–3161.
Moradpour, D., Gosert, R., Egger, D., Penin, F., Blum, H.E., Bienz, K., 2003. Membrane association of hepatitis C virus non-structural proteins and identification of the membrane alteration that harbors the viral replication complex. Antiviral Res. 60, 103–109.
Moradpour, D., Penin, F., Rice, C.M., 2007. Replication of hepatitis C virus. Nature Rev. Microbiol. 5, 453–463.
Mosimann, S.C., Cherney, M.M., Sia, S., Plotch, S., James, M.N., 1997. Refined X-ray crystallographic structure of the poliovirus 3C gene product. J. Mol. Biol. 273, 1032–1047.

Neddermann, P., Quintavalle, M., Di Pietro, C., Clementi, A., Cerretani, M., Altamura, S., Bartholomew, L., De Francesco, R., 2004. Reduction of hepatitis C virus NS5A hyperphosphorylation by selective inhibition of cellular kinases activates viral RNA replication in cell culture. J. Virol. 78, 13306–13314.

Nieva, J.L., Carrasco, L., 2015. Viroporins: structures and functions beyond cell membrane permeabilization. Viruses 7, 5169–5171.

Nieva, J.L., Madan, V., Carrasco, L., 2012. Viroporins: structure and biological functions. Nature Rev. Microbiol. 10, 563–574.

Novelli, G., D'Apice, M.R., 2012. Protein farnesylation and disease. J. Inherit Metabol. Dis. 35, 917–926.

Ollis, D.L., Kline, C., Steitz, T.A., 1985. Domain of *E. coli* DNA polymerase Ishowing sequence homology to T7 DNA polymerase. Nature 313, 818–819.

OuYang, B., Chou, J.J., 2014. The minimalist architectures of viroporins and their therapeutic implications. Biochim. Biophys. Acta 1838, 1058–1067.

Paliwal, D., Panda, S.K., Kapur, N., Varma, S.P.K., Durgapal, H., 2014. Hepatitis E virus (HEV) protease: a chymotrypsin like enzyme that processes both non-structural (pORF1) and capsid (pORF2) protein. J. Gen. Virol. 95, 1689–1700.

Patil, V.M., Gupta, S.P., Samanta, S., Masand, N., 2011. Current perspective of HCV NS5B inhibitors: A review. Curr. Med. Chem. 18, 5564–5597.

Paulmann, D., Magulski, T., Schwarz, R., Heitmann, L., Flehmig, B., Vallbracht, A., Dotzauer, A., 2008. Hepatitis A virus protein 2B suppresses beta interferon (IFN) gene transcription by interfering with IFN regulatory factor 3 activation. J. Gen. Virol. 89, 1593–1604.

Pawlotsky, J.-M., 2013. NS5A inhibitors in the treatment of hepatitis C. J. Hepatol 59, 375–382.

Pontrelli, G., Boccia, D., Di Renzi, M., Massari, M., Giugliano, F., Celentano, L.P., Taffon, S., Genovese, D., Scalise, F., Di Pasquale, S., Rapicetta, M., Croci, L., Salmaso, S., 2008. Epidemiological and virological characterization of a large community-wide outbreak of hepatitis A in southern Italy. Epidem. Infect. 136, 1027–1034.

Qin, W., Luo, H., Nomura, T., Hayashi, N., Yamashita, T., Murakami, S., 2002. Oligomeric interaction of hepatitis C virus NS5B is critical for catalytic activity of RNA-dependent RNA polymerase. J. Biol. Chem. 277, 2132–2137.

Quezada, E.M., Kane, C.M., 2009. The hepatitis C virus NS5A stimulates NS5B during in vitro RNA synthesis in a template specific manner. Open Biochem. J. 3, 39–48.

Raney, K.D., Sharma, S.D., Moustafa, I.M., Cameron, C.E., 2010. Hepatitis C virus non-structural protein 3 (HCV NS3): a multifunctional antiviral target. J. Biol. Chem. 285, 22725–22731.

Ryan, M.D., Flint, M., 1997. Virus-encoded proteinases of the picornaviral super-group. J. Gen. Virol. 78, 699–723.

Schaffer, F.L., Ehresmann, D.W., Fretz, M.K., Soergel, M.I., 1980. A protein, vpg, covalently linked to 36s calicivirusrna. J. Gen. Virol. 47, 215–220.

Schinazi, R.F., Bassit, L., Gavegnano, C., 2010. HCV drug discovery aimed at viral eradication. J. Viral Hepatol. 17, 77–90.

Schultheiss, T., Sommergruber, W., Kusov, Y., Gauss-Müller, V., 1995. Cleavage specificity of purified recombinant hepatitis A virus 3C proteinase on natural substrates. J. Virol. 69, 1727–1733.

Scott, C., Griffin, S., 2015. Viroporins: structure, function and potential as antiviral targets. J. Gen. Virol. 96, 2000–2027.

Seipelt, J., Guarne, A., Bergmann, E., James, M., Sommergruber, W., Fita, I., Skern, T., 1999. The structures of picornaviral proteinases. Virus Res. 62, 159–168.

Sharma, A., Gupta, S.P., Siddiqui, A.A., Sharma, N., 2017. HCV NS3/4A protease and its emerging inhibitors. J. Anal. Pharm. Res. 4, 1–9, 00108.
Sharma, A., Gupta, S.P., 2017. Fundamentals of viruses and their proteases. In: Gupta, S.P. (Ed.), Viral Proteases and Their Inhibitors. Elsevier, USA, pp. 1–24.
Sheridan, C., 2011. New Merck and Vertex drugs arise standard of care in hepatitis C. Nat. Biotech. 29, 553–554.
Shirota, Y., Luo, H., Qin, W., Kaneko, S., Yamashita, T., Kobayashi, K., Murakami, S., 2002. Hepatitis C virus (HCV) NS5A binds RNA-dependent RNA polymerase (RdRP) NS5B and modulates RNA-dependent RNA polymerase activity. J. Biol. Chem. 277, 11149–11155.
Shukla, A., Dey, D., Banerjee, K., Nain, A., Banerjee, M., 2015. The C-terminal region of the non-structural protein 2B from Hepatitis A Virus demonstrates lipid-specific viroporin-like activity. Sci. Rep. 5, 15884.
Slonczewski, J.L., Foster, J.W., 2013. Microbiology: An Evolving Science, third ed. W. W. Norton & Company, pp. 192–195.
Soriano, V., Peters, M.G., Zeuzem, S., 2009. New therapies for hepatitis C virus infection. Clin. Infect. Dis. 48, 313–320.
Stanway, G., Hughes, P.J., Mountford, R.C., Minor, P.D., Almond, J.W., 1984. The complete nucleotide sequence of a common cold virus: human rhinovirus 14. Nucleic Acids Res. 12, 7859–7875.
Steitz, T.A., 1999. DNA polymerases: structural diversity and common mechanisms. J. Biol. Chem. 274 (25), 17395–17398.
Tantillo, C., Ding, J., Jacobo-Molina, A., Nanni, R.G., Boyer, P.L., Hughes, S.H., Pauwels, R., Andries, K., Janssen, P.A., Arnold, E., 1994. Locations of anti-AIDS drug binding sites and resistance mutations in the three-dimensional structure of HIV-1 reverse transcriptase. Implications for mechanisms of drug inhibition and resistance. J. Mol. Biol. 243, 369–387.
Tavis, J.E., Cheng, X., Hu, Y., Totten, M., Cao, F., Michailidis, E., Aurora, R., Meyers, M.J., Jacobsen, J., Parniak, M.A., Sarafianos, S.G., 2013. The hepatitis B virus ribonuclease H is sensitive to inhibitors of the human immunodeficiency virus ribonuclease H and integrase enzymes. PLoS Pathog. 9, e1003125.
Tedesco, R., Shaw, A.N., Bambal, R., Chai, D., Concha, N.O., Darcy, M.G., Dhanak, D., Fitch, D.M., Gates, A., Gerhardt, W.G., Halegoua, D.L., Han, C., Hofmann, G.A., Johnston, V.K., Kaura, A.C., Liu, N., Keenan, R.M., Lin-Goerke, J., Sarisky, R.T., Wiggall, K.J., Zimmerman, M.N., Duffy, K.J., 2006. 3-(1,1-Dioxo-2H-(1,2,4)-benzothiadiazin-3-yl)-4-hydroxy-2(1H)-quinolinones, potent inhibitors of hepatitis C virus RNA-dependent RNA polymerase. J. Med. Chem. 49, 971–983.
Tellinghuisen, T.L., Foss, K.L., Treadaway, J.C., Rice, C.M., 2008. Identification of residues required for RNA replication in domains II and III of the hepatitis C virus NS5A protein. J. Virol. 82, 1073–1083.
Tsai, C.H., Lee, P.Y., Stollar, V., Li, M.L., 2006. Antiviral therapy targeting viral polymerase. Curr. Pharm. Des. 12, 1339–1355.
Uversky, V., Longhi, S., 2011. Flexible Viruses: Structural Disorder in Viral Proteins. Wiley, ISBN 9781118135549, p. 4.
van Hemert, F.J., Berkhout, B., Zaaijer, H.L., 2014. Differential binding of tenofovir and adefovir to reverse transcriptase of hepatitis B virus. PLoS One 9, e106324.
Vives-Adrián, L., Garriga, D., Buxaderas, M., Frag, J., Pereira, P.J.B., Macedo-Ribeiro, S., Verdaguera, N., 2015. Structural Basis for host membrane remodeling induced by protein 2B of hepatitis A virus. J. Virol. 89, 3648–3658.
Wang, M., Ng, K.K.-S., Cherney, M.M., Chan, L., Yannopoulos, C.G., Bedard, J., Morin, N., Nguyen-Ba, N., Alaoui-Ismaili, M.H., Bethell, R.C., James, M.N.G., 2003.

Non-nucleoside analogue inhibitors bind to an allosteric site on HCV NS5b polymerase. J. Biol. Chem. 278, 9489–9495.

Wang, Y.C., Huang, C.R., Chao, M., Lo, S.J., 2009. The C-terminal sequence of the large hepatitis delta antigen is variable but retains the ability to bind clathrin. Virol. J. 6, 31.

Welbourn, S., Pause, A., 2007. The hepatitis C virus NS2/3 protease. Curr. Issues Mol. Biol. 9, 63–70.

Welbourn, S., Green, R., Gamache, I., Dandache, S., Lohmann, V., Bartenschlager, R., Meerovitch, K., Pause, A., 2005. Hepatitis C virus NS2/3 processing is required for NS3 stability and viral RNA replication. J. Biol. Chem. 280, 29604–29611.

Xu, X., Thai, H., Kitrinos, K.M., Xia, G., Gaggar, A., Paulson, M., Ganova-Raeva, L., Khudyakov, Y., Lara, J., 2016. Modeling the functional state of the reverse transcriptase of hepatitis B virus and its application to probing drug-protein Interaction. BMC Bioinform. 17 (Suppl. 8), 280.

Yin, J., Bergmann, E.M., Cherney, M.M., Lall, M.S., Jain, R.P., Vederas, J.C., James, M.N.G., 2005. Dual modes of modification of Hepatitis A virus 3C protease by a serine-derived β-lactone: selective crystallization and formation of a functional catalytic triad in the active site. J. Mol. Biol. 354 (4), 854–871.

Yu, W., Coburn, C.A., Yang, D.-Y., Meinke, P.T., Wong, M., Rosenblum, S.B., Chen, K.X., Njoroge, G.F., Chen, L., Dwyer, M.P., Jiang, Y., Nair, A.G., Selyutin, O., Tong, L., Zeng, Q., Zhong, B., Ji, T., Hu, B., Agrawal, S., Xia, E., Zhai, Y., Liu, R., Kong, R., Ingravallo, P., Asante-Appiah, E., Nomeir, A., Fells, J., Kozlowski, J.A., 2016. Discovery of fused tricyclic core containing HCV NS5A inhibitors with pan-genotype activity. Bioorg. Med. Chem. Lett 26, 3158–3162.

FURTHER READING

DeGoey, D.A., Randolph, J.T., Liu, D., Pratt, J., Hutchins, C., Donner, P., Krueger, A.C., Matulenko, M., Patel, S., Motter, C.E., Nelson, L., Keddy, R., Tufano, M., Caspi, D.D., Krishnan, P., Mistry, N., Koev, G., Reisch, T.J., Mondal, R., Pilot-Matias, T., Gao, Y., Beno, D.A., Maring, C.J., Molla, A., Dumas, E., Campbell, A., Williams, L., Collins, C., Wagner, R., Kati, W.M., 2014. Discovery of ABT-267, a pan-genotypic inhibitor of HCV NS5B. J. Med. Chem. 57, 2047–2057.

CHAPTER 4

Hepatitis Therapy: Present Status

1. INTRODUCTION

Before we discuss hepatitis therapy, let us first define again acute and chronic hepatitis. Acute hepatitis refers to short-term infection that may last for less than 6 months, while chronic hepatitis refers to the infection that may last for more than 6 months and over time may progress to liver failure or liver cancer. Acute hepatitis can sometimes resolve on its own, or it may progress to chronic hepatitis, which needs severe treatment. Hepatitis refers to an inflammatory condition of the liver, which is located in the upper-right area of the abdomen and which performs many critical functions that affect the metabolic system, such as bile production, breakdown of carbohydrates, fats, and proteins, activation of enzymes, and synthesis of blood proteins. According to a Centers for Disease Control and Prevention (CDC) survey, approximately 4.4 million Americans are currently living with chronic hepatitis B and C. Of all kinds of hepatitis infections, hepatitis A is always acute. Though hepatitis E is also usually acute, it may be dangerous in pregnant women. Others (B, C, and D) are usually chronic. Symptoms of chronic infection may not occur till the damage affects liver function, while those of acute infection appear quickly, and these may be fatigue, flu-like symptoms, dark urine, pale stool, abdominal pain, loss of appetite, unexplained weight loss, signs of jaundice (yellow skin and eyes), etc. These signs and symptoms may be too subtle to notice in chronic hepatitis, as its development is very slow.

Now to refresh the memory about the characters and infectivity of hepatitis viruses, a brief of them is given next.

1.1 Hepatitis A Virus

This is a nonenveloped picornavirus having ssRNA of 7.5 kb. It has only one serotype and can infect children and young adults through fecal-oral

Studies on Hepatitis Viruses
ISBN 978-0-12-813330-9
https://doi.org/10.1016/B978-0-12-813330-9.00004-1

transmission. Poor sanitation is the greatest risk, so its risk is very low in developed countries like the United States, Europe, and Australia.

1.2 Hepatitis B Virus

This is an enveloped hepadnavirus having dsDNA of 3.2 kb. It has seven polypeptides and a surface antigen (HBsAg), a predominant spike protein. It has unusual genome replication that involves the enzyme reverse transcriptase.

1.3 Hepatitis C Virus

This is a flavivirus, with plus-strand RNA of 9.4 kb, which can be transmitted through blood product or sexual intercourse. Most acute infections are subclinical, but most become chronic. About 25,000 people die from hepatitis C virus (HCV) infection every year in the United States.

1.4 Hepatitis D Virus

This is not a virus but a viroid, the smallest infectious pathogen, composed of a short strand of circular ssRNA without any protein coat. Hepatitis D virus (HDV) has a minus-strand RNA of 1.7 kb and infects only in the presence of hepatitis B virus (HBV) infection.

1.5 Hepatitis E Virus

It possesses a plus-strand RNA of 7.6 kb and infects through oral-fecal transmission. Pregnant women are more sensitive to hepatitis E virus (HEV) infection.

Hepatitis virus infections in humans target the liver, leading to focal necrosis and jaundice.

HBV and HCV lead to hepatocellular carcinomas. HBV can also cause rash, arthritis, vasculitis, and glomerulonephritis. Fatality rates associated with hepatitis A to C are as follows:

Hep A: $<0.5\%$ (increases after age 40)

Hep B: 1%–2% (chronic in 5%–10% of infections)

Hep C: 0.5%–1% (chronic in 70%–90% of infections)

For hepatitis A, childhood vaccination is now routine, and hepatitis A virus (HAV) infection results in life-long immunity. For hepatitis B, vaccination is also routine, but the means of immune response is not very clear. Healthcare workers face higher risk. Nothing is known about the host response to hepatitis C. Hepatitis C infection can be treated with antiviral drugs like interferon and ribavirin.

2. TREATMENT OF HEPATITIS

The hepatitis treatment depends upon the type of hepatitis, e.g., hepatitis A usually does not require any treatment because it is a short-term illness. However, hepatitis A vaccine is available that can be given to children between ages 12–18 months. A vaccine for adults is also available that can be combined with the hepatitis B vaccine. Acute hepatitis does not require any specific treatment, but chronic cases are treated with antiviral medication, which is continued for several months or years. Hepatitis B can be prevented with vaccination, which requires a series of three vaccines to be completed over the first 6 months in newborns (CDC recommendation).

For hepatitis C, there is no vaccine. Both acute and chronic forms of hepatitis C are treated by antiviral medications. A combination of antiviral therapies is used for the treatment of chronic hepatitis C. A liver transplant may be required if people develop cirrhosis or liver disease as a result of chronic hepatitis C. For hepatitis D, no antiviral medications are developed, but a drug called alpha interferon can be used for it, which only shows improvement in 25%–30% of people. This will be discussed later. However, since hepatitis D develops only in the presence of hepatitis B, it can be prevented by vaccination for hepatitis B. Like hepatitis D, there is no antiviral therapy for hepatitis E. Hepatitis E is often acute and can resolve itself, but pregnant women with this infection require close monitoring and care.

As said, prevention is better than cure. Therefore some tips have been suggested to prevent hepatitis infection. If you are travelling to a developing country, you can avoid hepatitis A and E infections by taking precautions in using local water, ice, raw or undercooked seafood, raw fruits, and vegetables. Some preventive measures to safeguard from hepatitis B, C, and D are to not share drug needles, razors, or toothbrushes and not touching spilled blood. Additionally, one should also avoid unsafe sex because hepatitis B and C can also be contracted through sexual intercourse. People with chronic hepatitis B and C are encouraged to avoid alcohol because it can accelerate liver disease and failure.

3. HEPATITIS THERAPIES

3.1 Hepatitis A

As already discussed, hepatitis A treatment generally involves supportive care. However, attempts have been made to find antiviral drugs for hepatitis

A that may target any replication step in HAV. In a study, Kanda et al. (2015) noted that amantadine and interferon-lambda 1 (IL-29) inhibit HAV internal ribosomal entry site (IRES)-mediated translation and HAV replication. Additionally, some Janus kinase inhibitors were observed to inhibit La protein expression, HAV IRES activity, and HAV replication. Ubiquitous in eukaryotic cells, the La protein associates with the 3′ termini of many newly synthesized small RNAs. Attempts are being made to find more direct acting drugs, which will be discussed later in detail.

3.2 Hepatitis B and C

Direct-acting antivirals (DAAs) against HBV and HCV have been extensively studied. We cite here in Tables 4.1 and 4.2 the FDA-approved

Table 4.1 FDA approved anti-hepatitis B virus agents

Brand name	Generic name	Manufacturer	Indication
Baraclude	Entecavir	Bristol-Myers Squibb	Chronic hepatitis B virus infection with evidence of active viral replication
Epivir HBV	Lamivudine	GlaxoSmithKline	Chronic hepatitis B associated with hepatitis B viral replication and active liver inflammation
Hepsera	Adefovir Dipivoxil	Gilead Sciences	Chronic hepatitis B in patients about12 years of age
Intron A	Interferon Alpha-2b	Schering	Chronic hepatitis B in patients 1 year of age or older with compensated liver disease
Pegasys	Pegylated Interferon	Roche	Treatment of adult patients with HBeAg-positive and HBeAg-negative chronic hepatitis B who have compensated liver disease and evidence of viral replication and liver inflammation
Tyzeka	Telbivudine	Novartis	Chronic hepatitis B in adult patients with evidence of viral replication and either evidence of persistent elevations in serum aminotransferases (ALT or AST) or histologically active disease
Viread	Tenofovir	Gilead Sciences	Chronic hepatitis B in adults

Table 4.2 FDA approved anti-HCV agents

Brand name	Generic name	Manufacturer	Indication
CoPegus	Ribavirin	Roche	Use in combination with Pegasys or with Roferon for the treatment of adults with chronic HCV infection who have compensated liver disease and have not been previously treated with interferon alpha
Daklinza	Daclatasvir	Bristol-Myers Squibb Company	An NS5A replication complex inhibitor is indicated for use with sofosbuvir for the treatment of patients with chronic HCV genotype 3 infection. SVR rates are reduced in HCV genotype 3–infected patients with cirrhosis receiving this regimen. The recommended dosage of DAKLINZA is 60 mg, taken orally, once daily in combination with sofosbuvir for 12 weeks. DAKLINZA may be taken with or without food. The optimal duration of DAKLINZA and sofosbuvir for patients with cirrhosis has not been established
Epclusa	Sofosbuvir, Velpatasvir	Gilead	A fixed-dose combination of sofosbuvir, an HCV nucleotide analog NS5B polymerase inhibitor, and velpatasvir, an HCV NS5A inhibitor, indicated for the treatment of adult patients with chronic HCV genotype 1, 2, 3, 4, 5, or 6 infection, without cirrhosis or with compensated cirrhosis or with decompensated cirrhosis for use in combination with ribavirin
Harvoni	Ledipasvir/Sofosbuvir	Gilead	A fixed-dose combination of ledipasvir, an HCV NS5A inhibitor, and sofosbuvir, an HCV nucleotide analog NS5B polymerase inhibitor, indicated for the treatment of CHC genotype 1 infection in adults

Continued

Table 4.2 FDA approved anti-HCV agents—cont'd

Brand name	Generic name	Manufacturer	Indication
Incivek	Telaprevir	Vertex Pharmaceuticals	In combination with peginterferon alfa and ribavirin, for the treatment of genotype 1 CHC in adult patients with compensated liver disease, including cirrhosis, who are treatment-naïve or who have been previously treated with interferon-based treatment, including prior null responders, partial responders, and relapsers
Infergen	Interferon aphacon-1	Three Rivers Pharma	Treatment of CHC in patients 18 years of age or older with compensated liver disease who have anti-HCV serum antibodies and/or HCV RNA
Intron A	Interferon alpha-2b	Schering	Treatment of CHC in patients 18 years of age or older with compensated liver disease who have a history of blood or blood-product exposure and/or are HCV antibody positive
Olysio	Simeprevir	Janssen Pharmaceuticals	For the treatment of CHC genotype 1 infection as a component of a combination antiviral treatment regimen
Pegasys	Pegylated interferon	Roche	Treatment of adults with chronic HCV infection who have compensated liver disease and have not been previously treated with interferon alpha
Pegintron	Pegylated interferon alpha-2b	Schering	In combination with Rebetol, is indicated for the treatment of CHC in patients 3 years of age and older with compensated liver disease use alone for the treatment of CHC in patients with compensated liver disease previously untreated with interferon alpha and who are at least 18 years of age and intolerant to ribavirin

Rebetol	Ribavirin	Schering	Use in combination with Pegintron for treatment of CHC in patients with compensated liver disease who have not been previously treated with interferon alpha and are at least 18 years of age
Roferon	Interferon alpha-2a	Roche	Treatment of CHC in patients 18 years of age or older
Sovaldi	Sofosbuvir	Gilead Sciences	For the treatment of CHC infection as a component of a combination antiviral treatment regimen
Technivie	Ombitasvir, Paritaprevir, and Ritonavir	AbbVie Inc.	A fixed-dose combination containing ombitasvir, an HCV NS5A inhibitor; paritaprevir, an HCV NS3/4A protease inhibitor; and ritonavir, a CYP3A inhibitor. The product is indicated in combination with ribavirin for the treatment of patients with genotype 4 chronic HCV infection without cirrhosis
Victrelis	Boceprevir	Merck & Co.	Treatment of CHC genotype 1 infection, in combination with peginterferon alfa and ribavirin, in adult patients (18 years of age and older) with compensated liver disease, including cirrhosis, who are previously untreated or who have failed previous interferon and ribavirin therapy
Viekira Pak	Ombitasvir, Paritaprevir, and Ritonavir tablets copackaged with dasabuvir tablets	AbbVie Inc.	Use with or without ribavirin for the treatment of patients with genotype 1 chronic HCV infection including those with compensated cirrhosis
Zepatier	Elbasvir, Grazoprevir	Merck Sharp Dohme	A fixed-dose combination product containing elbasvir, an HCV NS5A inhibitor, and grazoprevir, an HCV NS3/4A protease inhibitor, indicated with or without ribavirin for treatment of chronic HCV genotypes 1 or 4 infection in adults

CHC, chronic hepatitis C; *HCV*, hepatitis C virus; *SVR*, sustained virologic response.

drugs that can be used to treat hepatitis B and C, respectively. Structures of some of the compounds of Tables 4.1 and 4.2 are given in Figs. 4.1 and 4.2, respectively. The lists may not be complete, but they cover the drugs approved up to the year 2017 (https://www.fda.gov/forpatients/illness/hepatitisbc/ucm408658.htm).

3.3 Hepatitis D

So far, there is no FDA-approved anti-HDV drug. Peginterferon alfa-2a (PEG-IFNα2a), which is an anti-HBV drug, has shown some anti-HDV activity (Bahcecioglu and Sahin, 2017; Petersen et al., 2016; Abeywickrama-Samarakoon et al., 2016). The use of this compound with or without adefovir, an anti-HBV drug, over 48 weeks was found to result in sustained HDV RNA clearance in approximately one-fourth of patients (Wedemeyer et al., 2011). In another study, PEG-IFN was found to

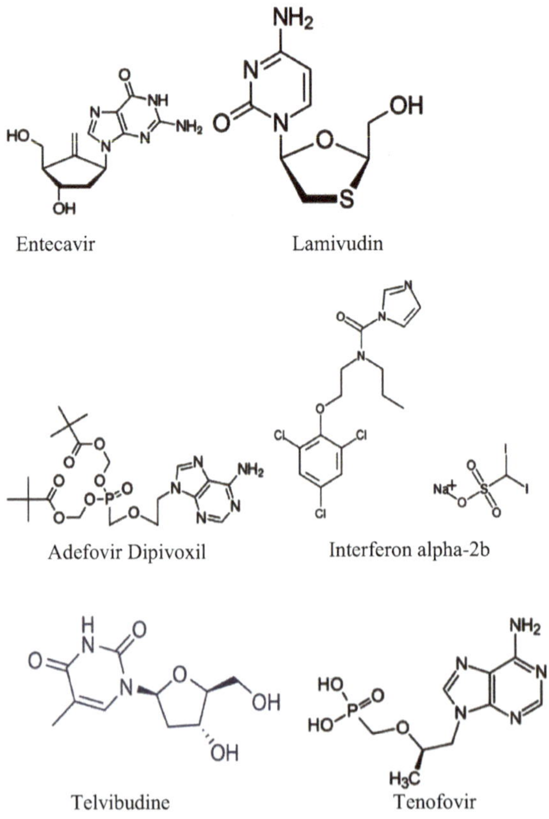

Figure 4.1 Structures of some of the compounds given Table 4.1.

achieve sustained virologic response (SVR) and remission in only 29.4% of patients (Samiullah et al., 2012). Thus, PEG-IFN has low rates of SVR and clinical improvement (Abeywickrama-Samarakoon et al., 2016). Interferon-based therapy, because of its low efficacy and serious side effects, has not been very successful (Bahcecioglu and Sahin, 2017), so attempts are

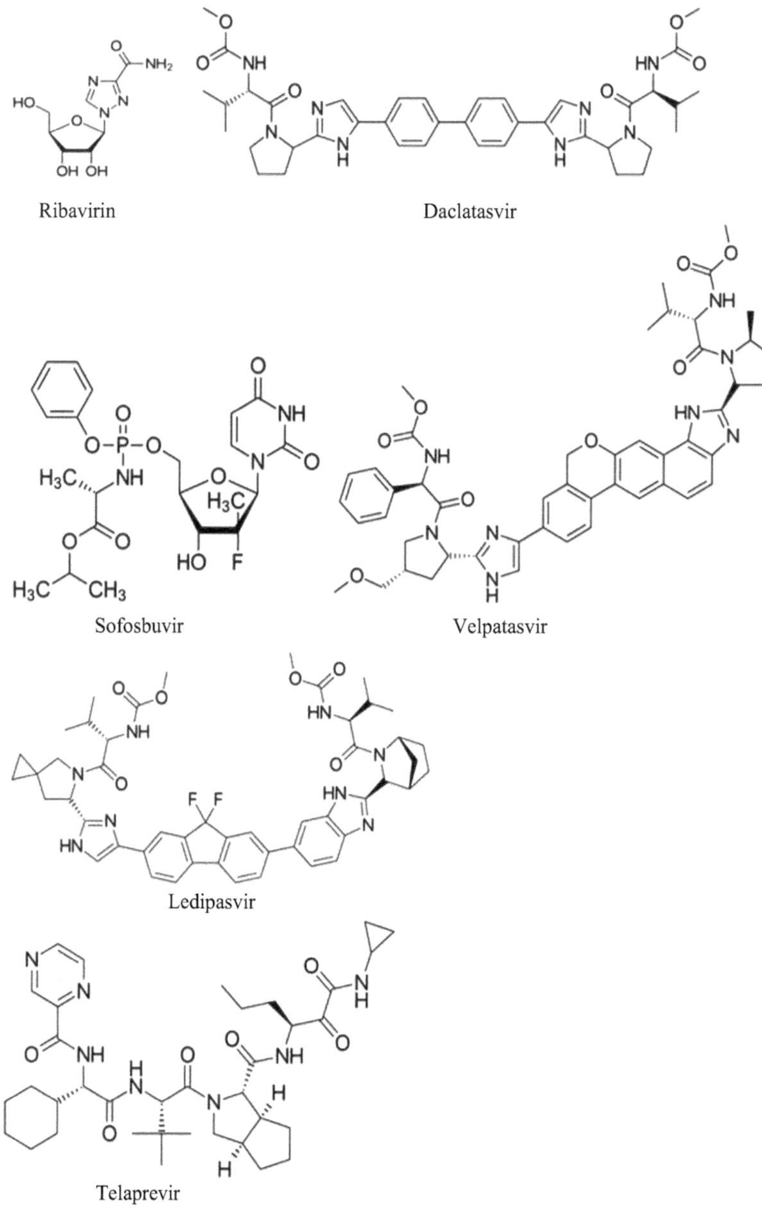

Figure 4.2 Structures of compounds given in Table 4.2.

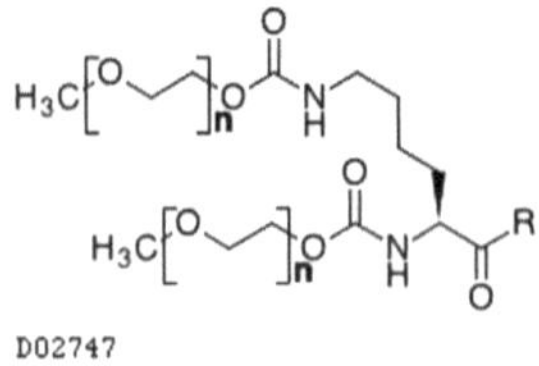

Pegasys (Peginterferon alfa-2a)

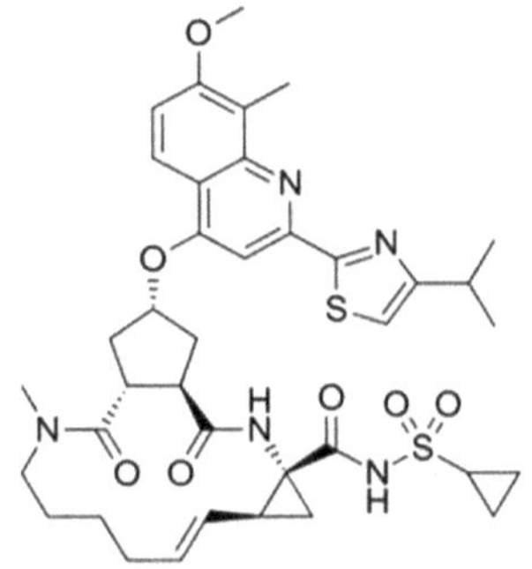

Simeprevir

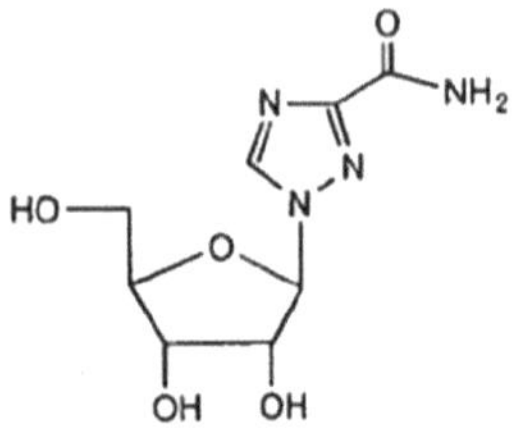

Pegintron

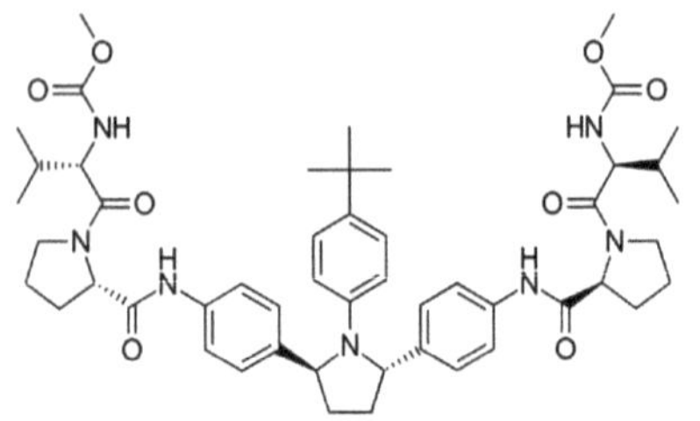

Omvitasvir

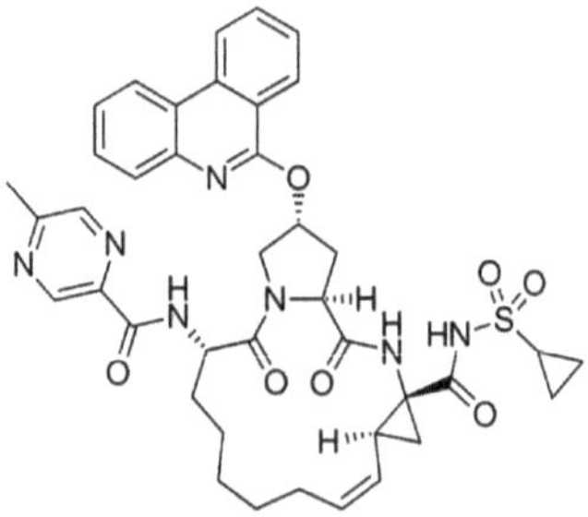

Paritaprevir

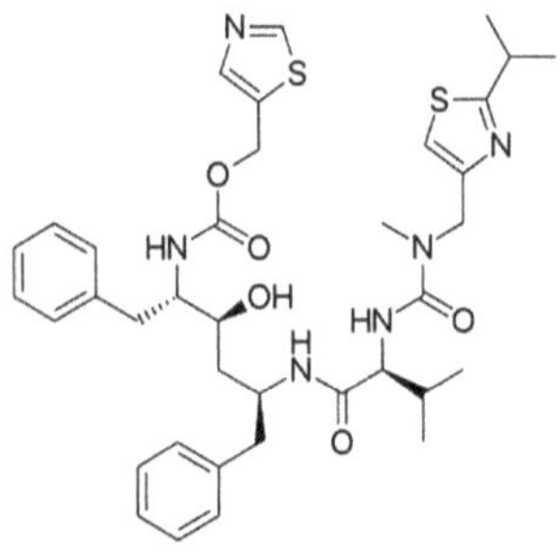

Ritonavir

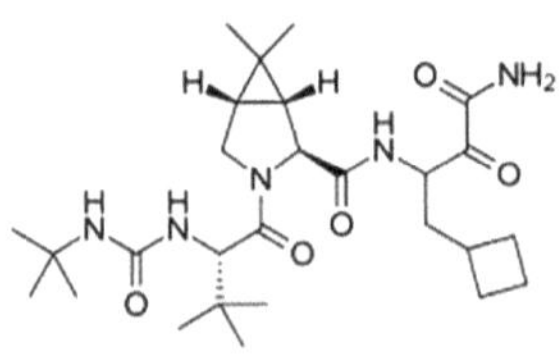

Boceprevir

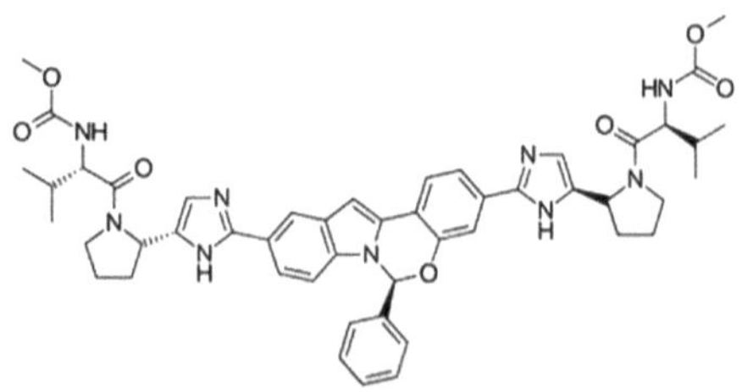

Elbasvir

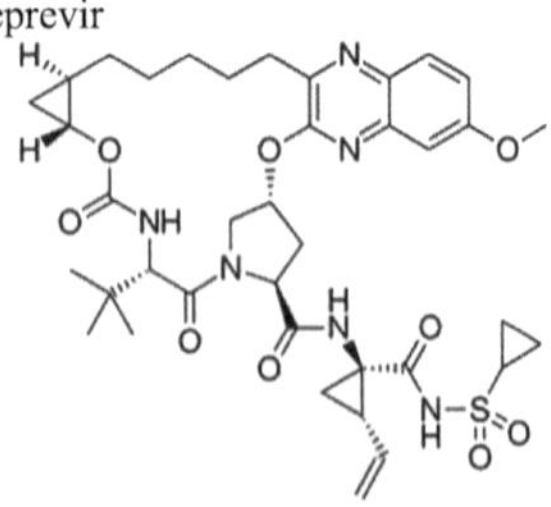

Grazoprevir

Figure 4.2 Cont'd

being made to find new drugs, targeting some important steps in the life cycle of HDV/HBV such as prenylation in HDV and HBsAg-release in HDV and HBV (Bahcecioglu and Sahin, 2017).

3.4 Hepatitis E

For hepatitis E, only prevention is the best cure. No effective antiviral therapy has been developed so far. In the name of antiviral therapy, a nucleotide analog, sofosbuvir, has been found to inhibit RNA replication of HEV genotype 3 in vitro. Also, this compound has been observed to have an additive antiviral effect in combination with ribavirin (Thi et al., 2016). Though this combination appeared promising, further studies are needed for the effect of this combination in chronic HEV infection, especially in patients in whom ribavirin monotherapy fails to achieve HEV elimination (Debing et al., 2015). Ribavirin monotherapy (600–1000 mg/day) for at least 3 months is practiced as the first treatment option for patients with chronic hepatitis E. Ribavirin monotherapy inhibits the replication of HEV in vivo and might induce a sustained virologic response in patients with chronic HEV infections (Kamar et al., 2010). In a patient with severe acute hepatitis E, significant improvement of liver enzymes and functions was reported when treated with ribavirin for 21 days (Gerolami et al., 2011). Further studies are required to determine the optimal duration of ribavirin therapy.

4. CONCLUSIONS

As far as hepatitis therapy is concerned, it can be conclusively said that it depends upon the type of hepatitis, e.g., hepatitis A usually does not require any treatment because it does not last long. Hepatitis B can be prevented with vaccination, and DAA drugs are also available for it, while both acute and chronic hepatitis C are treated by antiviral medications. For hepatitis D and E, no antiviral drugs have been developed. Lastly, it is said that prevention is better than cure, and those suffering from chronic hepatitis B and C are advised to keep away from alcohol.

REFERENCES

Abeywickrama-Samarakoon, N., Cortay, J.C., Deny, P., 2016. The hepatitis D satellite virus of hepatitis B virus: half-opening a new era to control viral infection? Curr. Opin. Infect. Dis. 29, 645–653.

Bahcecioglu, I.H., Sahin, A., 2017. Treatment of delta hepatitis: today and in the future - a review. Infect. Dis. 49, 241–250.

Debing, Y., Moradpour, D., Neyts, J., Gouttenoire, J., 2015. Update on hepatitis E virology: implication for critical practice. J. Hepatol. 15, 200–212.

Gerolami, R., Borentain, P., Raissouni, F., Motte, A., Solas, C., Colson, P., 2011. Treatment of severe acute hepatitis E by ribavirin. J. Clin. Virol. 52, 60–62.

Kamar, N., Rostaing, L., Abravanel, F., Garrouste, C., Lhomme, S., Esposito, L., Basse, G., Cointault, O., Ribes, D., Nogier, M.B., Alric, L., Peron, J.M., Izopet, J., 2010. Ribavirin therapy inhibits viral replication on patients with chronic hepatitis E virus infection. Gastroenterology 139, 1612–1618.

Kanda, T., Nakamoto, S., Wu, S., Nakamura, M., Jiang, X., Sasaki, R., Yokosuka, O., 2015. Direct-acting antivirals and host-targeting agents against the hepatitis A virus. J. Clin. Trans. Hepatol. 3, 205–210.

Petersen, J., Thompson, A.J., Levrero, M., 2016. Aiming for cure in HBV and HDV infection. J. Hepatol. 65, 835–848.

Samiullah, S., Bikharam, D., Nasreen, 2012. Treatment of chronic hepatitis delta virus with peg-interferon and factors that predict sustained viral response. World J. Gastroent. 18, 5793–5798.

Thi, V.L.D., Debing, Y., Wu, X., Rice, C.M., Neyts, J., Moradpour, D., Gouttenoire, J., 2016. Sofosbuvir inhibits hepatitis E virus replication in vitro and results in an additive effect when combined with ribavirin. Gastroenterology 150, 82–85.

Wedemeyer, H., CihanYurdaydìn, C., Dalekos, G.N., Erhardt, A., Çakaloğlu, Y.C., Değertekin, H., SelimGürel, M.D., Zeuzem, S., Zachou, K., Bozkaya, H., Koch, A., Bock, T., Dienes, H.P., Manns, M.P., 2011. Peginterferon plus Adefovir versus either drug alone for hepatitis delta. New Eng. J. Med. 364, 322–331.

CHAPTER 5

The Medicinal Chemistry of Antihepatitis Agents I: Anti-HAV, Anti-HDV, and Anti-HEV Agents

1. INTRODUCTION

During the last few decades, there has been a revolution in the drug industry due to development of medicinal chemistry and studies on the basic cause of development of different diseases. The development of medicinal chemistry has been rewarding (Bowen and Zhong, 2016). It has greatly facilitated the development of drugs and understanding the mechanism of drug actions. In its usual sense, medicinal chemistry simply meant the application of organic chemistry to drug synthesis, but now, under its umbrella come various other aspects of drug design, such as identification of metabolites, studies of enzymes responsible for different diseases, computational chemistry, study of three-dimensional structures of molecules and enzymes, study of physicochemical properties of drug molecules, topology, development of quantitative structure-activity relationships, molecular modeling, X-ray crystallography, nuclear magnetic resonance (NMR), and molecular dynamics (Gupta, 2011). Thus all research directed to design and develop drugs can be included in medicinal chemistry. Medicinal chemistry has been organic chemistry in the chemistry departments, where the chemists used to only find out ways of synthesis of a particular drug. Now it has become a valuable subject of pharmacy departments with the name pharmaceutical chemistry, where the pharmacists focus on medicinal chemistry to understand and predict what changes in molecular structures will lead to enhance the biologic activity vis-à-vis a medicinal chemist who viewed medicinal chemistry as strictly application oriented. Thus, medicinal chemistry is a complex science that has gone beyond the hands of classical organic chemists. Now a career in medicinal chemistry has become rewarding but more challenging, too. Today, a medicinal chemist is

Studies on Hepatitis Viruses
ISBN 978-0-12-813330-9
https://doi.org/10.1016/B978-0-12-813330-9.00005-3

expected to not only have the knowledge of organic chemistry but also cause of diseases. At the biologic interface, medicinal chemistry combines to form a set of highly interdisciplinary sciences encompassing synthetic chemistry, computational chemistry, spectroscopy, enzymology, virology, microbiology, toxicology, etc., and thus medicinal chemistry is now playing a significant role in design and development of drugs. In this chapter, we present both experimental and theoretical roles of medicinal chemistry in designing antihepatitis drugs. The virology has provided enough information regarding the life cycles of different life-threatening viruses and the crucial steps necessary for their replication. These crucial steps involve different polymerases and proteases that could be found as important targets for the design of potent antiviral drugs. Now, we describe here the exploitation of such polymerases and proteases that could be successfully utilized for the design of drugs against all five hepatitis viruses: hepatitis A virus (HAV) to hepatitis E virus (HEV). However, in this chapter, we have chosen only the discussion of drugs against HAV, hepatitis D virus (HDV), and HEV, as they are less studied compared to hepatitis B virus (HBV) and hepatitis C virus.

2. ANTI-HAV DRUGS STUDIES

2.1 Experimental

For designing anti-HAV drugs, the target that has been extensively exploited is its 3C protease (3Cpro). This HAV 3Cpro is a picornaviral cysteine protease that is essential for cleavage of the initially synthesized viral polyprotein precursor to mature fragments and is therefore required for viral replication in vivo (Huang et al., 1999). It has been observed that this protease cleaves two peptides, corresponding to the viral polyprotein 2 B/2C and 2 C/3A junctions, most efficiently (Jewell et al., 1992). 3Cpro generally recognizes peptide substrates with L-glutamine at the P1 site. Cysteine proteases usually adopt a chymotrypsin-like fold (Allaire et al., 1994; Matthews et al.,1994) and display an active site configuration like those of serine proteases. Because of these specificities, mostly peptide-based compounds were studied as HAV 3Cpro inhibitors. However, so far, the following different types of 3Cpro inhibitors have been studied:

a. peptide aldehyde inhibitors (Malcolm et al., 1995)
b. azaglutamine and keto-glutamine derivatives (Huang et al., 1999; Ramtohul et al., 2002a)
c. azodicarboxamides (Hill and Vederas, 1999)

d. pseudoxazolones (Ramtohul et al., 2002b)
e. serine and threonine β-lactones (Lall et al., 1999, 2002)
f. triazolo[4,3-b]pyridazines (Shamroukh and Ali, 2008)
g. heteroaromatic esters (Huitema et al., 2008)

2.1.1 Peptide Aldehyde Inhibitors

The development of peptide aldehyde inhibitors was based on the modification of acetyl-Leu-Ala-Ala sequence to act as the competitors of the natural substrate Leu-Arg-Thr (Jewell et al., 1992; Malcolm et al., 1995). On the same basis, some monofluoromethylketones (FMKs) were also developed as 3Cpro inhibitors (Morris et al., 1997), and it was found that when an FMK was complexed with HAV 3Cpro, it had covalent bonding with the catalytic cysteine (Yin et al., 2006). The common feature of substrate required for recognition by all 3C proteases was found to be glutamine residue in P1 position (Krausslich and Wimmer, 1988) (Pn and Pn′ refer to the binding sites on the left and right sides of the scissile peptide bond, respectively, and the corresponding binding sites on the substrate are referred to by Sn and Sn′, respectively). Substrates with replacement of glutamine with an alanine were not found to act as 3Cpro substrates (Jewell et al., 1992). With various considerations, Malcolm et al. (1995) studied four peptide aldehydes as 3Cpro inhibitors. These inhibitors were acetyl N,N-dimethylglutaminal (Ac-Q′-H), acetyl leucinylalanylalanyl N,N′-dimethylglutaminal (Ac-LAAQ′-H), the thioester precursor (Ac-LAAQ′-SEt), and the related peptide aldehyde acetyl alanylalanylalanyl N,N-dimethylglutaminal (Ac-AAAQ′-H). Out of these compounds, acetyl leucinylalanylalanyl N,N′-dimethylglutaminal (Ac-LAAQ′-H) was found to be the most active. In this inhibitor, if leucine (L) residue was replaced with alanine residue (A), it resulted in a fivefold loss in activity.

2.1.2 AzaGlutamine and Keto-Glutamine Derivatives

Since HAV 3Cpro generally recognizes peptide substrates with L-glutamine at the P1 site, four types of analogs having an azaglutamine residue, as shown in Fig. 5.1 by A, B, C, and D, were chemically synthesized by Huang et al. (1999). These compounds were hydrazo-*o*-nitrophenyl-sulfenamides (A), frame-shifted hydrazo-*o*-nitrophenylsulfenamides (B), the azaglutamine sulfonamides (C), and haloacetyl azaglutamine analogs (D).

For generation of inhibitors of proteases, including cysteine proteases, incorporation of hydrazine functionality into peptide backbones (azapeptides) was shown to be crucial (Han and Janda, 1996; Fässler et al., 1996; Abeles and

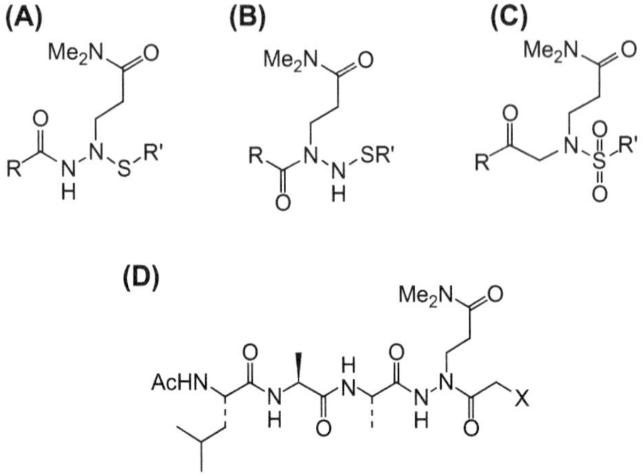

Figure 5.1 HAV 3Cpro inhibitors having azaglutamine residue: hydrazo-*o*-nitrophenylsulfenamides (A), frame-shifted hydrazo-*o*-nitrophenylsulfenamides (B), the azaglutamine sulfonamides (C), and haloacetyl azaglutamine analogs (D) (Huang et al., 1999).

Magrath, 1992). However, Huang et al. (1999) examined a number of sulfur-containing azaglutamine derivatives as potential inhibitors of HAV 3Cpro and provided synthetic access to four types of aza-amino acid peptide analogs, as shown in Fig. 5.1. Among all the groups of compounds, only one compound belonging to group B (R′ = Ac-Leu-Ala-Ala) and two compounds belonging to group D (X = Cl, and X = Br) were found to be active against HAV 3Cpro. These compounds were found to irreversibly react with the active site thiol (Cys-172) of the protease. It is therefore not unreasonable to expect that such substrate analogs may be developed to act as potent 3Cpro inhibitors.

Since the enzyme recognizes peptide substrates with a glutamine residue at the P1 site, a number of ketone-containing glutamine compounds were also synthesized and tested by Ramtohul et al. (2002a) for their inhibition activity against HAV 3Cpro. The crystal structure of HAV 3Cpro has been determined (Bergmann et al., 1997, 1999), and it has been found that it has a two-domain β-barrel fold, characteristic of chymotrypsin-like serine proteinases (Bergmann and James, 1999; Gorbalenya, and Snijder, 1996; Malcolm, 1995). In HAV 3Cpro an active site cysteine (Cys172) acts as a nucleophile with assistance from a histidine residue (His44), which behaves as a general acid-base catalyst. It leads to the formation of a tetrahedral intermediate being promoted by an electrophilic oxyanion hole (Bergmann and James, 1999; Gorbalenya, and Snijder, 1996). In the binding with the

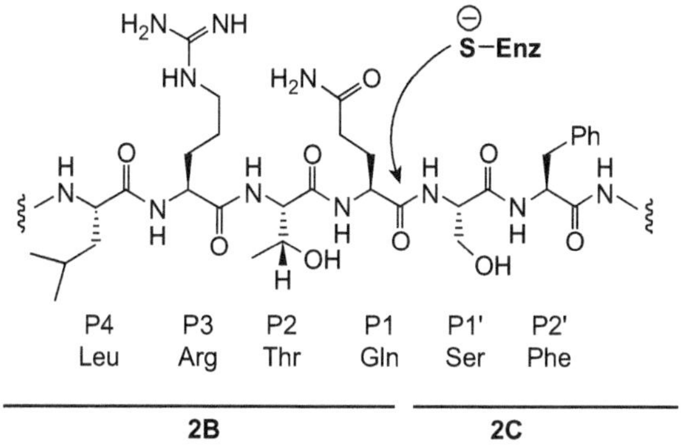

Figure 5.2 A schematic representation of HAV 3C protease showing the sequence of residues at the 2B–2C junction. *(Reprinted with permission from Ramtohul, et al., 2002a. J. Org. Chem. 67, 3169–3178. Copyright American Chemical Society 2002.)*

substrate, four to five residues (P5 to P1) preceding the scissile bond and two to three residues (P1′ to P3′) following it have been found to be involved (Ramtohul et al., 2002a; Schechter and Berger, 1967) (Fig. 5.2). Additionally, the enzyme-substrate interaction has also been shown to involve hydrogen bonding between the carbonyl oxygen of P1-glutamine of the enzyme and His191 of S1 in substrate (Bergmann and James, 1999; Gorbalenya, and Snijder, 1996; Malcolm, 1995). In search of potential therapeutic leads to develop potent 3Cpro inhibitors, a variety of compounds, as mentioned earlier in Section 2.1.2, were tried. A series of keto-glutamine analogs as 3Cpro Inhibitors were studied by Ramtohul et al. (2002a) to finally design a potent compound **1** having inhibition constant Ki equal to 9 μM. The activity of this compound was not affected in the presence of a 100-fold excess of dithiothreitol (DTT), which indicated that this type of phthalhydrazido system could be a specific inhibitor of HAV 3Cpro that would not react inadvertently with ubiquitous thiols (e.g., glutathione). Additionally, a compound **2** was suggested by these authors to be an interesting lead structure that could be further modified to generate drug candidates for picornaviral cysteine proteases.

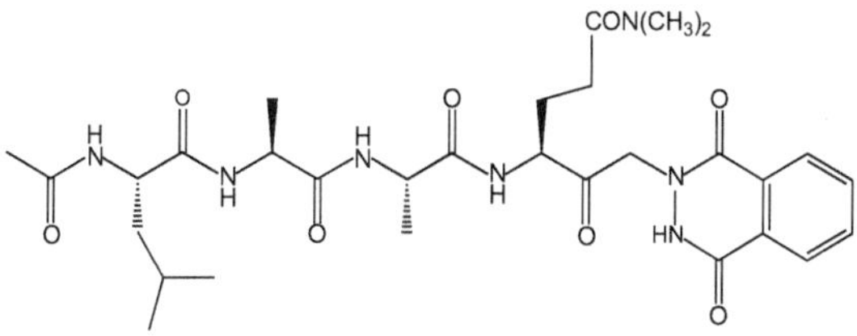

1

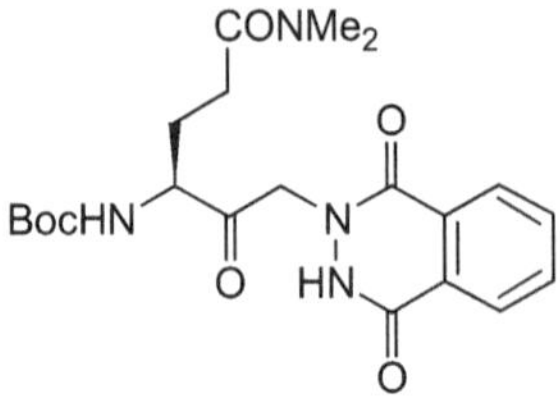

2

2.1.3 Azodicarboxamides

Hill and Vederas (1999) studied a group of azodicarboxamides as HAV 3Cpro inhibitors, among which a few compounds, as shown in Fig. 5.3, were found to have good inhibition activity. All four compounds, **3**−**6**, had their activity in the micromolar range as 23, 10, 32, and 62 μM, respectively. Hill and Vederas argued that these compounds inhibit the protease through the formation of a covalent complex with the enzyme. When the most active compound **4** was modeled into the active site of HAV 3Cpro based upon a cocrystal structure of HAV 3Cpro with a covalent addition of acetyl-Val-Phe-amide to Cys 172, it was observed that the carbonyl oxygen, which has the hydrogen bonding with the backbone nitrogen residue 170, neither appeared to fall into the oxyanion hole canonically nor to form any hydrogen bond with the nitrogen of Cys172 (Fig. 5.4). Also, the phenyl ring of the compound did not appear to fully enter the S2′ pocket because of the rigidity of the compound relative to the peptide. These

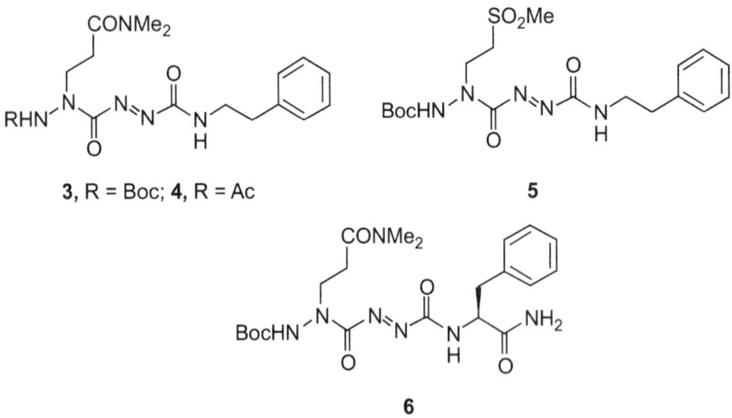

Figure 5.3 Some azodicarboxamides acting against HAV 3C proteases.

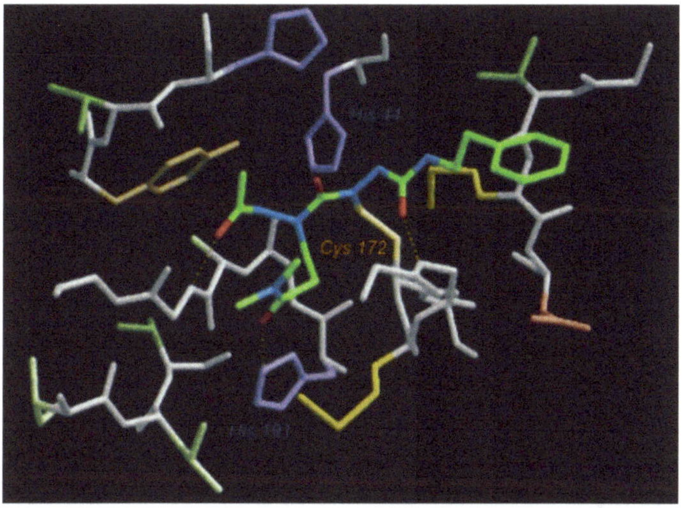

Figure 5.4 Molecular modeling of the most active compound (**4**) into the active site of HAV 3Cpro based upon a cocrystal structure of the protease with a covalent addition of acetyl-Val-Phe-amide to Cys172. The inhibitor molecule is shown in *green* with oxygen in *red* and nitrogen in *blue*. *(Reprinted with permission from Hill, Vederas, 1999. J.Org.Chem. 64, 9538–9546. Copyright American Chemical Society 1999.)*

azodicarboxamides were suggested to be potentially useful for cocrystallization studies, since recognition elements could be built onto either side of the azo moiety. This can span both the P and P′ sites of the active site, while usually cysteine protease inhibitors utilize recognition on only one side of the active site (Hill and Vederas, 1999).

2.1.4 Pseudoxazolones

Some pseudoxazolones, as represented by **7–10** in Fig. 5.5, were studied by Ramtohul et al. (2002b). In the entire series, these compounds were found to have the highest activity, having IC_{50} values of 6, 4, 3, and 4 μM, respectively. The enzyme inactivation by these compounds was suggested to involve covalent modification of the active site cysteine residue through nucleophilic addition of the thiolate to the imine moiety of the compounds. However, these pseudoxazolones were found to completely lose their activity in the presence of DTT. This indicated that added thiol reacts rapidly with these compounds, so such compounds were not assumed to be useful drug candidates without considerable modifications (Ramtohul et al., 2002b). However, they were suggested to be valuable biochemical tools to derive important information about enzyme-inhibitor interaction.

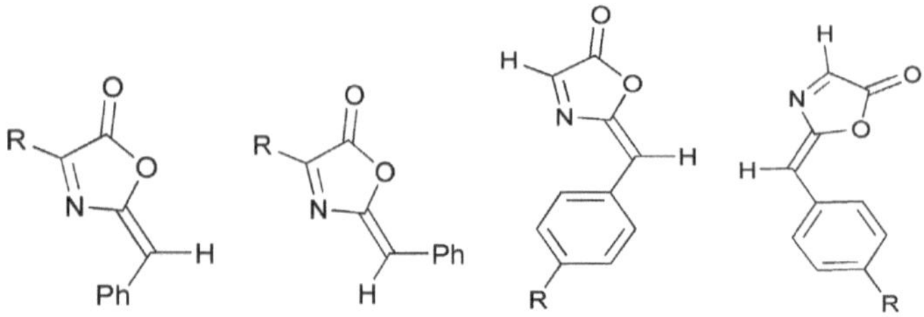

7, R = H 8, R = H 9, R = H 10, R = H

Figure 5.5 Some representative pseudoxazolones studied as HAV 3Cpro inhibitors.

2.1.5 Serine and Threonine β-Lactones

Lall et al. (1999) studied β-lactone functionality in the enantiomeric serine derivatives to find its effect on HAV 3Cpro inhibition and found that compounds such as **11** and **12** could display good irreversible and reversible inhibition, respectively, of this enzyme. This work was based on the fact that many of the naturally occurring β-lactones possess potent biologic activity (Lowe and Vederas, 1995; Pommier and Pons, 1995; Yang and Romo, 1999) and that the ability of the thiols to open the four-membered ring by nucleophilic attack at either the carbonyl or at β-position might be useful to make them irreversible inhibitors of cysteine proteases. These two compounds were, however, N-Cbz-serine β-lactones.

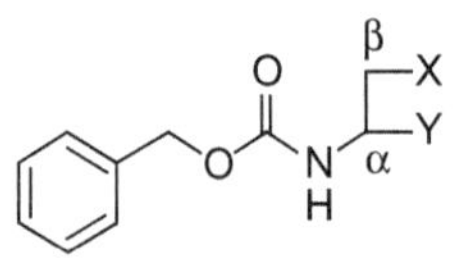

X-Y = O-CO

11, α-configuration S, IC_{50} = 35 μM

12, α-configuration R, IC_{50} = 06 μM

Despite the absence of a P1 glutamine side chain important for substrate recognition, compound **11** was found to be a good time-dependent irreversible inhibitor of HAV 3Cpro ($K_{inact} = 0.70\ min^{-1}$, $K_I = 1.84 \times 10^{-4}$ M, and $Kinact/K_I = 3800M^{-1}min^{-1}$) at an enzyme concentration of 0.1 μM, and the enantiomer **12** was found to be a competitive reversible inhibitor of HAV 3Cpro with K_I equal to 1.50 μM at a similar enzyme concentration. Thus, Lall et al. suggested that HAV 3Cpro active site may have different modes of binding for enantiomers **11** and **12**, so the former acts as an irreversible inhibitor and the latter as a reversible inhibitor.

Encouraged by this, Lall et al. later studied some threonine β-lactones (**13**) with different configurations at α and β positions and R = $SO_2(CH_n)_2Ph$ (n = 1 or 2) (Lall et al., 2002). Among threonine β-lactones, the two D-allo-threonine β-lactones, **14** and **15,** were found to be the most active against HAV 3Cpro with IC_{50} values of 5 and 12 μM, respectively. Thus both the communications of Lall et al. (1999, 2002) established the role of β-lactones in the inhibition of HAV 3C protease.

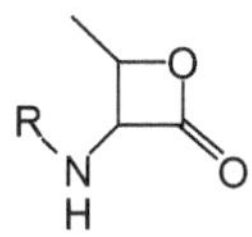

13

14, R=(E)-$SO_2(CH)_2Ph$; α, β-configuration = (R,R)

15, R= $SO_2(CH_2)_2Ph$; α, β-configuration = (R,R)

2.1.6 Heteroaromatic Esters

In an attempt to find broad-spectrum inhibitors of 3C and 3C-like (3CL) proteases of picornaviruses and coronaviruses, respectively, Huitema et al. (2008) screened a large library of heteroaromatic esters with some representative structures, as shown in Fig. 5.6. Three compounds in the library having

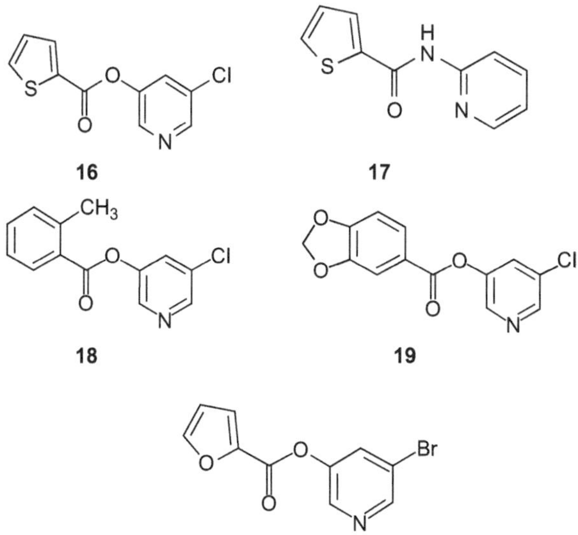

Figure 5.6 Some representative structures of heteroaromatic esters studied as $3C^{pro}$ and $3CL^{pro}$ inhibitors by Huitema et al. (2008).

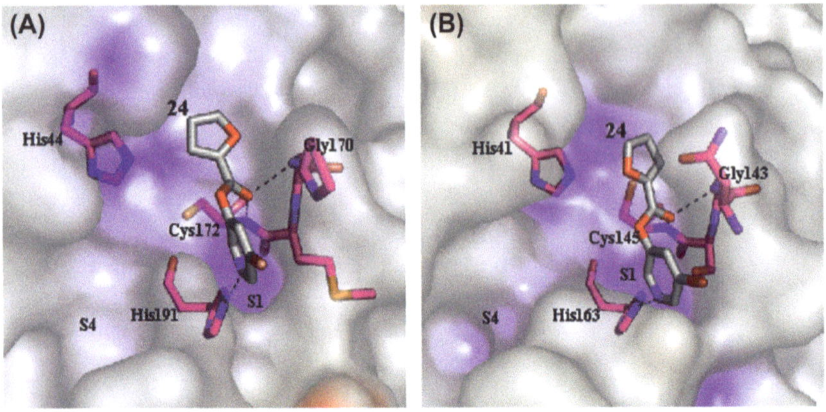

Figure 5.7 Docking of compound **20** in HAV 3Cpro (A) and SARS 3CLpro (B). In both the cases the compound is shown to have three hydrogen bonds with each protease and its halopyridinyl moiety to be engulfed into the S1 binding pocket. *(Reprinted with permission from Huitema et al., 2002a, 2008. Bioorg. Med. Chem. 16, 5761–5777. Copyright Elsevier 2008.)*

furan rings were found to inhibit HAV 3Cpro with K_I values between 120 and 240 nM. These compounds were reported to be among the most potent nonpeptidic 3Cpro inhibitors reported so far. The most potent compound, however, screened against HAV 3Cpro was **20** with an IC_{50} value of 53 ± 2 nM. A modeling study was performed on this compound to find that halopyridinyl moiety of the compound fits tightly into the S1-binding pocket and that its nitrogen is hydrogen-bonded to $N^{\varepsilon 2}$ of His191 (3.0 Å) and the halogen atom points out toward the solvent (Fig. 5.7A). Further, the carbonyl oxygen of the ester in the oxyanion hole of the enzyme forms two hydrogen bonds, one with nitrogen of Gly170 (3.1 Å) and other with nitrogen of Cys172 (2.7 Å). Similar interactions of this compound are shown with severe acute respiratory syndrome (SARS) 3CL protease (Fig. 5.7B).

2.2 Theoretical

No substantial theoretical studies are available on HAV inhibitors except the one discussed before in Section 2.1.6 for heteroaromatic esters.

3. ANTI-HDV DRUGS STUDIES

3.1 Experimental

As already discussed, HDV can cause severe liver disease, but so far, no specific therapy has been developed against this virus. In fact, in its life cycle, no crucial target has been found to be exploited for drug design. However, through some studies, it has been pointed out that the prenylation site of large delta antigen is a critical determinant of HDV particle assembly, so delta antigen prenylation can be pharmacologically exploited to develop anti-HDV drugs. In this context, a benzodiazepine peptidomimetic ($C_{27}H_{33}N_5O_5S_2$), BZA-5B, with IUPAC name of "Methyl (2S)-2-[[2-[3-[(2-amino-3-sulfanylpropanoyl)-methylamino]-2-oxo-5-phenyl-3H-1,4-benzodiazepin-1-yl]acetyl]amino]-4-methylsulfanylbutanoate" (**21**) was developed as an HDV prenylation inhibitor. This drug was originally synthesized as a specific prenyltransferase inhibitor and known to inhibit prenylation of the oncoprotein H-Ras^{V12} (Marsters et al., 1994). Glenn et al. (1998) later studied to establish that this BA-5B could be used as an anti-HDV drug.

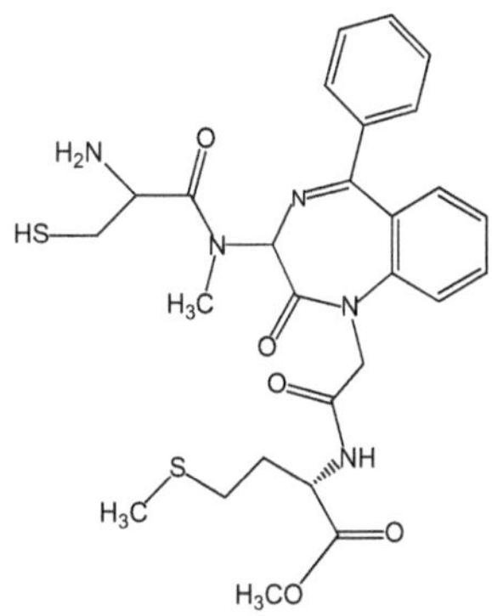

21, BZA-5B

Some authors have studied the bile acids (BAs) as anti-HDV agents (Yan et al., 2014; Pereira et al., 2015). BAs, with general structures as shown in Fig. 5.8, are steroid acids found predominantly in the bile of mammals and vertebrates. A polypeptide known as sodium-taurocholate cotransporting

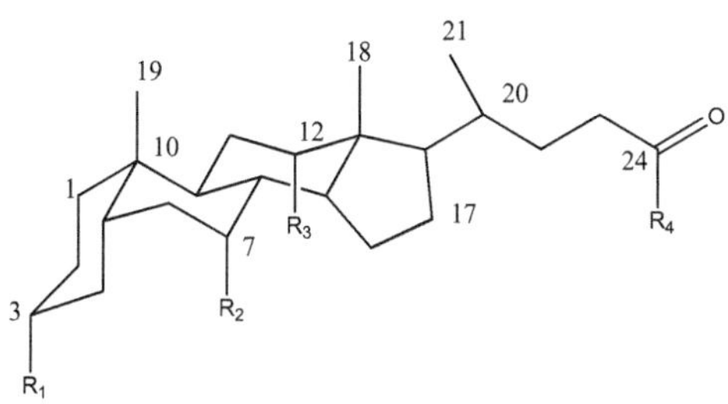

Figure 5.8 General structure of bile acids.

polypeptide (NTCP) is both a BA transporter mediating uptake of BA into hepatocytes and an essential receptor for HBV and HDV. Thus, it is of interest to study as to what extent and through what mechanism BAs can affect HDV cell entry. Inhibitory BAs can inhibit NTCP-mediated HDV entry into hepatocytes, suggesting that modulation of BA pool may affect HDV infection of hepatocytes. It is, however, reasonable to speculate that the conformation of NTCP favorable for substrate binding and that for virus binding may mutually affect each other. BAs act as substrate of NTPC. Working on a set of BAs, Pereira et al. (2015) demonstrated that BAs inhibit HDV cell entry, which occurs most likely through competition for binding to NTCP between BAs and the large HBV surface protein that has been shown to mediate binding of HBV and HDV to NTCP. However, with all efforts, there are still no approved drugs in the market or in advance clinical development stage against HDV.

3.2 Theoretical

Like HAV inhibitors, not many theoretical studies are available on HDV inhibitors. However, a compound, 3-(((4,7-Dichloro-1,3-benzoxazol-2-yl) methyl)amino)-5-ethyl-6-methylpyridin-2(1H)-one, (L-697661, **22**), was developed by Singh et al. (2011) as an anti-HDV drug through virtual screening and molecular docking studies. These authors selected a large set of HDV replication inhibitors from the National Centre for Biotechnology Information Pub-Chem compound database as ligand molecules having the ability to inhibit the replication of small delta antigen (S-HDAg) protein of HDV, which is required for HDV RNA replication in vivo (Kuo et al., 1989). For the docking study, the three-dimensional structure model of S-HDAg protein of HDV was retrieved from Protein Model Database (http://mi.caspur.it/PMDB/) as PM0075974 and used as a receptor. In the docking study, the binding modes and geometric orientation of all compounds were found to be almost identical, suggesting that all inhibitors occupied a common cavity in the receptor. However, compound **22** was considered to be a potential candidate drug for HDV inhibition.

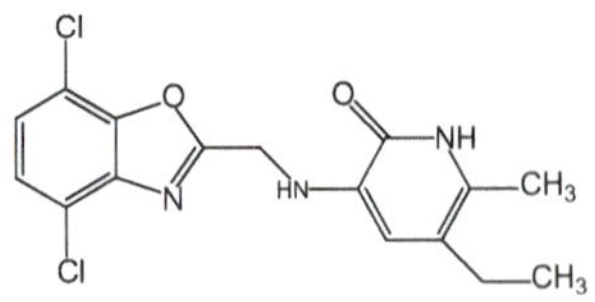

22, L-697661

4. ANTI-HEV DRUGS STUDIES

4.1 Experimental

It is now known that HEV can be responsible for chronic hepatitis in immunocompromised patients. Unfortunately, there is still no established therapy for this chronic infection, which can rapidly evolve to cirrhosis and can require liver transplantation (Haagsma et al., 2008; Gerolami et al., 2008; Kamar et al., 2008). It was, however, reported that a compound, ribavirin (**23**), may have antiviral effects (Mallet et al., 2010), but later the follow-up after ribavirin cessation was found to be very short. Although this compound has been shown to inhibit the replication of a wide range of RNA and DNA viruses (Sidwell et al., 1972; Patterson and Fernandez-Larsson, 1990), its sustained effect against HEV has hardly been reported. However, Kamar et al. (2010) reported that ribavirin monotherapy could inhibit the replication of HEV in vivo and might induce a sustained virologic response in patients with chronic HEV infections, but to determine the optimal duration of effect of ribavirin therapy, further studies are still required.

Some authors studied the effect of another antiviral compound, sofosbuvir (**24**), on HEV infection (Thi et al., 2016) and found that it can inhibit replication of HEV genotype 3 both in subgenomic replicon systems as well as full-length infectious clone, but they suggested that an additive antiviral effect may result if it is combined with ribavirin. Thus, sofosbuvir may be considered an add-on therapy to ribavirin for the treatment of chronic hepatitis E in immunocompromised patients.

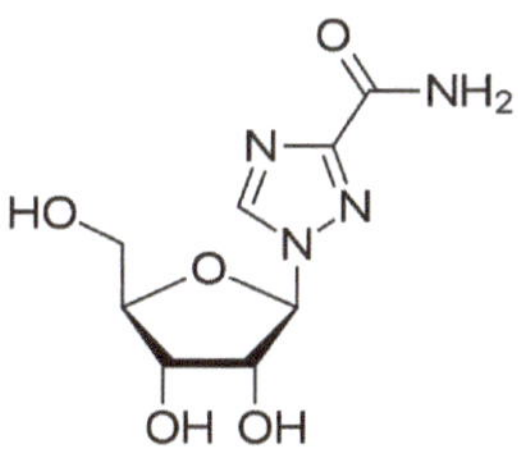

23, Ribavirin

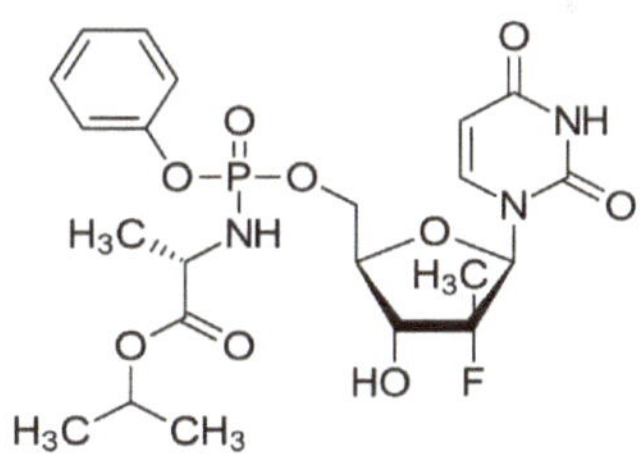

24, Sofosbuvir

In an effort to find a new category of anti-HEV drugs, Nan et al. (2015) tested a series of phosphorodiamidate morpholino oligomers (PMOs) for their ability to inhibit HEV replication in liver cells. PMOs are nuclease-resistant single-stranded DNA analogs containing a backbone of morpholine rings and phosphorodiamidate linkages (Summerton, 1999). A PMO can conjugate to an arginine-rich cell, penetrating the peptide to facilitate the entry into the cells (Abes et al., 2006; Summerton, 1999). These peptide-conjugated PMOs (PPMOs) are water soluble, so they can readily enter the cells. When these PPMOs were studied for HEV inhibition by Nan et al. (2015), it was found that some of them displayed potent inhibition of HEV genotype 1 replication. A PPMO that targeted start site region of HEV ORF1 (PPMO HP1) was observed to also effectively inhibit replication of genotype 3 Kernow-C1 strain in liver cells. The conserved nature of the target of PPMO HP1 and its overall efficacy suggested that this compound may be an HEV-specific inhibitor with antiviral activity across multiple HEV genotypes. In addition to PPMO HP1, three more PPMOs, HP2, HP3U, and HPN3, were found to be promising to be developed as anti-HEV agents. All these PPMOs are given in Table 5.1 with their specifications. It can be seen in Table 5.1 that PPMO HP1 and HP2 are complementary to a sequence in the 5′-end region of genomic and subgenomic RNA, respectively. On the other hand, HP3U is complementary to a sequence in the terminal region of the 3′ UTR, and HPN3 is the reverse complement of HP1 and was intended to interfere with the synthesis of positive-sense genomic RNA. PPMO HP1 and HP2 are complementary to a sequence in the 5′ end region of genomic and subgenomic RNA, respectively (Table 5.1). HP3U is complementary to a sequence in the terminal region of the 3′ UTR. HPN3 is the reverse complement of HP1 and was intended to interfere with the synthesis of positive-sense genomic RNA. A nonsense-sequence PPMO CP1 (Zhang et al., 2007), which has little agreement with HEV or human mRNA sequences, was used as a negative control PPMO. CP1 with fluorescein conjugated at its 3′ end (CP1−F) was used in the PPMO uptake assay.

Table 5.1 PPMOs designed against HEV Sar55 strain

PPMO	Sequence(5′-3′)	Target site in HEV genome (position)
HP1	GGGCCTCCATGGCATCGACC	Start site region of ORF1 (18−37)
HP2	CATGGGCGCAGCAAAAGACA	Start site region of ORF2 (5116−5135)
HP3U	GCGCGAAACGCAGAAAAGAG	Terminal region of 3′ UTR (7169−7188)
HPN3	GGTCGATGCCATGGAGGCCC	3′ terminal region of negative sense RNA

PPMOs were synthesized with an arginine-rich cell-penetrating peptide (P7) conjugated at the 5′ end at AVI BioPharma Inc (Corvallis, OR), as previously described (Abes et al., 2006).

4.2 Theoretical

No substantial theoretical work is available on anti-HEV drugs.

5. CONCLUSIONS

This chapter has presented theoretical and experimental studies on anti-HAV, anti-HDV, and anti-HEV drugs. For anti-HAV, the 3Cpro of the virus has been well exploited, and the following classes of inhibitors have been developed for its inhibition: peptide aldehydes, azaglutamine and keto-glutamine derivatives, azodicarboxamides, pseudoxazolones, serine and threonine β-lactones, triazolo[4,3-b]pyridazines, and heteroaromatic esters. A theoretical study has been conducted on heteroaromatic esters, where the most potent compound of the series (**20**) when docked into HAV 3Cpro was found to have its halopyridinyl moiety to completely fit into the S1-binding pocket of the enzyme and its nitrogen to have hydrogen bonding with $N^{\varepsilon 2}$ of His191 residue of the enzyme. Further, the carbonyl oxygen of the ester in the oxyanion hole of the enzyme was observed to form two hydrogen bonds, one with nitrogen of Gly170 (3.1 Å) and other with nitrogen of Cys172 (2.7 Å), and the halogen atom was found to point toward the solvent (Fig. 5.7A). For anti-HDV drugs, a delta antigen prenylation inhibitor such as **21** and some BAs were studied to inhibit NTCP-mediated HDV entry into hepatocytes. As far as theoretical study on anti-HDV drugs is concerned, a large set of HDV replication inhibitors was selected from the database to be docked in small delta antigen (S-HDAg) protein of HDV. The docking study had shown that the binding modes and geometric orientation of all compounds were almost identical, suggesting that all inhibitors occupied a common cavity in the receptor, but a compound **22** was considered to be a potential candidate drug for HDV inhibition. No more theoretical studies were available on anti-HDV drugs. For HEV therapy, a series of PMOs were studied. PMOs could conjugate to an arginine-rich cell, penetrating the peptide to facilitate the entry into the cells. Out of these peptide-conjugated PMOs (PPMOs), a few were found to be potent inhibitors of HEV genotype 1 replication. Further studies are not available on anti-HEV drugs, and theoretical studies are almost lacking.

REFERENCES

Abeles, R.H., Magrath, J., 1992. Cysteine protease inhibition by azapeptide esters. J. Med. Chem. 35, 4279–4283.

Abes, S., Moulton, H.M., Clair, P., Prevot, P., Youngblood, D.S., Wu, R.P., Iversen, P.L., Lebleu, B., 2006. Vectorization of morpholino oligomers by the (R-Ahx-R)4 peptide allows efficient splicing correction in the absence of endosomolytic agents. J. Control Release 116, 304–313.

Allaire, M., Chernaia, M., Malcolm, B.A., James, N.G., 1994. Picomaviral 3C cysteine proteinases have a fold similar to the chymotrypsin-like serine proteinases. Nature (London) 369, 72–77.

Bergmann, E.M., James, M.N.G., 1999. The 3C proteinases of picornaviruses and other positive-sense, single-stranded rna viruses. In: Von der Helm, K., Korant, B. (Eds.), Proteases as Targets for Therapy. Springer, Heiderlberg, pp. 117–143.

Bergmann, E.M., Mosimann, S.C., Chernaia, M.M., Malcolm, B.A., James, M.N.G., 1997. The refined crystal structure of the 3C gene product from hepatitis A virus: specific proteinase activity and RNA recognition. J. Virol. 71, 2436–2448.

Bergmann, E.M., Cherney, M.M., McKendrick, J., Frormann, S., Luo, C., Malcolm, B.A., Vederas, J.C., James, M.N.G., 1999. Crystal structure of an inhibitor complex of the 3C proteinase from hepatitis A virus (HAV) and implications for the polyprotein processing in HAV. Virology 265, 153–163.

Bowen, J.P., Zhong, H.A., 2016. The evolution in medicinal chemistry. Curr. Top. Med. Chem. 16, 897–898.

Fässler, A., Bold, G., Capraro, H.-G., Cozens, R., Mestan, J., Poncioni, B., Rösel, J., Tintelnot-Blomley, M., Lang, M., 1996. J. Med. Chem. 39, 3203–3216.

Gerolami, R., Moal, V., Colson, P., 2008. Chronic hepatitis E with cirrhosis in a kidney-transplant recipient. N. Engl. J. Med. 358, 859–860.

Glenn, J.S., Marsters Jr., J.C., Greenberg, H.B., 1998. Use of a prenylation inhibitor as a novel antiviral agent. J. Virol. 72, 9303–9306.

Gorbalenya, A.E., Snijder, E.J., 1996. Viral cystein proteases. Perspect. Drug Discov. Des. 6, 64–86.

Gupta, S.P., 2011. In: QSAR Studies and Molecular Modeling. Anamaya Publishers, New Delhi.

Haagsma, E.B., van den Berg, A.P., Porte, R.J., et al., 2008. Chronic hepatitis E virus infection in liver transplant recipients. Liver Transpl. 14, 547–553.

Han, H., Janda, K.D., 1996. Azatides: solution and liquid phase syntheses of a new peptidomimetic. J. Am. Chem. Soc. 118, 2539–2544.

Hill, R.D., Vederas, J.C., 1999. Azodicarboxamides: a new class of cysteine proteinase inhibitor for hepatitis A virus and human rhinovirus 3C enzymes. J. Org. Chem. 64, 9538–9546.

Huitema, C., Zhang, J., Yin, J., James, M.N.G., Vederas, J.C., Eltis, L.D., 2008. Heteroaromatic ester inhibitors of hepatitis A virus 3C proteinase: evaluation of mode of action. Bioorg. Med. Chem. 16, 5761–5777.

Huang, Y., Malcolmb, B.A., Vederas, J.C., 1999. Synthesis and testing of azaglutamine derivatives as inhibitors of hepatitis A virus (HAV) 3C proteinase. Bioorg. Med. Chem. 7, 607–619.

Jewell, D.A., Swietnicki, W., Dunn, B.M., Malcolm, B.A., 1992. Hepatitis A virus 3C proteinase substrate specificity. Biochemistry 31, 7862–7869.

Kamar, N., Mansuy, J.M., Cointault, O., et al., 2008. Hepatitis E virus-related cirrhosis in kidney- and kidney-pancreas-transplant recipients. Am. J. Transplant. 8, 1744–1748.

Kamar, N., Rostaing, L., Abravanel, F., Garrouste, C., Lhomme, S., Esposito, L., Basse, B., Cointault, O., Ribes, D., Nogier, M.B., Alric, L., Peron, J.M., Izopet, J., 2010. Ribavirin therapy inhibits viral replication on patients with chronic hepatitis E virus infection. Gastroenterology 139, 1612–1618.

Krausslich, H.G., Wimmer, E., 1988. Viral proteinases. Annu. Rev. Biochem. 57, 701–754.

Kuo, M.Y., Chao, M., Taylor, J., 1989. Initiation of replication of the human hepatitis delta virus genome from cloned DNA: role of delta antigen. J. Virol. 63, 1945–1950.

Lall, M.J., Karvellas, C., Vederas, J.C., 1999. β-Lactones as a new class of cysteine proteinase inhibitors: inhibition of hepatitis A virus 3C proteinase by *N*-Cbz-serine *β*-lactone. Org. Lett. 1, 803–806.

Lall, M.S., Ramtohul, Y.K., James, M.N.G., Vederas, J.C., 2000. Serine and threonine *β*-lactones: a new class of hepatitis A virus 3C cysteine proteinase inhibitors. J. Org. Chem. 67, 1536–1547.

Lowe, C., Vederas, J.C., 1995. Naturally occurring β-lactones: occurrence, syntheses and properties. Org. Prep. Proced. Int. 27, 305–346.

Malcolm, B.A., 1995. The picornaviral 3C proteinases: cysteine nucleophiles in serine proteinase folds. Protein Sci. 4, 1439–1445.

Malcolm, B.A., Lowe, C., Shechosky, S., McKay, R.T., Yang, C.C., Shah, V.J., Simon, R.J., Vederas, J.C., Santi, D.V., 1995. Peptide aldehyde inhibitors of hepatitis A virus 3C proteinase. Biochemistry 34, 8172–8179.

Mallet, V., Nicand, E., Sultanik, P., et al., 2010. Brief communication: case reports of ribavirin treatment for chronic hepatitis E. Ann. Intern. Med. 153, 85–89.

Matthews, D.A., Smith, W.W., Ferre, R.A., Condon, B., Budahazi, G., Sisson, W., Villafranca, J.E., Janson, C.A., McElroy, H.E., Gribskow, C.L., Worland, S., 1994. Cell 77, 761–771.

Marsters Jr., J.C., McDowell, R.S., Reynolds, M.E., Oare, D.A., Somers, T.C., et al., 1994. Benzodiazepine peptidomimetic inhibitors of farnesyltransferase. Bioorg. Med. Chem. 2, 949–957.

Morris, T.S., Frormann, S., Shechosky, S., Lowe, C., Lall, M.S., Gauss-Müller, V., Purcell, R.H., Emerson, S.U., Vederas, J.C., Malcolm, B.A., 1997. In vitro and ex vivo inhibition of hepatitis A virus 3C proteinase by a peptidyl monofluoromethyl ketone. Bioorg. Med. Chem. 5, 797–807.

Nan, Y., Ma, Z., Kannan, H., Stein, D.A., Iversen, P.I., Meng, X.-J., Zhang, Y.-J., 2015. Inhibition of hepatitis E virus replication by peptide-conjugated morpholino oligomers. Antiviral Res. 120, 134–139.

Patterson, J.L., Fernandez-Larsson, R., 1990. Molecular mechanisms of action of ribavirin. Rev. Infect. Dis. 12, 1139–1146.

Pereira, I.V.A., Buchmann, B., Sandmann, L., Sprinzl, K., Schlaphoff, V., Döhner, K., Vondran, F., Sarrazin, C., Manns, M.P., Souza de Oliveira, C.P.M., Sodeik, B., Ciesek, S., Thomas von Hahn, T., 2015. Primary biliary acids inhibit hepatitis D virus (HDV) entry into human hepatoma cells expressing the sodium-taurocholate cotransporting polypeptide (NTCP). PLoS One 1–16. https://doi.org/10.1371/journal.pone.0117152.

Pommier, A., Pons, J.M., 1995. The synthesis of natural 2-oxetanones. Synthesis 7, 729–744.

Ramtohul, Y.K., James, M.N.G., Vederas, J.C., 2002a. Synthesis and evaluation of keto-glutamine analogues as inhibitors of hepatitis A virus 3C proteinase. J. Org. Chem. 67, 3169–3178.

Ramtohul, Y.K., Martin, N.I., Silkin, L., James, M.N.G., Vederas, J.C., 2002b. Synthesis of pseudoxazolones and their inhibition of the 3C cysteine proteinases from hepatitis A virus and human rhinovirus-14. J. Chem. Soc. Perkin Trans. 1, 1351–1359.

Schechter, I., Berger, A., 1967. On the size of the active site in proteases. I. Papain. Biochem. Biophys. Res. Commun. 27, 157–162.

Shamroukh, A.H., Ali, M.A., 2008. Anti-HAV activity of some newly synthesized triazolo [4,3-b]pyridazines. Arch. Pharm. Chem. Life Sci. 341, 223–230.

Sidwell, R.W., Huffman, J.H., Khare, G.P., Allen, L.B., Witkowski, J.T., Robins, R.K., 1972. Broad-spectrum antiviral activity of virazole: 1-beta-D-ribofuranosyl-1,2,4-triazole-3-carboxamide. Science 177, 705–706.

Singh, S., Gupta, S.K., Nischal, A., Khattri, S., Nath, R., Pant, K.K., Seth, P.K., 2011. Identification and characterization of novel small-molecule inhibitors against hepatitis delta virus replication by using docking strategies. Hepat. Mon. 11, 803–809.

Summerton, J., 1999. Morpholino antisense oligomers: the case for an RNase H-independent structural type. Biochim. Biophys. Acta 1489, 141–158.

Thi, V.L.D., Debing, Y., Wu, X., Rice, C.M., Neyts, J., Moradpour, D., Gouttenoire, J., 2016. Sofosbuvir inhibits hepatitis E virus replication in vitro and results in an additive effect when combined with ribavirin. Gastroenterology 150, 82–85.

Yan, H., Peng, B., Liu, Y., Xu, G., He, W., Ren, B., Jing, Z., Sui, J., Li, W., 2014. Viral entry of hepatitis B and D viruses and bile saltstransportation share common molecular determinants on sodiumtaurocholate cotransporting polypeptide. J. Virol. 88, 3273–3284.

Yang, H.W., Romo, D., 1999. Methods for the synthesis of optically active β-lactones (2-oxetanones). Tetrahedron 55, 6403–6434.

Yin, J., Cherney, M.M., Bergmann, E.M., Zhang, J., Huitema, C., Pettersson, H., Eltis, L.D., Vederas, J.C., James, M.N., 2006. An episulfide cation (thiiranium ring) trapped in the active site of HAV 3C proteinase inactivated by peptide-based ketone inhibitors. J. Mol. Biol. 361, 673–686.

Zhang, Y.J., Wang, K.Y., Stein, D.A., Patel, D., Watkins, R., Moulton, H.M., Iverson, P.L., Matson, D.O., 2007. Inhibition of replication and transcription activator and latency-associated nuclear antigen of Kaposi's sarcoma-associated herpes virus by morpholino oligomers. Antiviral Res. 73, 12–23.

FURTHER READING

Marsters Jr., J.C., Greenberg, H.B., 1998. Use of a prenylation inhibitor as a novel antiviral agent. J. Virol. 72, 9303–9306.

CHAPTER 6

The Medicinal Chemistry of Antihepatitis Agents II: Anti-HBV Agents

1. INTRODUCTION

HBV infection has been more serious than HAV, HDV, and HEV infections, so more attention has been paid to develop anti-HBV drugs. However, currently approved drugs to treat chronic HBV infection include interferon-α and two nucleoside analogs, lamivudine and adefovir dipivoxil (Zhao et al., 2006), but the treatment with these drugs was not found very satisfactory because of side effects of interferon and the viral resistance to other two drugs. Therefore, there have been continuous efforts to find more useful anti-HBV drugs.

As already discussed in Chapter 3, the most important targets that have been mostly exploited to find the potent anti-HBV drugs have been its polymerase, three antigens, surface antigen (HBsAg), core antigen (HBcAg), and e-antigen (HBeAg). While HBV polymerase (reverse transcriptase) is required for the DNA synthesis necessary for the virus, the antigens are required for its replication. Recently, a review was presented by Kaushik et al. (2010) describing some efforts to find drugs against DNA synthesis. Here, we discuss in detail both the experimental and theoretical efforts made by various workers to design the potent anti-HBV drugs based on all possible targets.

2. EXPERIMENTAL STUDIES

2.1 Drugs Acting Against DNA Synthesis

Many different categories of compounds have been studied for the inhibition of DNA synthesis in HBV. Most important of them have been nucleosides, indoles, benzimidazoles, thiazolides, myristoyl analogs, and caudatin analogs. Studies on these compounds are discussed here.

Studies on Hepatitis Viruses
ISBN 978-0-12-813330-9
https://doi.org/10.1016/B978-0-12-813330-9.00006-5

2.1.1 Nucleosides

A series of nucleosides as represented by general structure **1** was synthesized and studied by Lee and Chu (1995), and it was found that a 5-fluoro analog (1, X = F) had the highest activity. The screening of activity was done in a human hepatoma cell line carrying HBV 2.2.15 cells, and it was observed that replacement of fluorine with any other substituent led to a significant decrease in the activity. Similarly, anti-HBV activities of two enantiomers of cis-5-fluoro-1-[2-hydroxymethyl)-1,3-oxathiolan-5-yl]cytosine (FTC; **2:** R = F) were studied by Furman et al. (1992), in which the (−) isomer was found to be a potent inhibitor of viral replication with an IC_{50} value of 10 nM (**2** shows the structure of (−) isomer), while the (+) isomer was found to be comparatively less active. However, both were found to possess minimal toxicity to HepG2 cells. Earlier, a series of open-chain acyclic sugar nucleoside analogs were studied for their anti-DNA synthesis activity by Kumar and Tyrrell (2001), of which two structurally similar analogs, 9-[(2-hydroxyethoxy)methyl]guanine (**3**, acyclovir) and 9-[2-hydroxy-1-(hydroxymrthyl) ethoxy)methylguanine (**4**, ganciclovir) were found to be active in tissue culture and in vivo studies in animal models. However, acyclovir is an antiviral medication, primarily used for the treatment of herpes simplex virus infections, chickenpox, and shingles (de Clercq and Field, 2005), and ganciclovir is an antiviral medication used to treat cytomegalovirus (Rossi, 2006). However, a prodrug of ganciclovir, famciclovir (**5**), was found to possess anti-HBV effect in clinical studies.

Later, Zhang et al. (2005) tried to investigate the anti-HCV activity of a number of ring-expanded nucleobases containing the imidazo[4,5-c][1,3] diazepine-1,8-dione ring system with general structure, as shown by **6**. In this series, only one compound, ZP-28 (**7**), was found to have extracellular anti-HBV virion DNA activity, having an EC_{50} value of (1.7 ± 0.2) μM. In fact, the nucleosides and nucleotides were most widely studied by different authors for anti-HBV activity, and usually all of them were found to exhibit their activity through the DNA synthesis. The most active compounds found in different series of nucleosides and nucleotides studied by various authors are described in Table 6.1.

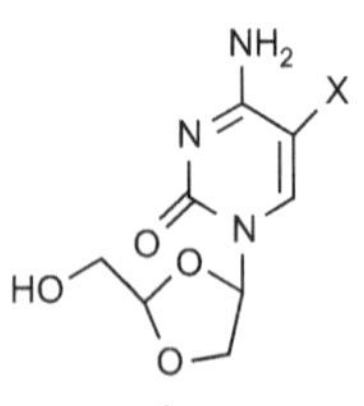

1

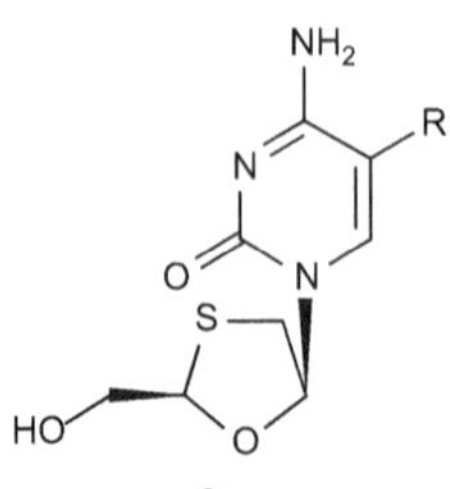

2

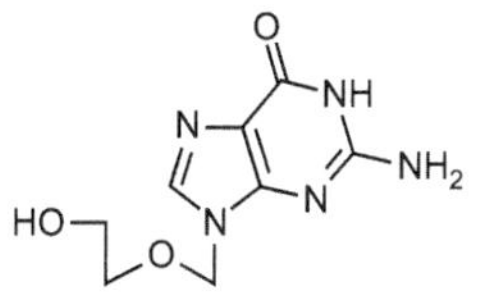

3, Acyclovir

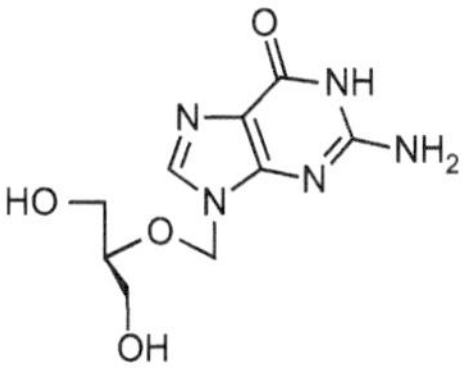

4, Ganciclovir

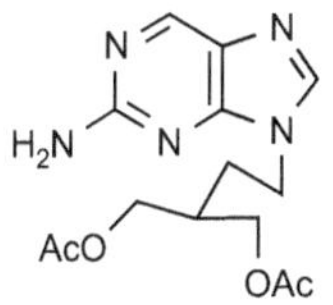

5, Famciclovir

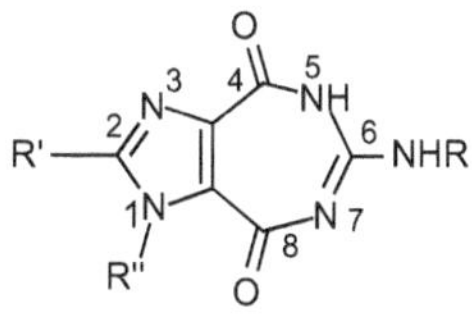

6

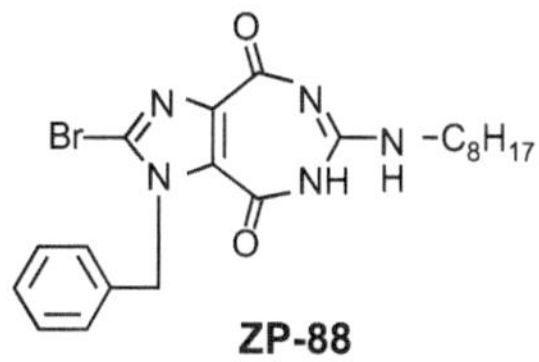

ZP-88

7

2.1.2 Indoles

Arbidol (**8**) and its derivatives were observed to have a variety of biologic effects, such as antiviral effects, immunostimulating effects, and interferon-induced actions (Leneva et al., 1994; Glushkov, 1992). Some structural modifications in this compound led Chai et al. (2006) to have a series of compounds with anti-HBV effects. This encouraged Zhao et al. (2006) that the efficacy of the compounds could be increased by making hydrophobic and electronic changes in the structures. With this thought, they made changes at the 1,2, 4, and 6 positions of the indole ring, and thus with the general structure as **9**, they studied a series of 20 compounds to find that Mannich basic functionalities at position 4 (R_3-substituents) had appreciable effect on the antiviral activity. Of them, heterocyclic groups, particularly pyrrolidinyl and imidazolyl groups, were more effective. Consequently, analogs **9a** and **9b** were found to have better IC_{50} and selectivity index for the replication of HBV DNA and analogs **9c** and **9d** to have significant

Table 6.1 Some nucleosides and nucleotides acting against DNA synthesis in HBV

No.	Name	Structure	Anti-HBV activity	References
1	L-Carbocyclic 2′,3′-didehydro-2′,3′-dideoxyadenosine	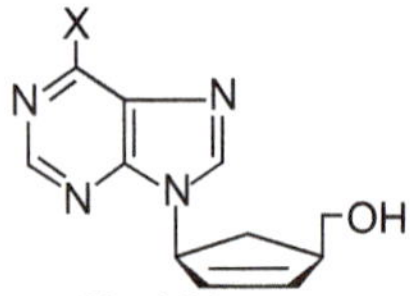 X = NH_2	$EC_{50} = 0.9\ \mu M$	Wang et al. (1999)
2	9-(2-Deoxy-2-fluoro-β-L-arabinoturanosyl) purine nucleosides, **A** and **B**	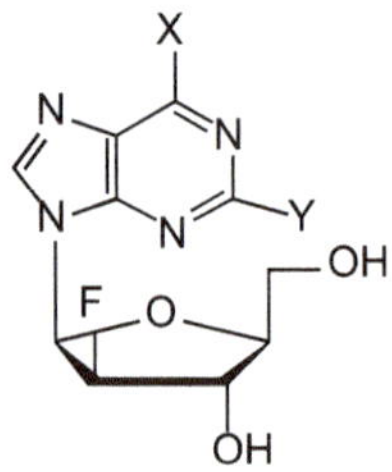 **A**: X = NH_2, Y = H **B**: X = OH, Y = H	**A** $EC_{50} = 1.5\ \mu M$ **B** $EC_{50} = 8\ \mu M$	Ma et al. (1997)
3	β −L-2′,3′-didehydro-2′,3′-dideoxyadenosine	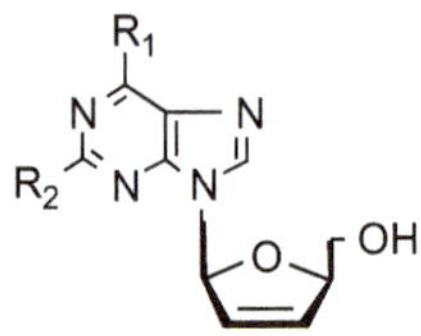 $R_1 = NH_2$, R_2 = H	$EC_{50} = 1.2\ \mu M$	Bolon et al. (1996)
4	2′-Fluoro-2′,3′-unsaturated L-nucleoside	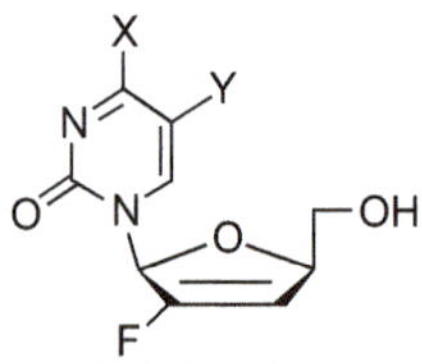 X = NH_2, Y = H	$EC_{50} = 0.18\ \mu M$	Lee et al., 1999

anti-HBV effects even better than lamivudine. These two compounds also had a better effect on replication of HBV DNA.

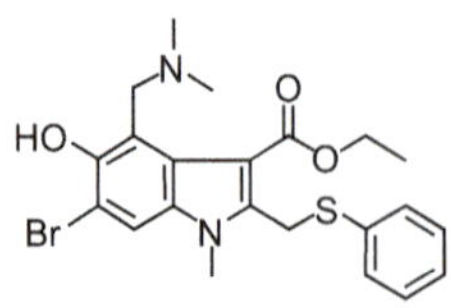

8, Arbidol

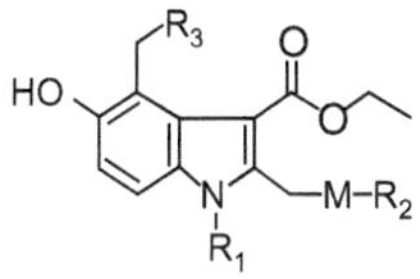

9

9a: R_1 = cyclopropyl, **M** = sulfonyl, **R_2** = 2-furanylmethyl, **R_3** = pyrrolidinyl

9b: R_1 = cyclopropyl, **M**,= sulfonyl, **R_2** = p-methylphenyl, **R_3** = imidazolyl

9c: R_1 = cyclopropyl, **M** = sulfinyl, **R_2** = phenyl, **R_3** = 2-methylimidazolyl

9d: R_1 = methyl, **M**,= sulfonyl, **R_2** = m-methoxyphenyl, **R_3** = pyrrolidinyl

2.1.3 Benzimidazoles

A series of isopropylsulfonyl-2-amine benzimidazole derivatives, based on a lead compound (**10**), were synthesized and evaluated for their anti-HBV activity by Li et al. (2007). Compound **10**, the N-(2-amino-1-isopropylsulfonyl)-1H-benzo[d]imidazole-6-yl)-3-(trifluoromethyl) benzenesul-fonamide, with an IC_{50} of 4 μM to inhibit HBV DNA replication and a selectivity index 27 in vitro, was identified in a communication by Li et al. (2006). In the present study, the compounds synthesized by these authors (Li et al., 2007) and their anti-HBV activity and cytotoxicity are given in Table 6.2. These compounds belong to two different series, **11** and **12**. As can be seen in Table 6.2, except a few, all the compounds had IC_{50} values less than 5.0 μM and selectivity indices greater than 40. Out of these compounds, **11d** and **j** had the potential to be developed as potent anti-HBV agents.

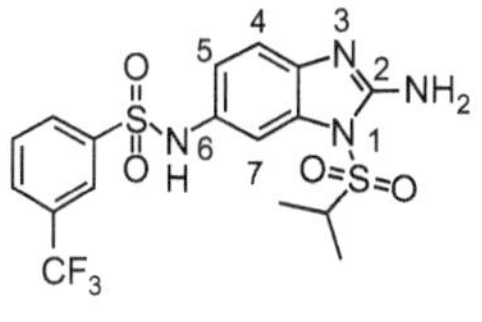

10

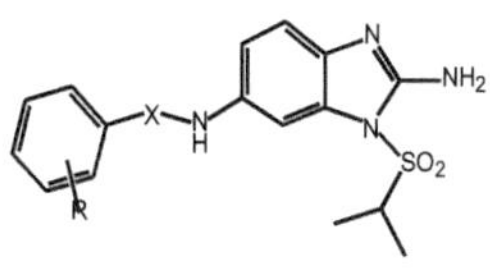

11

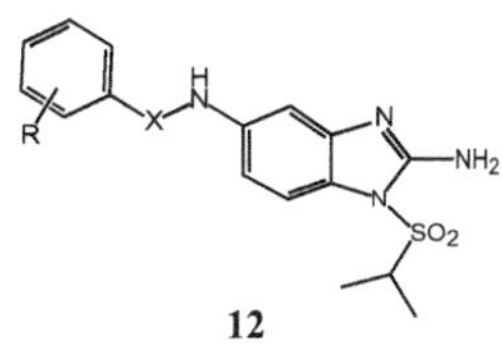

12

Table 6.2 A series of 1-isopropylsulfonyl-2-amine benzimidazole analogs and their anti-HBV activity and cytotoxicity (Li et al., 2007)

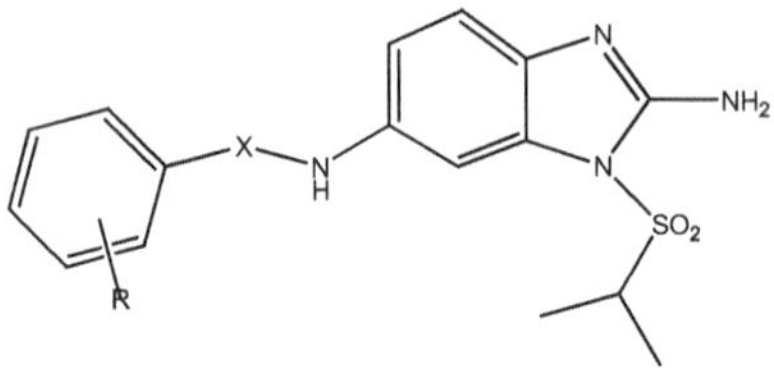

11a-k

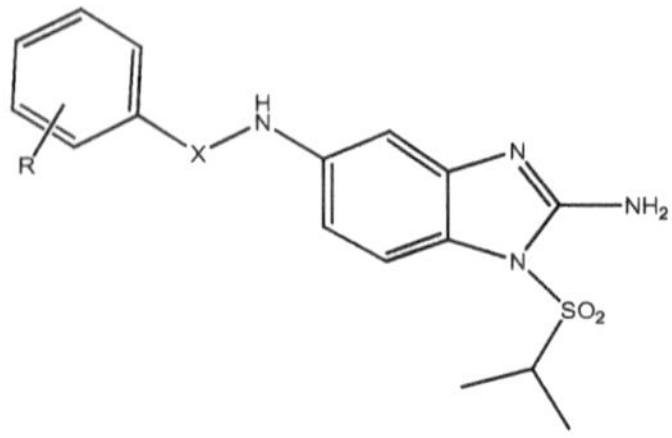

12a-c

Compound	X	R	IC_{50}(μM)	CC_{50}(μM)	SI
11a	CO	H	33.7	>500	>15
11b	CO	2-F	0.94	48	51
11c	CO	4-F	1.2	67	56
11d	CO	2,6-diF	0.70	192	274
11e	CO	2-Cl	0.82	88	107
11f	CO	4-CH_3	4.20	>500	>119
11g	SO_2	H	1.52	62	41
11h	SO_2	4-CH_3	4.80	80	17
11i	SO_2	3-NO_2	10.5	57	5
11j	–	–	0.70	86	123
11k	–	–	NA	98	–
12a	CO	H	2.0	>500	>250
12b	CO	4-F	3.7	34	9
12c	SO_2	4-CH_3	16	151	9
Lamivudine	–	–	0.38	>1000	>2632
Adefovir	–	–	1.7	57	34

2.1.4 Thiazolides

A series of thiazolides (2-hydroxyaroyl-N-(thiazol-2-yl)-amides) with general structures as **13** were synthesized and studied for their activity against HBV replication by Stachulski et al. (2011). Of all the analogs, only one analog, **13a**, with a chlorine at R_4 position, was found to be worthy of further development.

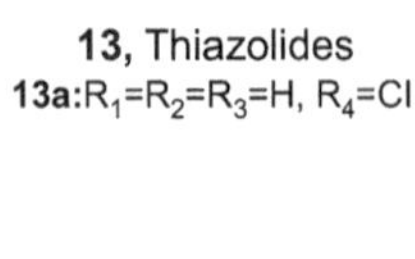

13, Thiazolides
13a:R_1=R_2=R_3=H, R_4=Cl

2.1.5 Matijing-Su Derivatives

For anti-HBV drug development, natural products with their enormous structural diversity have been a great source, e.g., Matijing-Su [MTS, N-(N-benzoyl-L-phenylalanyl-O-acetyl-L-phenylalnol)], a dipeptide derivative isolated from a Chinese ethnic drug Matijing (*Dichondra repens* Forst) was found to have anti-HBV activity (Liang et al., 2002). The result of MTS encouraged Liang group to make further research on MTS to find more potent anti-HBV drugs (Xu et al., 2009). Consequently, a series of MTS were synthesized and evaluated for their anti-HBV activity by Xu et al. (2009). Of the total series of compounds, belonging to the basic structure **14**, three compounds, **14a–c**, were found to have high potency against the HBV DNA replication with $IC_{50} < 4\ \mu M$ (1.40, 2.33, and 2.36 μM, respectively) and selectivity indices (SI) > 20 (28.27, 28.29, and 45.93, respectively). Thus, as compared to positive control lamivudine, the IC_{50} values of all the three compounds were much less and the SI of **14c** was much higher (lamivudine, IC50 = 82.42 μM, SI = 41.59).

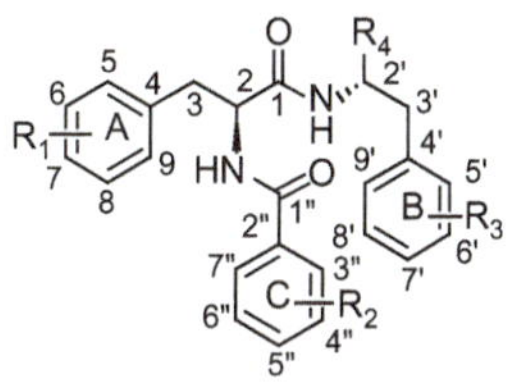

14

14a: R_1 = H, R_2 = 5″-Fluoro, R_3 = H, R_4 = Methyl carboxylate
14b: R_1 = H, R_2 = 5″-Choro, R_3 = H, R_4 = Hydroxymethyl
14c: R_1 = 7-Nitro, R_2 = H, R_3 = 7′-Hydroxy, R_4 = Methyl carboxylate

2.2 Anti-HBsAg and Anti-HBeAg Agents

2.2.1 Lignan Derivatives

The lignans are a large group of polyphenols found in plants. They have great health value. Several C_{18} (possessing 18-carbon atom skeleton) dibenzocyclooctadiene lignans isolated from the plants of *Schisandra* have been found to possess many therapeutic effects such as antioxidant, antihepatitis, antihepatotoxic, and antilipid peroxidative effects (Bao et al., 1979; Lu and Liu, 1992; Hikino et al., 1984; Yang et al., 1992a,b). Several of C_{18} dibenzocyclooctadiene lignans were also found to have anti-HBsAg and anti-HBeAg effects. These results encouraged Kuo et al. (2005) to further study these lignans to find potential anti-HBV agents. A member of this family of lignans, +-Gomisin K_3 (**15**), was taken as a lead compound. +-Gomisin K_3 was isolated from *Schisandra arisanensis*. It was found that substitution at C_{14}-position had appreciable effect on HBsAg and HBeAg inhibition and toxicity effects of the compounds, so some compounds, as shown in Table 6.3, were found to possess good anti-HBsAg and anti-HBeAg potency and least toxicity.

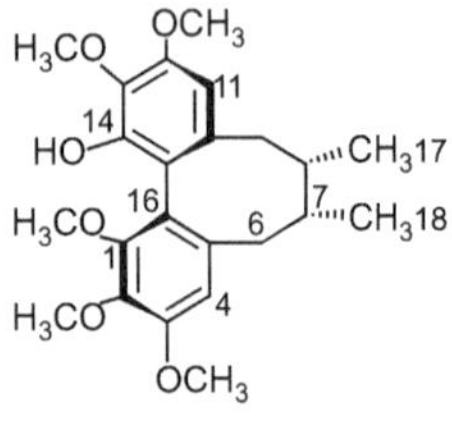

15, +-Gomisin K_3

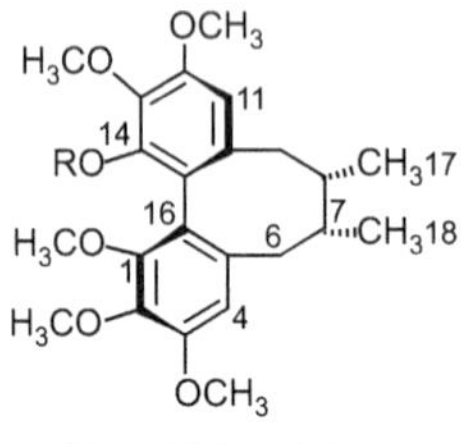

15a-i (Table 6.3)

Table 6.3 Some potent derivatives of +-Gomisin K_3 (**15**) acting as anti-HBsAg and anti-HBeAg agents (Kuo et al., 2005)

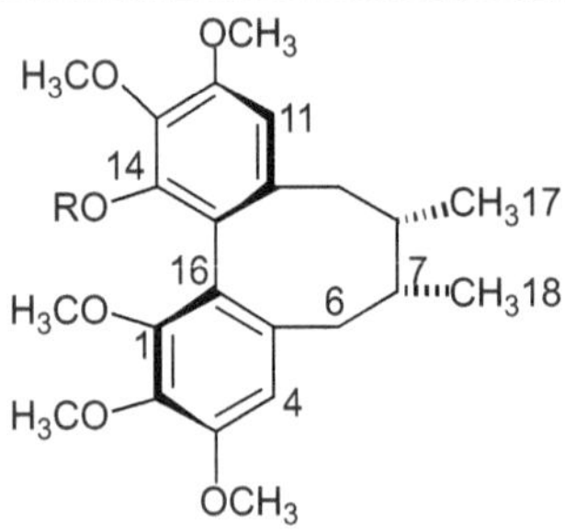

15a-i

Compound	R
15a	Ac (acetyl)
15b	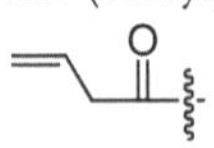
15c	Bz (benzyl)
15d	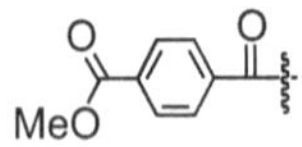
15e	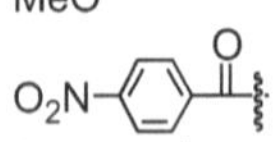
15f	2-Furoyl
15g	2-Thenoyl
15h	Mesyl (methanesulfonyl)
15i	Tosyl (*p*-toluenesulfonyl)

2.2.2 Alisol A Derivatives

Alisol A (**16**), which is a protostane-type triterpene, is also a natural product that comprises 0.05%–0.1% of rhizomes of Alisma orientalis Juzep (Murata et al., 1970). It carries four hydroxyl groups, providing the opportunity to have a large number of derivatives. Using this facility, Zhang et al. (2008) synthesized a large series of **16**, out of which a compound, **17**, was found to be highly potent against HBsAg and HBeAg with IC_{50} values of 0.024 and 0.028 mM, respectively, which were much lower than the positive control drug lamivudine ($IC_{50\ HBsAg} = 12$ mM, $IC_{50HBeAg} = 26$ mM). This compound also had a very high selectivity index for both antigens (>108 for HBsAg and >93 for HBeAg). This study on alisol A derivatives concluded the following:

1. Acylation of hydroxy group at positions 11, 23, and 24 would decrease cytotoxicity, and an epoxy group at C-13 may be an important feature to confer relatively low toxicity.
2. A hydroxyl group at C-25 may be important for lead optimization, and a carbonyl function at C-3 would hardly play any role in biologic activity.

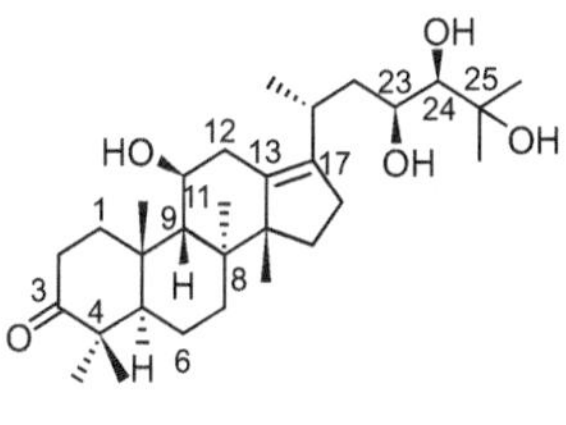

16, Alisol A

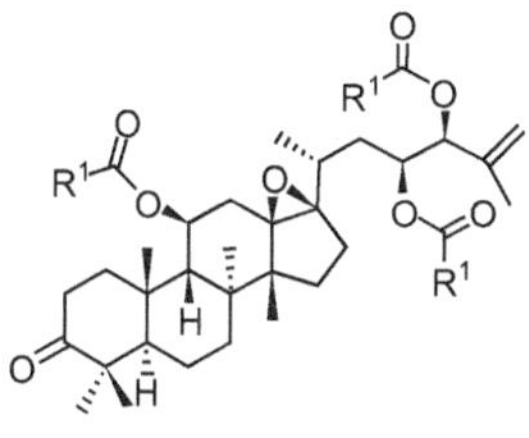

17, $R_1 = CH_3$

2.2.3 Coumarins

Coumarins are a class of natural compounds that are widely distributed and that possess a number of biologic activities. Some pyranocoumarins, extracted from the roots of *Clausena excavate* (Rutaceae) (Huang et al., 1997), were used as traditional medicines to treat several diseases such as hepatitis, cough, asthma, fever, headache, dermatological conditions, and gastrointestinal diseases (Sasaki, 1924; Kongkathip et al., 2005; Sunthitikawinsakul et al., 2003). Su et al. (2009) synthesized a series of pyranocoumarins in which the two compounds **18** and **19** were found to be most potent against HBV with EC_{50} values of 1.14 and 1.34 μM, respectively. The strong anti-HBV activity and ease of preparation of these compounds make pyranocoumarins worth trying for development of still stronger anti-HBV drugs.

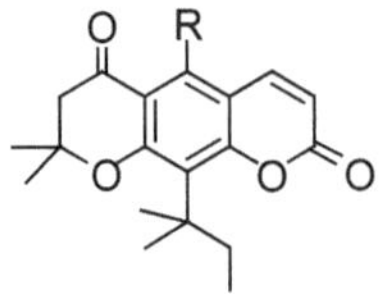

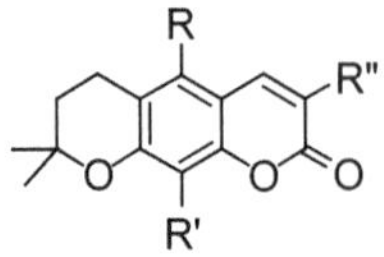

18, R = OH **19, R = OH, R' =**  **, R" = H**

2.2.4 Quinolines

In the previous Section 2.1.2, indoles were discussed for their potency against HBV DNA replication by some authors (Chai et al., 2006; Zhao et al., 2006a,b). To explore the SAR surrounding the indole region, the same group of authors expanded the indole ring to a quinoline ring and targeted the positions 2 and 5 of the quinoline ring (Liu et al., 2008). Thus, taking the basic structure **20,** a new series of compounds were tried with arylthiomethyl or arylsulfonylmethyl groups at position 2 (X = −S− or −(S=O)−; R_1 = H, F, alkyl or alkoxyl) and aliphatic aminomethyl or N-heteroaromatic methyl functionality at position 5 (R_2 = aliphatic amino or N-heteroaromatic) for obtaining new anti-HBV drugs. All the compounds synthesized were evaluated for their activity against HBsAg and HBeAg, as well as HBV DNA replication. Most of the compounds studied were found to have high potency against HBsAg, as well as HBeAg, and a compound, which had an imidazolyl group at the 5-position and a 4-fluorophenylsulfinyl group at the 2-position (**20a**), had exceptionally high potency against HBV DNA replication with an IC_{50} value of 4.7 μM, a 66-fold improvement over lamivudine (IC_{50} = 311.2 μM). However, a few more compounds had also shown potency against DNA replication.

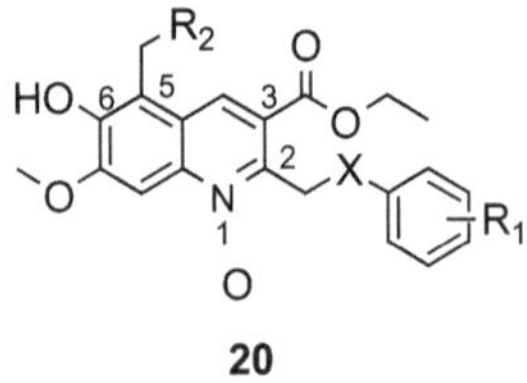

20

20a: X= -(S=O)-, R_1 = 4-F, R_2 = 1H-imidazolyl

Further innovation by this group of authors was made by replacing the quinoline ring with a 6H-[1]benzothiopyrano[4,3-b]quinoline ring, while retaining some of the structure features of previous quinoline derivatives (Jia et al., 2009). Thus the basic structure taken was as **21.** They examined the effect of substituents on the D-ring (the last ring with numbers 1−4), taking particularly the methyl group and F atom at its different position. However, the compounds that were found to be active against HBsAg were those having a methyl or dimethyl group at the D-ring and a dimethylamino at A-ring (R_2-substituent). Thus only three compounds, **21a−c**, exhibited potency against HBsAg with IC_{50} values 9.3, 13.8, and 22.1 μM, respectively, but three other compounds, **21d−f**, were found to be quite potent against HBV DNA replication with IC_{50} values of 4.2, 2.0, and 2.4 μM, respectively. From these results, it appeared that at R_1-position 2- or 2,4-diMe group and at R_2-position dimethylamino group were more favorable for anti-HBV activity of the compounds.

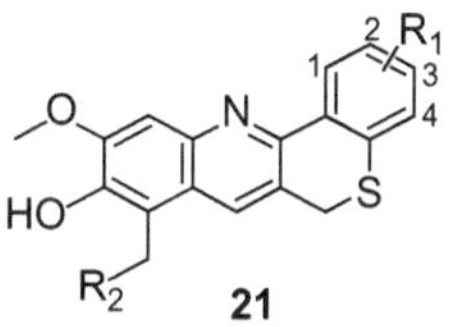

21

21a: R_1 = H, R_2= Dimethylamino; **21b:** R_1= 2,4-diMe, R_2= Dimethylamino
21c: R_1 = 2-Me, R_2= Dimethylamino; **21d:** R_1= 2,4-diMe, R_2= Piperidin-1-yl
21e: R_1 = 2-Me, R_2= Diethylamino; **21f:** R_1= 3-F, R_2= Diethylamino

2.2.5 Caudatin Analogs

Some authors have found that caudatin, a natural product extracted from *Cynanchum auriculatam,* a Chinese herb called Bai-Shou-Wu, possesses anti-HBV activity. They therefore made extensive study on its derivatives (Wang et al., 2012a, 2012b, 2015) to investigate therapeutically useful anti-HBV agents. The structure of caudation is as shown by **22**. In one of their communications, Wang et al. (2012a) studied a series of 3-O-substituted analogs of caudatin (**23**) to find that six of their compounds, as shown in Table 6.4, were highly potent against DNA replication, and a few had high potency against HBsAg and HBeAg as well. As can be seen, all six compounds were active against DNA replication, and except for one (**23f**), they were quite potent against HBsAg. However, only two of these compounds (**23a** and **b**) were active against HBeAg. In a recent communication, however, Wang et al. (2015) extended their work on further substitutions at C-3 position of the caudatin and also tried to explore the effect of substituting the C-17 position of the caudation. They ultimately

Table 6.4 Some caudatin analogs possessing high potency against HBV HBsAg, HBeAg, and DNA replication (Wang et al., 2012a, 2015)

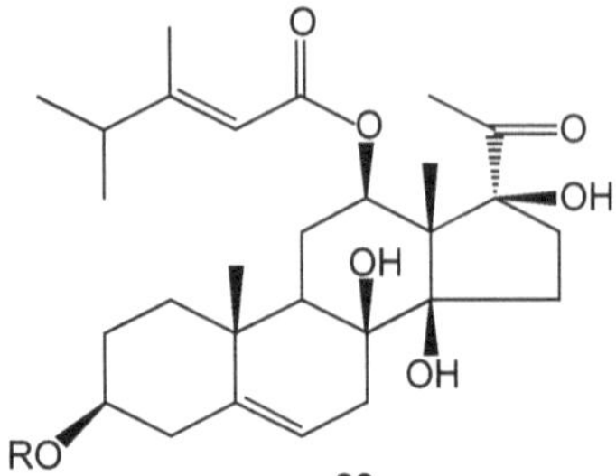

23

Compound[a]	R	HBsAg IC_{50} (μM)	HBeAg IC_{50} (μM)	DNA Repl. IC_{50} (μM)
23a	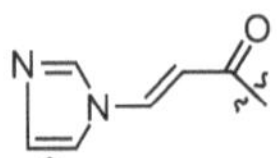	18.68	13.16	7.48
23b	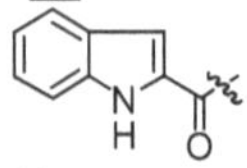	21.71	33.73	3.63
23c	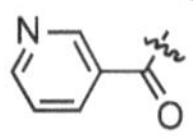	20.84	>1038.05	4.06
23d	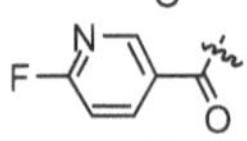	15.59	>687.36	4.97
23e	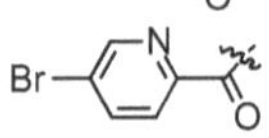	14.52	>720.99	2.82
23f	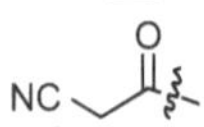	100.48	>161.49	7.32
23g	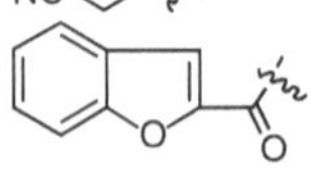	63.02	204.80	24.55
23h	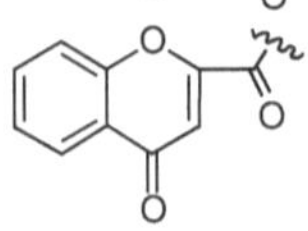	52.81	173.51	5.69
23i	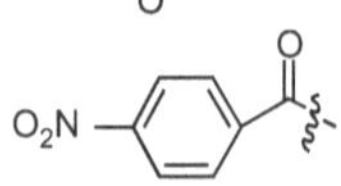	56.08	70.39	8.23

[a]Compounds **23a–f** were reported by Wang et al. (2012a) and **23g–i** by Wang et al. (2015).

found that the modification at the C-3 position was dominant; the conclusions regarding the SAR were as follows:

1. The activity and toxicity would be enhanced with the incorporation of C=C and C≡C bond in the structure.
2. An increase in the length of the introduced fatty chain to eight carbon atoms would increase the activity.

3. Introduction of heterocyclic rings containing the oxygen and sulfur atoms could largely enhance the activity.
4. Hydroxyl groups were important features in maintaining antiviral activity.
5. Hydroxyl group at C-3 position was a good target for further lead optimization by introducing the rational substituents.

In this study, these authors reported that compounds **23g–i**, as added in Table 6.4, had excellent activities against the secretion of HBsAg (IC_{50} = 63.02, 52.81, and 56.08 μM, respectively) and HBeAg (IC_{50} = 204.80, 173.51, and 70.39 μM, respectively), as well as DNA replication (IC_{50} = 24.55, 5.69, and 8.23 μM, respectively) (Table 6.4). However, a comparison shows that these compounds were in no way better than those reported by these authors earlier (compounds **23a–h**) (Wang et al., 2012a).

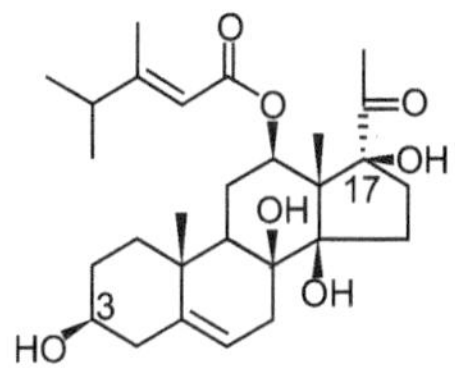

22, Caudatin

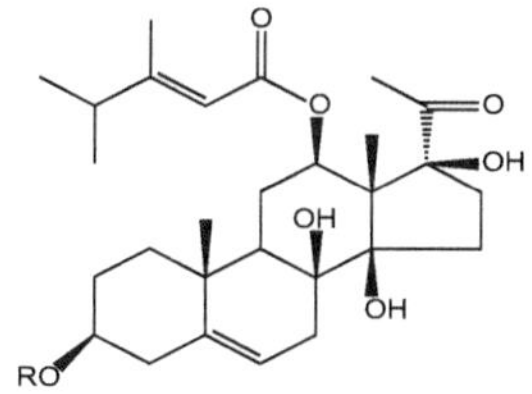

23

In another communication, these authors studied the effect of combination of caudatin with cinnamic acids on anti-HBV activity (Wang et al., 2012b), as cinnamic acid analogs were found to possess a number of biologic activities, including antiviral activity (Zhuang et al., 2009; Lee et al., 2007). Thus, a derivative of **24** with three methoxy groups at 3-,4-, and 5-positions of the aryl ring was found to possess significant activity against HBsAg, HBeAg, and HBV DNA replication with IC_{50} values of 5.52, 5.52, and 2.44 μM, respectively, and they had a good safety index (LD50 > 1250 mg/kg).

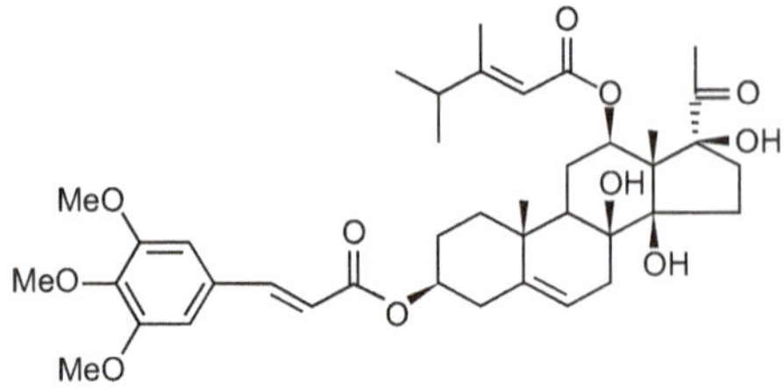

24

2.3 Human Hepatoblastoma Cell Lines (HepG2/2.2.15 Cells)

Human hepatoblastoma cell line (HepG2) is a perpetual cell line that was derived from the liver tissue of a 15-year-old Caucasian American male with a well-differentiated hepatocellular carcinoma. The HepG2 cell lines have been found easy to use for the study of the effects of anti-HBV drugs, as the level of HBV gene expression and replication can be easily achieved in HBV baculovirus-infected HepG2 cells. HepG2 cells can be successfully grown in a large-scale cultivation system. HBV replication in HepG2 cells is readily detectable by 1 day postinfection with baculovirus and persists up to at least 11 days. In place of HepG2 cells, HepG2-2.2.15 cells are also used, which are derived from HepG2 cells by stable transfection with full-length HBV plasmid. The drug efficacy can be measured by studying the level of inhibition of virus replication in these cells. The antiviral activity of nucleoside analogs in HepG2/2.2.15 cells was evaluated as a means of development and qualification of single-plate assay.

2.3.1 Purine and Pyrimidine Nucleosides

Purine and pyrimidine nucleosides have been widely studied for the development of a range of antiviral drugs to prevent viral replication in infected cells. They have been particularly exploited to develop drugs against HBV, HCV, herpes simplex virus, and HIV. Once they are phosphorylated, they work as antimetabolites by being similar enough to be incorporated into growing DNA strands. They can act as chain terminators and inhibit viral DNA polymerase. A variety of nucleoside analogs have been found to act as reverse transcriptase inhibitors (NRTIs) as well as inhibitors of other polymerases. Some nucleoside analogs have been found to act both as NRTIs as well as polymerases of other viruses. Extensive studies have been done to exploit purine and pyrimidine analogs to develop anti-HBV and anti-HCV drugs. For the development of anti-HBV drugs, Ku group (Kim et al., 1992; Lee and Chu, 1995; Xiang et al., 1995; Bolon et al., 1996; Ma et al., 1997; Wang et al., 1999; Lee et al., 1999) performed several studies, as described subsequently.

In a study on a dioxolane cytosine nucleosides, Kim et al. (1992) had found a compound, (−)-β-L-dioxolane-cytosine [**25**, (−)-OddC], that had exhibited extremely potent anti-HBV activity in 2.2.15 cells, but its in vitro cellular cytotoxicity precluded its use as an useful drug to treat HBV infection. Therefore, in another communication, Lee and Chu (1995) performed a study on D-dioxolane cytosine derivatives (**26**), varying the substituents (particularly F, Cl, Br, and I) at the 5-position. In this study, they found that the fluoro derivative (**26**a) was the most potent with an EC_{50} value of 0.02 μM and selectivity index >1500. Any other modification than 5-fluoro group at 5-position of the pyrimidine ring was found to reduce the activity. Following this work, Xiang et al. (1995) studied the corresponding L-nucleosides and reported the synthesis of 1-or-9-(2,3-dideoxy-2-fluoro-β-L-threo-pentofuranosyl)-cytosine and −adenine and their anti-HBV activities, in which a compound with structure **27** was found to have just modest anti-HBV activity but was considered a lead compound to search for better compounds.

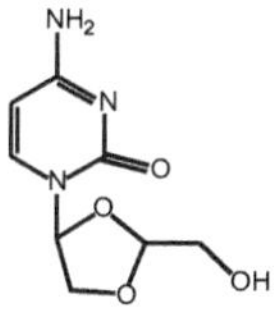

25, (-)-OddC

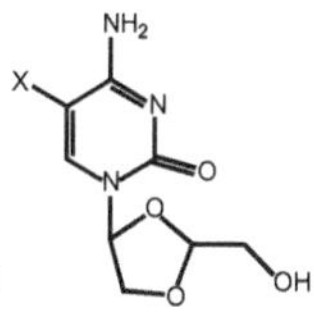

26, (**32a**: X = F)

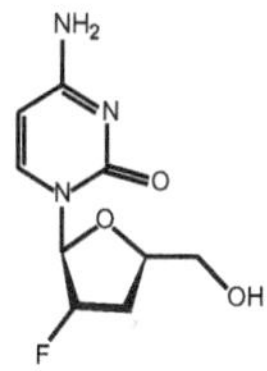

27

A series of β-L-2′,3′-dideoxyadenosine and β-L-2′,3′-didehydro-2′, 3′-dideoxyadenosine analogs were synthesized in a stereoselective manner and studied for their anti-HIV and anti-HBV activities by Bolon et al. (1996). However, in this study, only one compound, β-L-2′,3′ -didehydro-2′,3′ -dideoxyadenosine (β-L-d4A, **28**), was found to exhibit significant anti-HBV activity with an EC_{50} value of 1.2 μM in 2.2.15 cells. Another compound, **29** (β-L-d2A), was also found to be potent against HBV, but its EC_{50} value (6.0 μM) was slightly higher than that of **28**. Continuing their study on adenosine nucleosides for anti-HBV drugs, Chu group studied a series of 9-(2-deoxy-2- β-L-arabinofuranosyl)purine nucleosides (**30**) to find that only two compounds of this series, adenine derivative (**30a**) and hypoxanthine derivative (**30b**), had good potency against HBV when evaluated in HepG2/2.2.15 cells, with EC_{50} values of 1.5 and 8 μM, respectively, without any significant toxicity up to 200 μM (Ma et al., 1997).

Two years later, Ku group (Wang et al., 1999) performed an asymmetric synthesis of L-carbocyclic 2′,3′-didehydro-2′,3′-dideoxy- and 2′,3′-dideoxypyrimidine and purine nucleoside analogs and evaluated their anti-HIV and HBV activities. In this study, they found only one compound, L-carbocyclic 2′,3′-didehydro-2′,3′dideoxyadenosine (**31**), to exhibit potent anti-HBV activity, having an EC_{50} value of 0.09 μM, without any toxicity up to 100 μM. In the next study, Lee et al. (1999) studied a series of L-nucleoside analogs containing 2′-vinylic fluoride (**32**) for the activity against HIV-1 and HBV. In this study, they found two compounds that contained pyrimidine ring (**32a** and **32b)** and one compound that contained purine ring (**33)** had appreciable anti-HBV activity with EC_{50} values of 0.18, 0.225, and 1.7 μM, respectively, without any significant toxicity up to 100 μM.

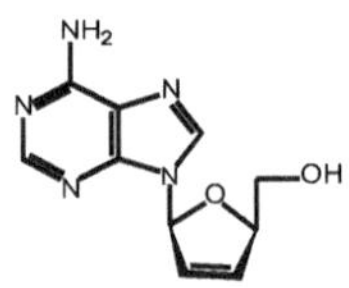

28, β-L-d4A

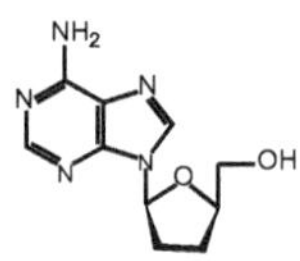

29, β-L-d2A

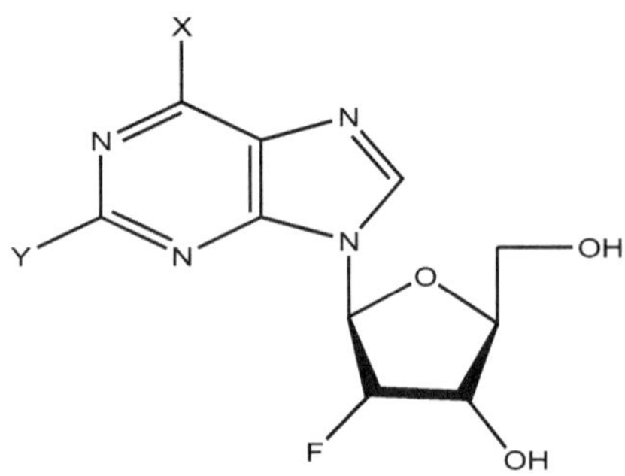

30

30a: X = NH_2, Y = H

30b: X = OH, Y = H

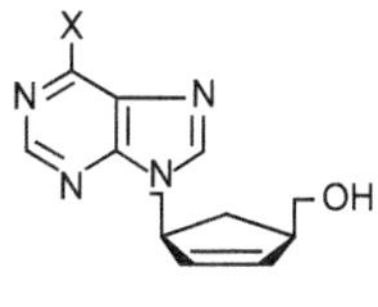

31: X = NH_2

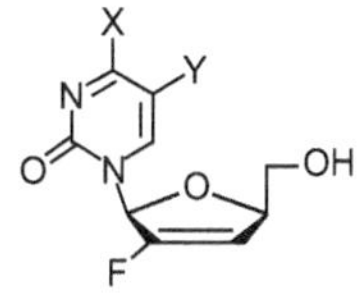

32

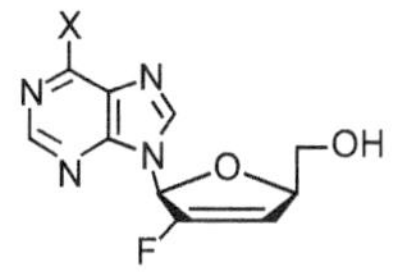

33: X = NH_2

32a: X = NH_2, Y = H
32b: X = NH_2, Y = F

A novel carbocyclic 2′-deoxyguanosine analog, BMS-200,475 (**34**), having potent and selective anti-HBV activity with $EC_{50} = 3$ nM and relatively low toxicity ($CC_{50} = 30$ μM) in HepG2/2.2.15 cell lines, was investigated by Bisacchi et al. (1997). The enantiomer of this compound as well as adenine, thymine, and iodouracil analogs were found to be much less active than this compound. A few novel 5-vinyl pyrimidine nucleosides (**35**) were attempted for anti-HBV activity by Kumar and Tyrrell (2001). In their study, these authors found only two compounds, **35a** and **35b**, to be potent against DHBV (duck HBV) when measured in cell culture.

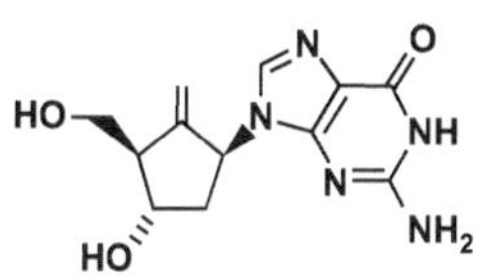

34, BMS-200475

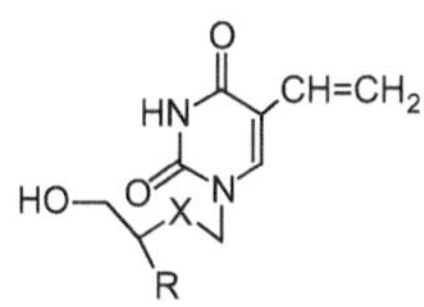

35

35a: X = O, R = CH_2OH
35b: X = C, R = CH_2OH

The current HBV therapy depends primarily on nucleoside/nucleotide analogs. They effectively inhibit virus replication but fail to completely eliminate the virus. In developing countries where the infected population is large, adopting lifelong therapy for patients may not be preferred. Thus, nucleosides or nucleotides may not be final choice to treat HBV infection, but developing new anti-HBV drugs is a difficult and complex project. The final FDA-approved nucleoside/nucleotide analogs and their active forms, as cited by Kang et al. (2015) and Huang et al. (2016), are given in Table 6.5. The structures of the compounds given in this table can be seen in Fig. 6.1. A brief description of these nucleosides is given next.

Table 6.5 Marketed nucleoside analogs to treat HBV infection (these drugs act through inhibition of HBV DNA polymerase) (Huang et al., 2016)

No.	Generic name	Structure	Brand name	Company	Year of FDA approval
1	Lamivudine	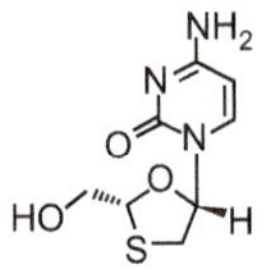	Zeffix, Heptovir, Epivir, Epivir-HBV	GlaskoSmith Kline	1998 (for adults), 2000 (for children), USA
2	Adefovir	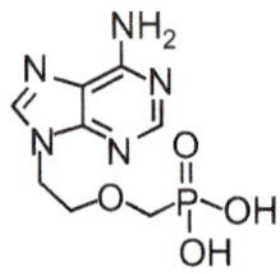	Preveon, Hepsera	Gilead	2002, USA
3	Entecavir	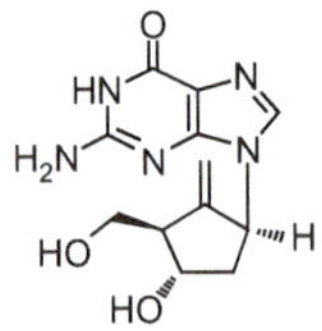	Baraclude	Bristol-Myers-Squib	2005, USA
4	Telbivudine	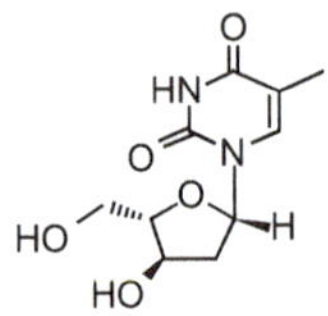	Sebivo (Europe) Tyzeka (USA)	Idenix, Novartis	2006, USA
5	Tenofovir	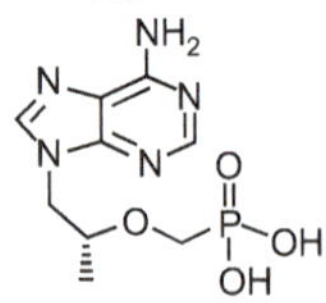	Viread	Gilead Sciences	2008, USA
6	Clevudine	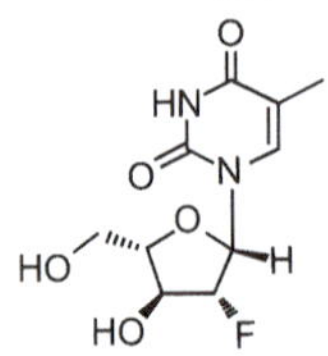	Levovir, Revovir	Bukbang Pharm	2006, Korea

2.3.1.1 Lamivudine

Lamivudine, a pyrimidine nucleoside, a prodrug, was the first approved drug to treat chronic HBV infection. It is first converted to its triphosphate after entering the cell. As a reverse transcriptase inhibitor, it lowers HBV DNA synthesis and inhibits HBV replication. Lamivudine needs a long-term course to avoid reversal of HBV infection.

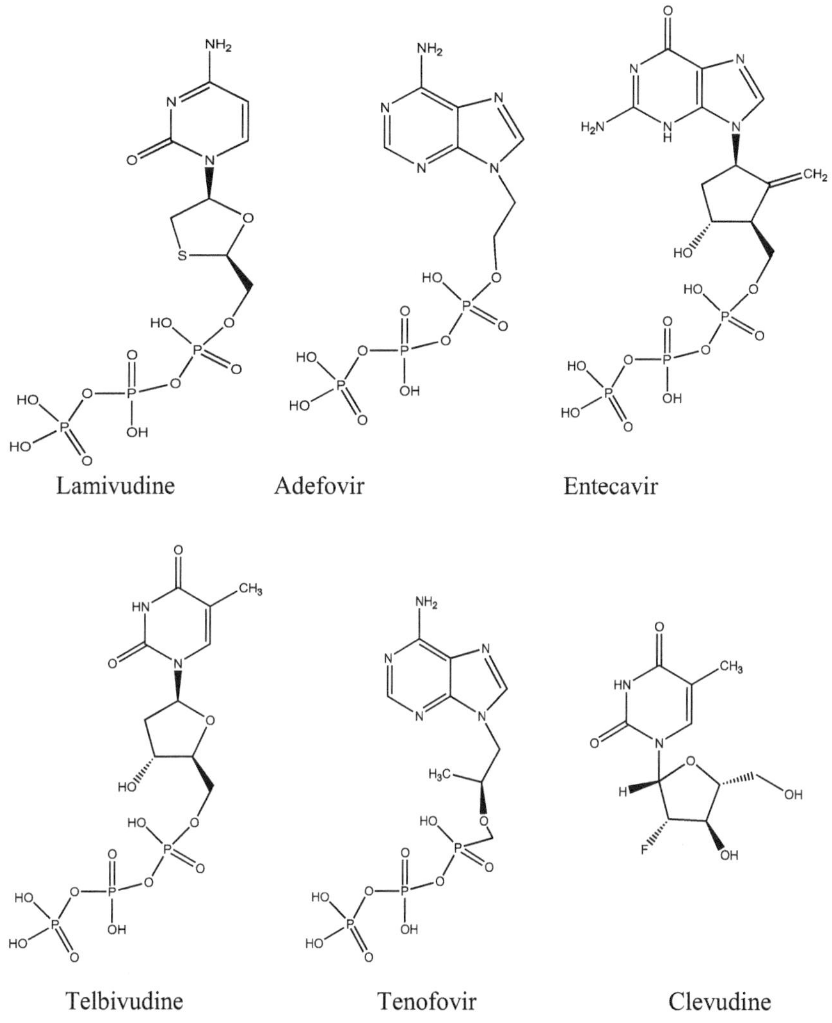

Figure 6.1 The active forms of nucleosides given in Table 6.5.

2.3.1.2 Adefovir

Adefovir, a purine nucleoside prodrug, is first converted to its metabolite adefovir diphosphate after entering the cell, which substitutes for normal substrates of dATP. Once incorporated into viral DNA chain, it stops the elongation of viral DNA synthesis, required for viral replication.

2.3.1.3 Entecavir

Entecavir is a guanine nucleoside. It is also a prodrug, which is first converted to its metabolite, an active triphosphate, in the cells. It inhibits all three functions of viral polymerase, initiation, synthesis of minus strand DNA from the pregenomic RNA, and synthesis of viral DNA chain.

2.3.1.4 Telbivudine

Telbivudine, a purine nucleoside, is a specific, selective, and oral drug to treat chronic HBV infection by inhibiting the synthesis of both the strands of viral DNA. Its efficacy is stronger than that of lamivudine (Hou et al., 2008; Liaw et al., 2009), and it is the only drug that can be safely used for treating pregnant female patients.

2.3.1.5 Tenofovir

Tenofovir, an acyclic adenine nucleotide, is also a prodrug, which is first converted to tenofovir diphosphate on entering the cell. It inhibits both HBV and HIV replication at the reverse transcription stage (WHO guidelines, 2015). It leads to potent inhibition, offers a high barrier to drug resistance, and maintains overall balance between benefits and risks, which is why it is preferably recommended for the treatment of the viral infection.

2.3.1.6 Clevudine

Like other nucleoside analogs, clevudine also inhibits HBV polymerase by competing with the natural substrate thymidine. It was approved by Korea and the Philippines, but its phase III clinical trial was terminated due to some myopathy cases in patients, and because of this, its approval in South Korea was also revoked.

2.3.2 Adverse Effects of Nucleoside/Nucleotide Analogs

While nucleoside/nucleotide analogs have been preferred drugs for chronic HBV infection treatment, they are not free from adverse effects. In the long course of treatment, they produce nephrotoxic and myopathic effects (Fung and Lok, 2004; Fleischer and Lok, 2009). Nephrotoxic effect leads to gradual increase in serum creatine and decreases in serum phosphorus due to inhibition of kidney function, while myopathy leads to muscle pain, muscle weakness, or muscle tenderness (Lewis et al., 2003; Seok et al., 2009)). Viruses also develop resistance to these drugs. Like other polymerases, HBV-RT polymerase lacks proofreading activity on the newly synthesized viral genome, which results in the introduction of random mutations into the progeny HBV DNA.

As pointed out by Zoulim (2004), the major features of the HBV replication cycle relevant to antiviral therapy are the following: (1) HBV replication does not induce a cytopathic effect (structural changes) in infected cells, which in turn is one of the factors involved in viral persistence, (2) covalently closed circular DNA (cccDNA) is the transcriptionally active form of the viral genome and has a long half-life in infected cells, and (3) viral genome replication occurs via a reverse transcription that leads to the production of the viral DNA genome within nucleocapsids, which are then enveloped prior to virion release.

The HBV-RT activity also generates viral mutants at each replication cycle. Moreover, viral nucleocapsids may also be recycled back to the nucleus to initially amplify and maintain a stable pool of viral cccDNA and whose complete clearance from infected cells is almost impossible by the present anti-HBV drugs; it has been shown that none of the currently available polymerase inhibitors can inhibit the de novo formation of cccDNA in a newly infected cell (Delmas et al., 2002; Kock et al., 2003; Seignères et al., 2003). It is suggested that during antiviral therapy, residual circulating virions may still be capable of infecting new cells.

The molecular mechanism of resistance to these drugs and its clinical importance for chronic HBV have been reviewed (Ghany and Liang, 2007; Zoulim and Locarnini, 2009). Attempts have been made to understand the molecular basis of such drug resistance by HBV, for which a homology model based on HIV-RT has been used, as three-dimensional (3D) X-ray structure of HBV polymerase is not available. A detailed study by Kim et al. (2010) showed that most of the drug-resistant mutations alter the conformation of at least one of the following sites: dNTP (deoxynucleotide) binding site, primer binding site, dNTP incoming region, dNTP binding pocket, or magnesium ions binding site, as shown in Fig. 6.2 for lamivudine, for example. The HBV-RT model structure was constructed and refined as previously reported by Daga et al. (2010).

Thus, because of adverse effects and viral resistance, attempts to find foolproof drugs are still ongoing. Kim et al. (2010) compiled a series of new nucleosides/nucleotides, as shown in Table 6.6, that were at various stages of clinical trials. Kim et al. (2010) also compiled a list of nonnucleoside analogs, most of which were found to be active against viral replication (Table 6.7).

2.3.3 Chinese Herbs

As already discussed, Chinese herbs have been fruitfully studied for anti-HBV therapy. Chen et al. (2016) recently studied the herb *Hypericum japonica* and found that its extract had the highest activity in inhibiting

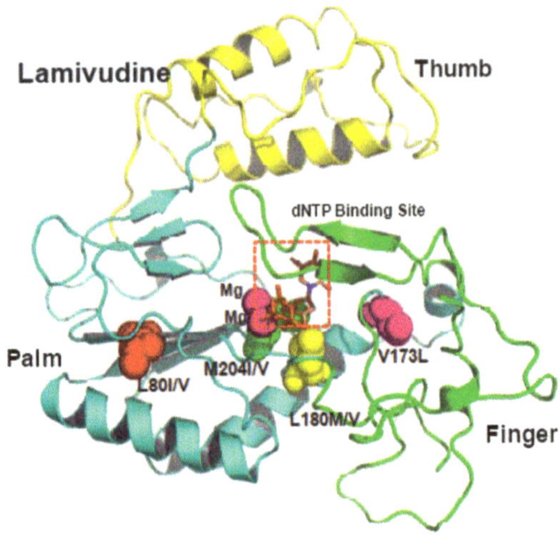

Figure 6.2 The ribbon diagram of homology model of HBV-RT, showing, for example, the location of mutations to lamivudine. HBV-RT consists of three subdomains: fingers (amino acids 1–19 and 90–72 in *green*), palm (amino acids 50–89 and 173–267 in *cyan*), and thumb (amino acids 268–351 in *yellow*). The locations of the mutations are indicated with a sphere model. *(Reprinted from Kim, K.-Y., Kim, N.D., Seong, B.-L., 2010. Discovery and development of anti-HBV agents and their resistance. Molecules 15, 5878–5908.)*

secretion of HBeAg in HepG2-2.2.15 cells. Four compounds were extracted from the ethyl acetate fraction of *H. japonica*. These compounds, as shown in Fig. 6.3, were vincetoxicoside B, (2R, 3R) double hydrogen quercetin-7-O-α-L-rhamnoside, quercitrin, and 1,3,6,7-tetrahydroxy-9H-xanthen-9-one, of which the last one was found to possess the highest activity for the inhibition of secretion of HBeAg in HepG2-2.2.15 cells with an IC_{50} value of 40.25 μg/mL and 67.66% inhibition rate.

3. THEORETICAL STUDIES

Currently, HBV DNA polymerase has drawn great attention to be exploited as a target for designing potential anti-HBV agents. Therefore, study on its 3D structure has been of great importance. Its structure may be exploited to understand the molecular basis of many of its properties, such as mechanism of its polymerization, inhibition, drug resistance, and interpretation of clinical and biochemical data. If 3D X-ray structure of any enzyme is not available, mutation studies have been found to provide valuable information about structure and functions and useful clues to

Table 6.6 New nucleoside/nucleotide analogs under development

Compound (class)	Structure	Company	Phase, year
Lagociclovir valactate (nucleoside), prodrug	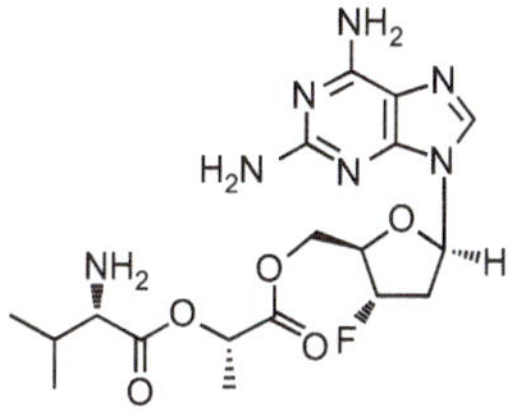	Medvir AB	Phase II 2005, Sweden
Elvucitabine (nucleoside)	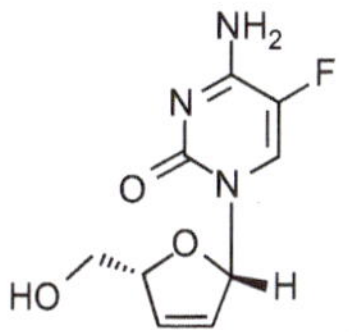	Achillion Pharmaceuticals	Phase II 2001, USA
B-80380 (nucleotide)	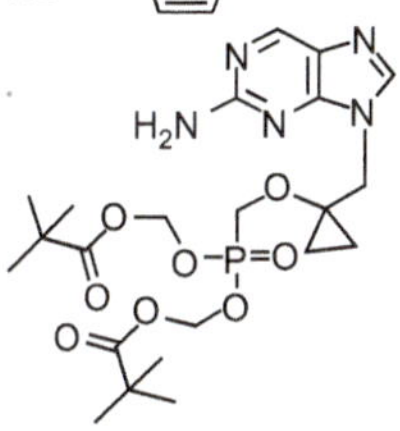	LG Chem Ltd	Phase II 2003, Korea
Radofovir (nucleotide), prodrug	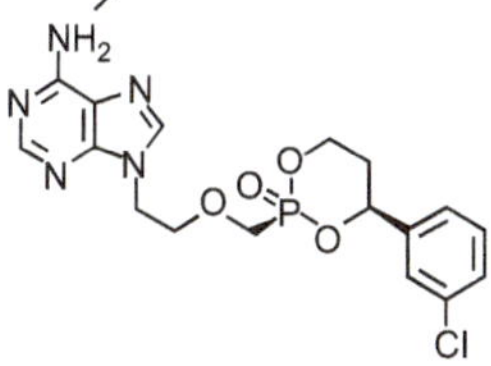	Metabasis Therapeutics	Phase II 2007, USA
Altorcitabine (nucleoside), prodrug	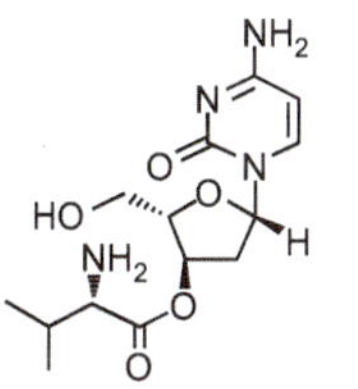	Idenix Pharmaceuticals/ Novartis	Phase II 2003, USA

All drugs reported were tested against HBV-RT (Kim et al., 2010).

develop a homology model. To develop a homology model for HBV polymerase based on HIV-1 RT 3D structure, mutation studies in HIV-1 RT were found to be of great importance. Several mutations present in HIV-1 RT were reported to be present in corresponding positions of HBV-RT. Using such information, homology models of catalytic domains of HBV DNA polymerase (HDP), based on 3D X-ray structure of HIV-1 RT, were developed by various authors (Das et al., 2001; Bartholomeusz

Table 6.7 Some nonnucleoside inhibitors targeting HBV replication (Kim et al., 2010)

No.	Structure	Activity, IC_{50}	Mechanism of action	Reference
1	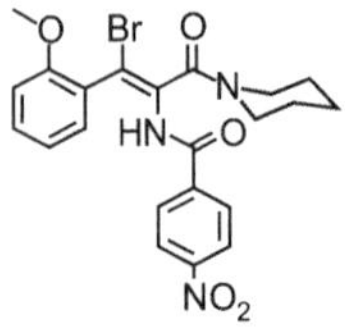	2.4 μM in HepG2 cells	Replication inhibition by blocking RNA packaging	Delaney (2002), King et al. (1998), and Feld et al. (2007)
2	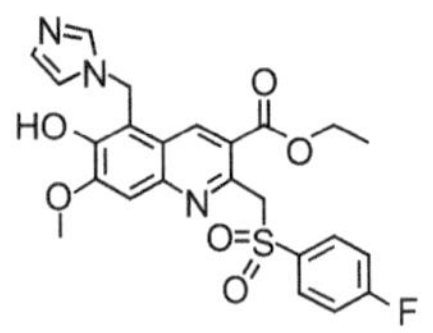	4.7 μM (replication) 26.2 μM (HBsAg) 98.1 μM (HBeAg) in HepG2-2.2.15 cells	Replication inhibition, HBsAg and HBeAg secretion	Liu et al. (2008)
3	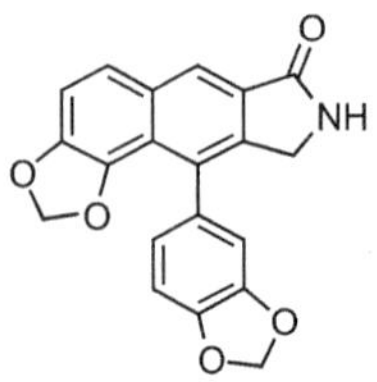	0.08 μM in HepG2-2.2.15 cells	Replication inhibition by down-regulation of HNF-3 and 4	Li et al. (2005) and Ying et al. (2007)
4	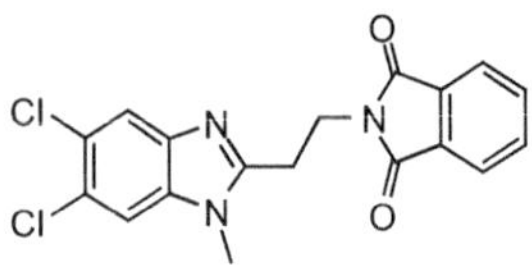	0.9 μM in HepG2 cells	Replication inhibition	Li et al. (2006)
5	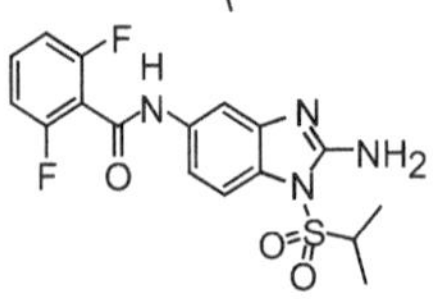	0.7 μM in HepG2-2.2.15 cells	Replication inhibition	Li et al. (2007)
6	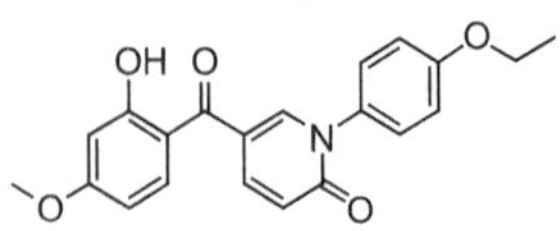	0.206 μM in HepG2 cells	Replication inhibition	Lv et al. (2010)
7	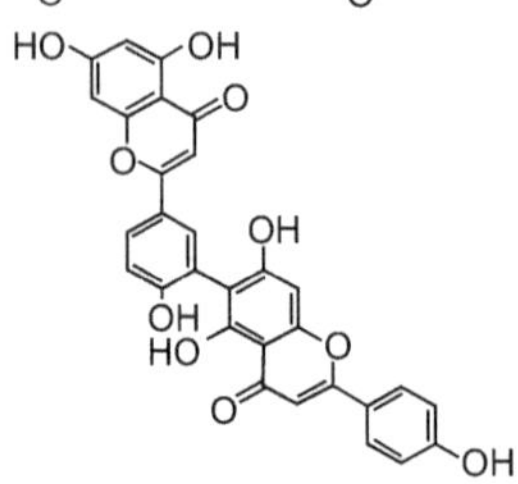	0.25 μM in HepG2-2.2.15 cells	Replication inhibition	Zembower et al. (1996)
8	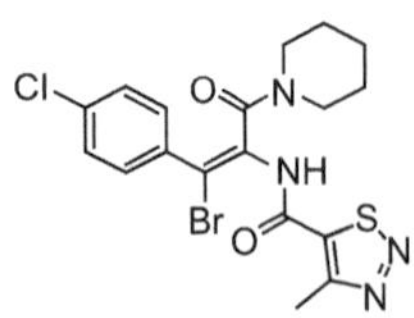	3.59 μg/mL in HepG2-2.2.15 cells	Replication inhibition	Dong et al. (2010)

Table 6.7 Some nonnucleoside inhibitors targeting HBV replication (Kim et al., 2010)—cont'd

No.	Structure	Activity, IC_{50}	Mechanism of action	Reference
9	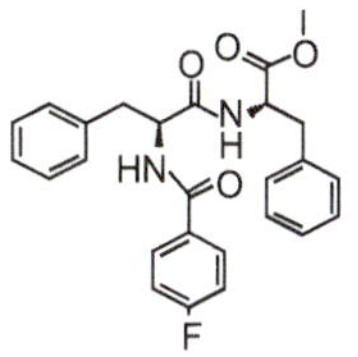	1.40 μM in HepG2-2.2.15 cells	Replication inhibition, HBsAg and HBeAg secretion	Xu et al. (2009)
10	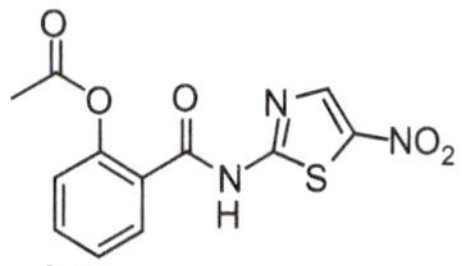	EC_{50} = 0.12 μM in HepG2-2.2.15 cells	Replication inhibition	Korba et al. (2008)
11	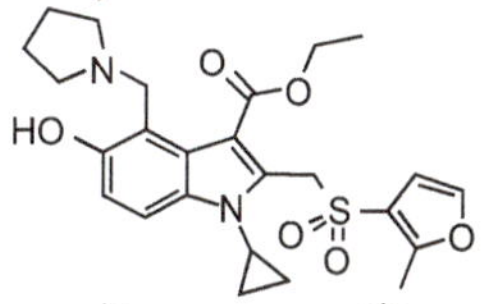	1.52 μg/mL in HepG2-2.2.15 cells	Replication inhibition	Zhao et al. (2006)
12	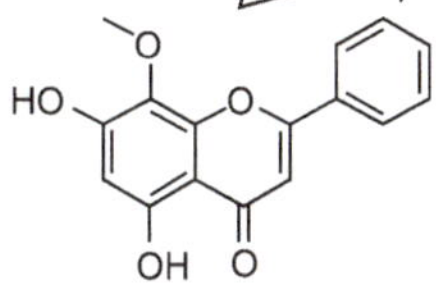	0.57 μg/mL in DHBV replication 4.0 μg/mL (HBsAg, HBeAg)	Replication inhibition, HBsAg and HBeAg secretion	Guo et al. (2007)
13	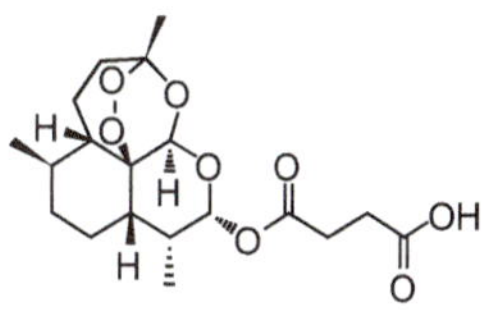	2.3 μM (HBsAg) 0.5 μM (Replication)	Replication inhibition and HBsAg secretion	Romero et al. (2005)
14	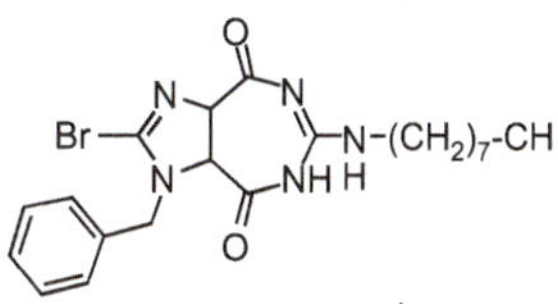	EC_{50} = 1.7 μM in HepG2-2.2.15 cells	Replication and virion secretion inhibition	Zhang et al. (2005)
15	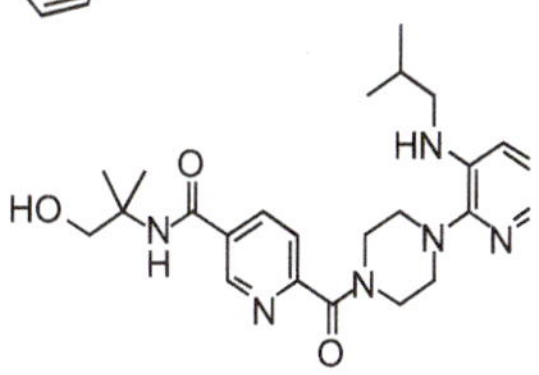	~0.01 μg/mL	RT inhibition	Lee et al. (2002)
16	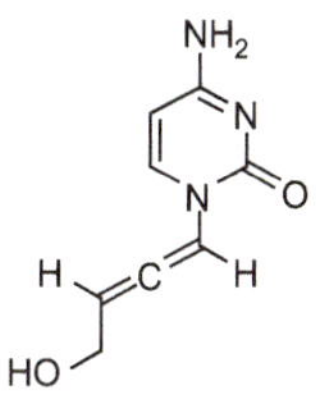	0.08 μM in HepG2-2.2.15 cells	Replication inhibition	Zhu et al. (1997)

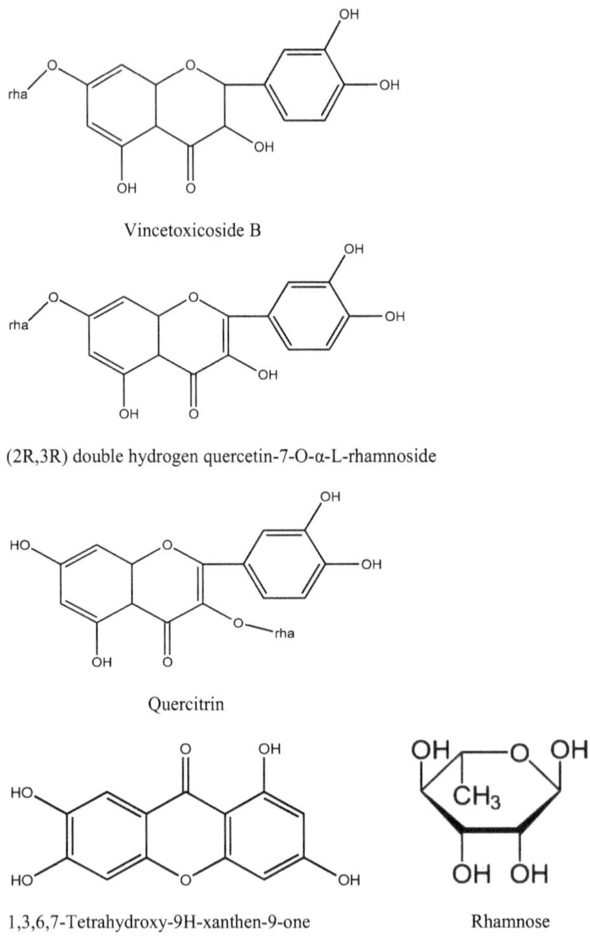

Figure 6.3 Extracts of *H. japonica* acting as inhibitors of secretion of HBeAg in HepG2-2.2.15 cells (rha: rhamnoside).

et al., 2004; Langley et al., 2007), which were found to be useful to provide insights about the binding mode of various HBV inhibitors and development of resistance to them. The sequence alignment model of HDP designed by Das et al. (2001) based on a protein data bank code 1RTD of X-ray structure of HIV-1 RT was used to reveal the mechanism of HDP resistance to lamivudine and emtricitabine and then later by Chu and co-workers for molecular dynamics studies to explain the mechanism of action of clevudine and the sensitivity of adefovir and drug resistance (Chong and Chu, 2002; Yadav and Chu, 2004; Sharon and Chu, 2008).

Daga et al. (2010) reported a different homology model of HDP based on a better alignment and a more recent HIV-1 RT X-ray structure, which could be used to explain all mutations in HDP reported to date. With this model, Daga et al. made two different docking studies to find in both very similar docking poses for four substrates and 10 inhibitors. In a molecular dynamics study, the conformational changes were found to lead to the formation of a small hydrophobic pocket near the nucleotide binding site, which was occupied by exocyclic alkene moiety of entecavir and the sulfur atom of lamivudine, explaining the better binding affinity of these inhibitors than natural substrates.

A docking study of two herbal compounds, bentysrepinine (Y101, **42**) and repensine (**43**), was performed with a homology model of HDP by Meng et al. (2015) to find that except the 2 Mg^{2+} ions coordinated by residues Asp418 and Asp540, the active pocket was mainly made by hydrophobic residues, such as Val419, Ala421, Ala422, Phe423, Tyr424, Met506, and Asn571. The hydrogen bond interactions of both the compounds may be with Arg376. These findings may be useful for designing potent anti-HBV drugs.

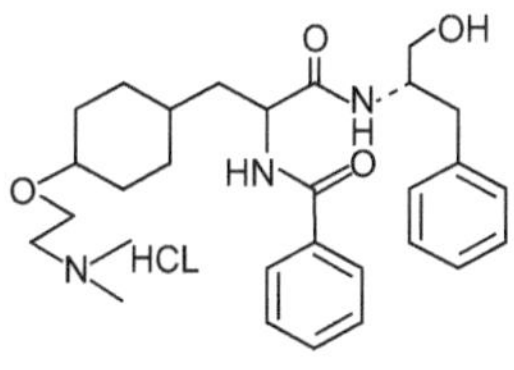

36, Y1o1

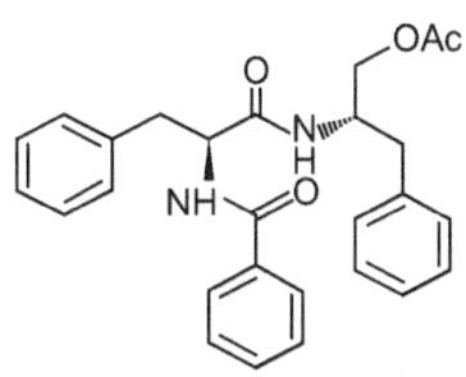

37, Repensine

A recent attempt to find a homology of HBV-RT was made by Xu et al. (2016). Till today, no experimentally derived HBV-RT structure could be available. However, HIV and HBV-RTs have been recognized as homologous (van Hemert et al., 2014), which possess similar multiple catalytic activities. However, the structural model of HBV-RT with full confidence could still not be obtained. Since nucleoside/nucleotide reverse transcriptase inhibitors (NRTIs) are still the drugs of choice to treat chronic HBV infections and since NRTIs are also potent HIV-1 inhibitors, they have been preferentially used to theoretically study the 3D structure of HBV-RT based on the structure of HIV-1 RT. Tenofovir disoproxil fumarate (TDF) is an oral prodrug of nucleotide analog tenofovir (TFV) and is easily converted to the latter following absorption and readily catalyzed to the active diphosphate form. This TDF is a nonselective potent

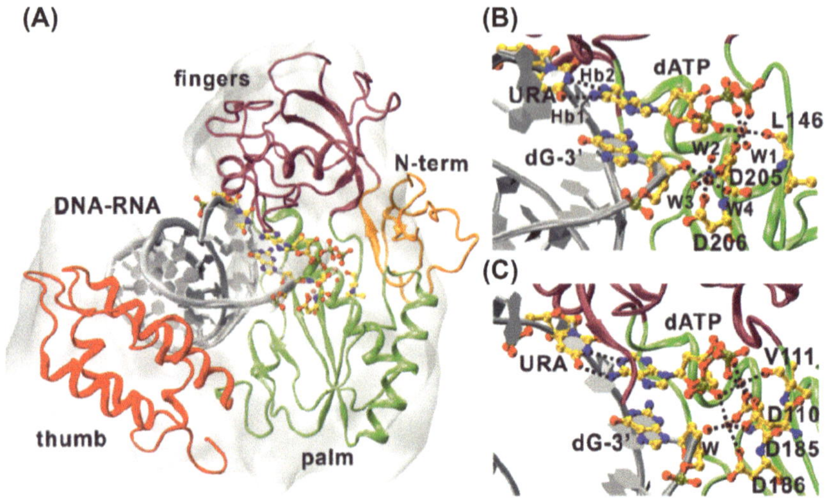

Figure 6.4 Comparative structures of HBV-RT4/DNA-RNA/dATP and HIV-RT4/DNA-RNA/dATP crystal structures with Mg^{2+} as active site: (A) HBV-RT4/DNA-RNA/dATP structure is shown with its three domains in different colors and the interactions between its substrate and residues in stick and ball representations. Structurally aligned to this HBV model is shown the crystal structure of HIV-RT4/DNA-RNA/dATP (PBD code:4PQU, chain A, residues 1–320) by *gray surface* representation. (B) The active site configuration of HBV-RT4/DNA-RNA/dATP model, exhibiting the nonbonded interactions represented by *black dashed lines* as the two hydrogen bonds (Hb1 and Hb2), the ionic bonds formed between the Mg^{2+} ions (in *red ball* representations without labeling), and the chelating oxygen atoms. The four water oxygen atoms participating in chelation are labeled as W1 to W4. (C) The active site configuration of HIV-RT4/DNA-RNA/dATP crystal structure with only one water molecule (W) involved and interacting with Mg^{2+}. The D110 has no counterpart in HBV-RT model. Hydrogen atoms have not been shown for clarity. *(Reprinted from Xu, X., Thai, H., Kitrinos, K.M., Xia, G., Gaggar, A., Paulson, A., Ganova-Raeva, L., Khudyakov, Y., Lara, J., 2016. Modeling the functional state of the reverse transcriptase of hepatitis B virus and its application to probing drug-protein interaction. BMC Bioinform. 17, 583–592. Open Access publication.)*

inhibitor of HIV-1 RT (Robbins et al., 1998; Miller et al., 2001; Durand-Gasselin et al., 2009), as well as an effective drug to treat chronic HBV infection (Perry and Simpson, 2009; Michailidis et al., 2012; Jones and Hu, 2013; Jung et al., 2015) with no evidence of resistance (Kitrinos et al., 2014). As HBV-RT is homologous to HIV-RT (van Hemert et al., 2014), it can be assumed that a similar mechanism may be involved in the differential treatment response to TDF. Therefore, Xu et al. (2016) utilized the crystal structure of HIV-RT to model the atomic structure of HBV-RT in its functional state to delineate in detail the protein-drug interactions between TFV-DP (tenofovir diphosphate) and active sites of genetically distinct variants of HBV-RT. In models of NRTIs bound to HBV-RT, Xu et al. (2016) assumed that an effective drug binds to the active site, as if it were a native substrate in position to be further covalently incorporated, similar to HIV-RT/DNA-DNA/TFV-DP complex (Tuske et al., 2004).

Out of several models designed for HBV-RT, based on variant HIV-1 RT models, a model HBV-RT4 was chosen for further refinement using multinanosecond molecular dynamics simulation with the native hybrid DNA-RNA duplex and with dATP (deoxy adenosine triphosphate) into the active site, as their original orientation in the HIV-RT/DNA-RNA/dATP crystal structure. A comparison of this model is shown in Fig. 6.4. Another model, as shown in Fig. 6.5, was designed with replacement of dATP by TFV-DP. These figures show the comparative interactions of substrates with the HBV model and 3D structure of HIV-1 (PBD code:4PQU).

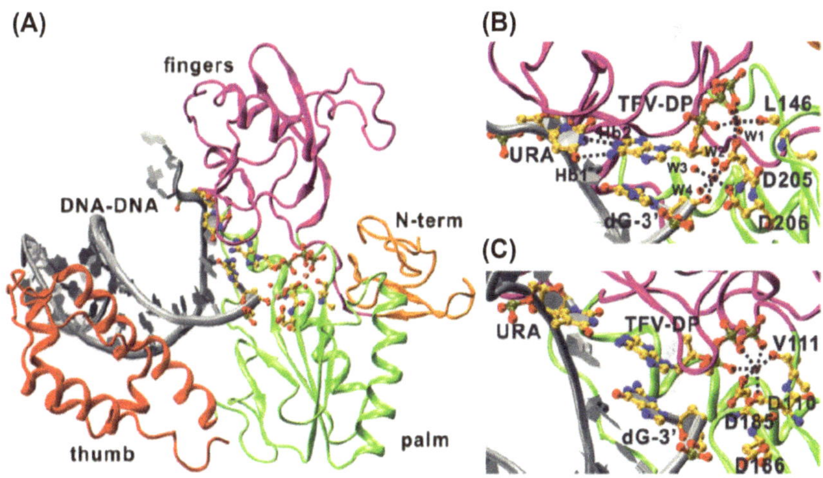

Figure 6.5 Comparative structures of HBV-RT4/DNA-RNA/TFV-DP and HIV-RT4/DNA-RNA/TFV-DP crystal structures with Mg^{2+} as active site: (A) HBV-RT4/DNA-RNA/TFV-DP structure is shown with all its domains in different colors and the interactions between its substrate and residues in stick and ball representations. (B) The active site configuration of HBV-RT4/DNA-RNA/dATP model, exhibiting the nonbonded interactions represented by *black dashed lines* as the two hydrogen bonds (Hb1 and Hb2) and the ionic bonds formed between the Mg^{2+} ions (in *red ball* representations without labeling), and the chelating oxygen atoms. The four water oxygen atoms participating in chelation are labeled as W1 to W4. (C) The active site configuration of HIV-RT4/DNA-RNA/TFV-DP crystal structure with only 1 Mg^{2+} identified in the active site and no water molecule involved. Hydrogen atoms have not been shown for clarity. *(Reprinted from Xu, X., Thai, H., Kitrinos, K.M., Xia, G., Gaggar, A., Paulson, A., Ganova-Raeva, L., Khudyakov, Y., Lara, J., 2016. Modeling the functional state of the reverse transcriptase of hepatitis B virus and its application to probing drug-protein interaction. BMC Bioinform. 17, 583–592. Open Access publication. Open Access publication.)*

4. CONCLUSIONS

In conclusion, presently there are only five drugs—lamivudine, adefovir, entecavir, telbivudine, and tenofovir—that have been approved by the FDA to treat the chronic HBV infection. These are nucleoside analogs. Nucleoside and nucleotide analogs have been extensively studied for antiviral therapy, but they soon develop resistance from viruses. However, some other classes of drugs such as indoles, benzimidazoles, thiazolides, myristoyl analogs, and caudatin analogs have also been extensively studied for their anti-HBV effects. These chemicals have been studied particularly for the inhibition of HBV polymerase (reverse transcriptase), its three antigens, surface antigen (HBsAg), core antigen (HBcAg), and e-antigen (HBeAg), or human hepatoma cell lines (HepG2/2.2.15 cells). Since polymerase inhibitors have been of great interest and since 3D structure of HBV is still not available, attempts have been made to develop its homology model based on 3D X-ray structure of HIV-1 reverse transcriptase. Molecular modeling studies have shown that its active sites are almost identical to those of HIV-1 RT.

REFERENCES

Bao, T.T., Xu, G.F., Liu, T.G., Sun, R.H., Song, Z.Y., 1979. A comparison of the pharmacological actions of seven constituents isolated from fructus schizadrae (author's transl). Acta Pharm. Sinica 14, 1–7.

Bartholomeusz, A., Tehan, B.G., Chalmers, D.K., 2004. Comparison of the HBV and HIV polymerase, and anti-viral resistance mutations. Antiviral Ther. 9, 149–160.

Bisacchi, G.S., Chao, S.T., Bachard, C., Daris, J.P., Innaimo, S., Jacobs, G.A., Kocy, O., Lapointe, P., 1997. BMS-200475, a novel carbocyclic 2′-deoxyguanosine analog with potent and selective anti-hepatitis B virus activity in vitro. Bioorg. Med. Chem. Lett. 7, 127–132.

Bolon, P.J., Wang, P., Chu, C.K., Gosselin, G., Boudou, V., Pierra, C., Mathe, C., Imbach, J.L., Faraj, A., Abdelaziz el Alaoui, M., Sommadossi, J.P., Balakrishnapai, S., Zhu, Y.L., Lin, J.S., Cheng, Y.C., Schinazi, R.F., 1996. Anti-human immunodeficiency and anti-hepatitis B virus activity of beta-l-2′,3′-dideoxy purine nucleosides. Bioorg. Med. Chem. Lett. 6, 1657–1662.

Chai, H., Zhao, Y., Zhao, C., Gong, P., 2006. Synthesis and in vitro anti-hepatitis B virus activities of some ethyl 6-bromo-5-hydoxy-1H-indole-3-carboxylates. Bioorg. Med. Chem. Lett. 14, 911–917.

Chen, G.Y., Liu, M., Yu, M.B., Zheng, C., Wei, S., 2016. Study on the chemical constituents and anti-HBV activity of hypericum japonica. MATEC Web Conf. 60, 1–5.

Chong, Y., Chu, C.K., 2002. Understanding the unique mechanism of L-FMAU (Clevudine) against hepatitis B virus: molecular dynamics studies. Bioorg. Med. Chem. Lett. 12, 3459–3462.

Daga, P.R., Duan, J., Doerksen, R.J., 2010. Computational model of hepatitis B virus DNA polymerase molecular dynamics and docking to understand resistant mutations. Protein Sci. 19, 796–807.

Das, K., Xiong, X., Yanh, H., Westland, C.E., Gibbs, C.S., Sarafianos, S.G., Arnold, E., 2001. Molecular modeling and biochemical characterization reveal the mechanism of hepatitis B virus polymerase resistance to lamivudine (3TC) and emtricitabine (FTC). J. Virol. 75, 4771–4779.

De Clercq, E., Field, H.J., 2005. Antiviral prodrugs – the development of successful prodrug strategies for antiviral chemotherapy. Br. J. Pharmacol. 147, 1–11.

Delaney 4th, W.E., Edwards, R., College, D., Shaw, T., Furman, P., Painter, G., Locarnini, S., 2002. Phenylpropenamide derivatives AT-61 and AT-130 inhibit replication of wild-type and lamivudine-resistant strains of hepatitis B virus. Antimicrob. Agents Chemother. 46, 3057–3560.

Delmas, J., Schorr, O., Jamard, C., Gibbs, C., Trepo, C., Hantz, O., Zoulim, F., 2002. Inhibitory effect of adefovir on viral DNA synthesis and covalently closed circular DNA formation in duck hepatitis B virus-infected hepatocytes in vivo and in vitro. Antimicrob. Agents Chemother. 46, 425–433.

Dong, W.L., Liu, Z.X., Liu, X.H., Li, Z.M., Zhao, W.G., 2010. Synthesis and antiviral activity of new acrylamide derivatives containing 1, 2, 3-thiadiazole as inhibitors of hepatitis B virus replication. Eur. J. Med. Chem. 45, 1919–1926.

Durand-Gasselin, L., Van Rompay, K.K., Vela, J.E., Henne, I.N., Lee, W.A., Rhodes, G.R., Ray, A.S., 2009. Nucleotide analogue prodrug tenofovir disoproxil enhances lymphoid cell loading following oral administration in monkeys. Mol. Pharm. 6, 1145–1151.

Feld, J.J., Colledge, D., Sozzi, V., Edwards, R., Littlejohn, M., Locarnini, S.A., 2007. The phenylpropenamide derivative AT-130 blocks HBV replication at level of viral RNA packaging. Antiviral Res. 76, 168–177.

Fleischer, R.D., Lok, A.S., 2009. Myopathy and neuropathy associated with nucleos(t)ide anolog therapy for hepatitis B. J. Hepatol. 51, 787–791.

Fung, S.K., Lok, A.S., 2004. Drug insight: nucleoside and nucleotide analog inhibitors for hepatitis B. Nat. Clin. Pract. Gastroenterol. Hepatol. 1, 90–97.

Furman, P.A., Davis, M., Liotta, D.C., Paff, M., Frick, L.W., Nelson, D.J., Dornsife, R.E., Wurster, J.A., Wilson, A.J., Fyfe, J.A., Tuttle, J.B., Miller, W.H., Condreay, L., Averett, D.R., Schinazi, R.F., Painter, G.R., 1992. The anti-hepatitis B virus activities, cytotoxicities, and anabolic profiles of the (-) and (+) enantiomers of cis-5-fluoro-1-[2-hydroxymethyl)-1, 3-oxathiolan-5-yl]cytosine. Antimicrob. Agents Chemother. 36, 2686–2692.

Ghany, M., Liang, T.J., 2007. Drug targets and molecular mechanisms of drug resistance in chronic hepatitis B. Gastroenterology 132, 1574–1585.

Glushkov, R.G., 1992. Arbidol Antiviral, immunostimulant, interferon inducer. Drug Future 17, 1079–1081.

Guo, Q., Zhao, L., You, Q., Yang, Y., Gu, H., Song, G., Lu, N., Xin, J., 2007. Anti-hepatitis B virus activity of wogonin in vitro and in vivo. Antiviral Res. 74, 16–24.

Hikino, H., Kiso, Y., Taguchi, H., Ikeya, Y., 1984. Antihepatotoxic actions of lignoids from Schizandra chinensis fruits. Planta Med. 50, 213–218.

Hou, J., Yin, Y.K., Xu, D., Tan, D., Niu, J., Zhou, X., Wang, Y., Zhu, L., He, Y., Ren, H., 2008. Telbivudine versus lamivudine in Chinese patients with chronic hepatitis B. Hepatology 47, 447–454.

Huang, C., Lin, H.H., Wan, J.-B., He, C., Hu, Y., 2016. R&D of hepatitis B drugs: an analysis based on technology flows measured by patent citations. PLoS One 11, e0164328.

Huang, S.C., Wu, P.L., Wu, T.S., 1997. Two coumarins from the root bark of Clausena excavate. Phytochemistry 44, 179–181.

Jia, W., Liu, Y., Li, W., Liu, Y., Zhang, D., Zhang, P., Gon, P., 2009. Synthesis and in vitro anti-hepatitis B virus activity of 6H-[1]benzothiopyrano[4,3-b]quinolin-9-ols. Bioorganic Med. Chem. 17, 4569–4574.

Jones, S.A., Hu, J., 2013. Hepatitis B virus reverse transcriptase: diverse functions as classical and emerging targets for antiviral intervention. Emerg. Microb. Infect. 2, e56.

Jung, S.K., Kim, K.A., Ha, S.Y., Lee, H.K., Kim, Y.D., Lee, B.H., Paik, W.H., Kim, J.W., Bae, W.K., Kim, N.H., Lee, J.S., Jwa, Y.J., 2015. Tenofovir disoproxyl fumarate monotherapy for necleos(t)ide analogue naïve and necleos(t)ide analogue experienced cronic hepatitis B patients. Clin. Mol. Hepatol. 21, 41–48.

Kang, L., Pan, J., Wu, J., Hu, J., Sun, Q., Tang, J., 2015. Anti-HBV drugs: progress, unmet needs and new hope. Viruses 7, 4960–4977.

Kaushik, S., Gupta, S.P., Sharma, P.K., 2010. Design and development of hepatitis B virus agents. Curr. Med. Chem. 17, 3377–3392.

Kim, H.O., Ahn, S.K., Alves, A.J., Beach, J.W., Jeong, L.S., Choi, B.G., Van Roey, P., Schinazi, R.F., Chu, C.K., 1992. Antiviral optically pure dioxolane purine nucleosides. J. Med. Chem. 35, 1987.

Kim, K.-Y., Kim, N.D., Seong, B.-L., 2010. Discovery and development of anti-HBV agents and their resistance. Molecules 15, 5878–5908.

King, R.W., Ladner, S.K., Miller, T.J., Zaifert, K., Perni, R.B., Conway, S.C., Otto, M.J., 1998. Inhibition of human hepatitis B virus replication by AT-61, a phenylpropenamide derivative, alone and in combination with (-) beta-L-2′,3′-dideoxy-3′thiacytidine. Antimicrob. Agents Chemother. 42, 3179–3186.

Kitrinos, K.M., Corsa, A., Liu, Y., Flaherty, J., Snow-Lampart, A., Marcellin, P., Borroto-Esoda, K., Miller, M.D., 2014. No detectable resistance to tenofovir disoproxil fumarate after 6 years of therapy in patients with cronic hepatitis B. Hepatology 59, 434–442.

Kock, J., Baumert, T.F., Delaney, W.E.I.V., Blum, H.E., Von Weizsacker, F., 2003. Inhibitory effect of adefovir and lamivudine on the initation of hepatitis B virus infection in primary tupaia hepatocytes. Hepatology 38, 1410–1418.

Kongkathip, B., Kongkathip, N., Sunthitikawinsakul, A., Napaswat, C., Yoosook, C., 2005. Anti-HIV-1 constituents from Clausena excavata: Part II. carbazoles and a pyranocoumarin. Phytother Res. 19, 728–731.

Korba, B.E., Montero, A.B., Farrar, K., Gaye, K., Mukerjee, S., Ayers, M.S., Rossignol, J.F., 2008. Nitazoxanide, tizoxanide and other thiazolides are potent inhibitors of hepatitis B virus and hepatitis C virus replication. Antiviral Res. 77, 56–63.

Kumar, R., Tyrrell, D.L.J., 2001. Novel 5-vinyl pyrimidine nucleosides with potent anti-hepatitis B virus activity. Bioorg. Med. Chem. Lett. 11, 2917–2920.

Kuo, Y.-H., Wu, M.-D., Hung, C.-C., Huang, R.L., Yang Kuo, L.-M., Shen, Y.-C., Ong, C.-W., 2005. Syntheses of C_{18} dibenzocyclooctadiene lignan derivatives as anti-HBsAg and anti-HBeAg agents. Bioorganic Med. Chem. 13, 1555–1561.

Langley, D.R., Walsh, A.W., Baldick, C.J., Eggers, B.J., Rose, R.E., Levine, S.M., Kapur, A.J., Colnno, R.J., Tenney, D.J., 2007. Inhibition of hepatitis B virus polymerase by entecavir. J. Virol. 81, 3992–4001.

Lee, J., Shim, H., Park, Y., Park, S., Shin, J., Yang, W., Lee, H., Park, W., Chung, Y., Lee, S., 2002. 2, 5-Pyridinedicarboxylic acid derivatives as non-nucleosidic reverse transcriptase inhibitors of hepatitis B virus. Bioorg. Med. Chem. Lett. 12, 2715–2717.

Lee, K., Choi, Y., Gullen, E., Schluter-Wirtz, S., Schinazi, R.F., Cheng, Y.-C., Chu, C.K., 1999. Synthesis and anti-HBV activities of 2′-fluoro-2′,3′-unsaturated L-nucleosides J. Med. Chem. 42, 1320–1328.

Lee, M., Chu, C.K., 1995. Dioxolane cytosine nucleosides as anti-hepatitis B agents. Bioorg. Med. Chem. Lett. 5, 2011–2014.

Lee, S.U., Shin, C.G., Lee, C.K., Lee, Y.S., 2007. Bioactivity and structure-activity relationship of cinnamic acid esters and their derivatives as potential antifungal agents for plant protection. Eur. J. Med. Chem. 42, 1309–1315.

Leneva, I.A., Fadeeva, N.I., Fedykina, I.T., 1994. In: Proceedings of the 7th International Conference on Antiviral Research, Abstract 187.
Lewis, W., Day, B.J., Copeland, W.C., 2003. Mitochondrial toxicity of NRTI antiviral drugs: an integrated cellular perspective. Nat. Rev. Drug Discov. 2, 812–822.
Li, Y.-F., Wang, G.-F., He, P.-L., Huang, W.-G., Zhu, F.-H., Gao, H.-Y., Tang, W., Luo, Y., Feng, C.-L., Shi, L.-P., Ren, Y.-D., Lu, W., Zuo, J.-P., 2006. Synthesis and anti-hepatitis B virus activity of novel benzimidazole derivatives. J. Med. Chem. 49, 4790–4794.
Li, Y.-F., Wang, G.-F., Luo, Y., Huang, W.-G., Tang, W., Feng, C.-L., Shi, L.-P., Ren, Y.-D., Lu, W., Zuo, J.-P., 2007. Identification of 1-isopropylsulfonyl-2-amine benzimidazoles as a new class of inhibitors of hepatitis B virus. Eur. J. Med. Chem. 42, 1358–1364.
Li, Y., Fu, L., Yeo, H., Zhu, J.L., Chou, C.K., Kou, Y.H., Yeh, S.F., Gullen, E., Austin, D., Cheng, Y.C., 2005. Inhibition of hepatitis B virus gene expression and replication by helioxanthin and its derivatives. Antivir. Chem. Chemother. 16, 193–201.
Liang, G.Y., Liu, Y.M., Xu, B.X., 2002. Patent 02160309.X.
Liaw, Y.-F., Gane, E., Leung, N., Zeuzem, S., Wang, Y., Lai, C.L., Heathcote, E.J., Manns, M., Bzowej, N., Niu, J., Han, S.-H., Huang, S.G., Cakaloglu, Y., Tong, M.J., Papatheodoridis, G., Chen, Y., Brown, N.A., Albanis, E., Galil, K., Naoumov, N.V., 2009. 2-Year globe trial results: telbivudine is superior to lamivudine in patients with chronic hepatitis B. Gastroenterology 136, 486–495.
Lu, H., Liu, G.T., 1992. Anti-oxidant activity of dibenzocyclooctene lignans isolated from Schisandraceae. Planta Med. 58, 311–313.
Lv, Z., Sheng, C., Wang, T., Zhang, Y., Liu, J., Feng, J., Sun, H., Zhong, H., Niu, C., Li, K., 2010. Design, synthesis, and antihepatitis B virus activities of novel 2-pyridone derivatives. J. Med. Chem. 53, 660–668.
Ma, T., Lin, J.S., Newton, M.G., Cheng, Y.C., Chu, C.K., 1997. Synthesis and anti-hepatitis B virus activity of 9-(2-Deoxy-2-fluoro-β-L-arabinoturanosyl)purine nucleosides. J. Med. Chem. 40, 2750–2754.
Meng, F.C., Xu, W.R., Li, Y.Z., Hunag, Z.M., Liang, G.Y., Liu, C.X., 2015. In silico molecular docking study of repensine and bentysrepinine against HBV DNA polymerase. Chinese Herb. Med. 7, 39–44.
Michailidis, E., Kirby, K.A., Hachiya, A., Yoo, W., Hong, S.P., Kim, S.O., Folk, W.R., Sarafianos, S.G., 2012. Antiviral therapies: focus on hepatitis B reverse transcriptase. Int. J. Biochem. Cell Biol. 44, 1060–1071.
Miller, M.D., Margot, N.A., Hertogs, K., Larder, B., Miller, V., 2001. Antiviral activity of tenofovir (PMPA) against nucleoside-resistant clinical HIV samples. Nucleos. Nucleot. Nuclei. Acids 20, 1025–1028.
Murata, T., Imai, Y., Hirata, T., Miyamoto, M., 1970. Biological-active triterpenes alismatis rhizoma. Chem. Pharm. Bull. 18, 1347–1353.
Perry, C.M., Simpson, D., 2009. Tenofovir disoproxil fumarate in chronic hepatitis B. Drugs 69, 2245–2256.
Robbins, B.L., Srinivas, R.V., Kim, C., Bischofberger, N., Fridland, A., 1998. Anti-human immunodeficiency virus activity and cellular metabolism of a potential prodrug of the acyclic nucleoside phosphonate 9-R-(2-phosphonomethoxypropyl) adenine (PMPA), bis(isopropyloxymethylcarbonyl)PMPA. Antimicrob. Agents Chemother. 42, 612–617.
Romero, M.R., Efferth, T., Serrano, M.A., Castano, B., Macias, R.I., Briz, O., Marin, J.J., 2005. Effect of artemisinin/artesunate as inhibitors of hepatitis B virus production in an "in vitro" replicative system. Antiviral Res. 68, 75–83.
Rossi, S. (Ed.), 2006. Australian Medicines Handbook. Australian Medicines Handbook Pty Ltd.

Sasaki, S., 1924. In: Khoyo Taiwan Minkan, Yakyo Shokubustu Shi. Khobunkan, Taipei, p. 36.

Seignères, B., Martin, P., Werle, B., Schorr, O., Jamard, C., Rimsky, L., Trépo, C., Zoulim, F., 2003. Effects of pyrimidine and purine analog combinations in the duck hepatitis B virus infection model. Antimicrob. Agents Chemother. 47, 1842–1852.

Seok, J.I., Lee, D.K., Lee, C.H., Park, M.S., Kim, S.Y., Kim, H.S., Jo, H.Y., Kim, D.S., 2009. Long-term therapy with clevudine for chronic hepatitis B can be associated with myopathy characterized by depletion of mitochondrial DNA. Hepatology 49, 2080–2086.

Sharon, A., Chu, C.K., 2008. Understanding the molecular basis of HBV drug resistance by molecular modeling. Antiviral Res. 80, 339–353.

Stachulski, A.V., Pidathala, C., Row, E.C., Sharma, R., Berry, N.G., Iqbal, M., Bentley, J., Allman, S.A., Edwards, G., Helm, A., Hellier, J., Korba, B.E., Semple, J.E., Rossignol, J.F., 2011. Thiazolides as novel antiviral agents. 1. inhibition of hepatitis B virus replication. J. Med. Chem. 54, 4119–4132.

Su, C.R., Yeh, S.F., Liu, C.M., Damu, A.G., Kuo, T.-S., Chiang, P.C., Basto, K.F., Lee, K.H., Wu, T.S., 2009. Anti-HBV and cytotoxic activities of pyranocoumarin derivatives. Bioorg. Med. Chem. 17, 6137–6143.

Sunthitikawinsakul, A., Kongkathip, N., Kongkathip, B., et al., 2003. Coumarins and carbazoles from Clausena excavata exhibited antimycobacterial and antifungal activities. Planta Med. 69, 155–157.

Tuske, S., Sarafianos, S.G., Clark Jr., A.D., Ding, J., Naeger, L.K., White, K.L., Miller, M.D., Gibbs, C.S., Boyer, P.L., Clark, P., Wang, G., Gaffney, B.L., Jones, R.A., Jerina, D.M., Hughes, S.H., Arnold, E., 2004. Structures of HIV-1 RT-DNA complexes before and after incorporation of the anti-AIDS drug tenofovir. Nat. Struct. Mol. Biol. 11, 469–474.

van Hemert, F.J., Berkhout, B., Zaaijer, H.L., 2014. Differential binding of tenofovir and adefovir to reverse transcriptase of hepatitis B virus. PLoS One 9, 1–9.

Wang, L.J., Chen, H., Ma, Y.-B., Huang, X.-Y., Geng, C.-A., Hang, X.-M., Chen, J.-J., 2015. Design, synthesis and biological evaluation of caudatin analogs as potent hepatitis B virus inhibitors. Med. Chem. 11, 165–179.

Wang, L.J., Geng, C.A., Ma, Y.B., Huang, X.Y., Luo, J., Chen, H., Guo, R.H., Zhang, X.M., Chen, J.J., 2012a. Synthesis, structure–activity relationships and biological evaluation of caudatin derivatives as novel anti-hepatitis B virus agents. Bioorganic Med. Chem. 20, 2877–2888.

Wang, L.-J., Geng, C.-A., Ma, Y.-B., Luo, J., Huang, X.-Y., Chen, H., Zhou, N.J., Zhang, X.-M., Chen, J.-J., 2012b. Design synthesis, and molecular hybrids of caudatin and cinnamic acids as novel anti-hepatitis B virus agents. Eur. J. Med. Chem. 54, 352–365.

Wang, P., Gullen, B., Newton, M.G., Cheng, Y.C., Schinazi, R.F., Chu, C.K., 1999. Asymmetric synthesis and antiviral activities of L-carbocyclic 2′,3′-didehydro-2′,3′-dideoxy and 2′,3′-dideoxy nucleosides. J. Med. Chem. 42, 3390–3399.

WHO guidelines, 2015. For the Prevention, Care and Treatment of Person with Chronic Hepatitis B Infection. Available online:http://www.who.int/hiv/pub/hepatitis/hepatitis-b-guidlines/en/.

Xiang, Y., Cavalcanti, S., Schinazi, R.F., Pai, S.B., Zhu, Y.-L., Chu, C.K., 1995. Anti-human immunodeficiency virus type-1 (HIV-1) and anti-hepatitis B virus (HBV) activities of (2,3-dideoxy-2-fluoro-β-L-threo-pentofuranosyl) nucleosides. Bioorg. Med. Chem. Lett. 5, 877–880.

Xu, B., Huang, Z., Liu, C., Cai, Z., Pan, W., Cao, P., Hao, X., Liang, G., 2009. Synthesis and biological evaluation of Matijing-Su derivatives as potent anti-HBV agents. Bioorganic Med. Chem. 17, 3118–3125.

Xu, X., Thai, H., Kitrinos, K.M., Xia, G., Gaggar, A., Paulson, A., Ganova-Raeva, L., Khudyakov, Y., Lara, J., 2016. Modeling the functional state of the reverse transcriptase of hepatitis B virus and its application to probing drug-protein interaction. BMC Bioinform. 17, 583–592.

Yadav, V., Chu, C.K., 2004. Molecular mechanisms of adefovir sensitivity and resistance in HBV polymerase mutants: a molecular dynamics study. Bioorg. Med. Chem. Lett. 14, 4313–4317.

Yang, X.W., Hattori, M., Namba, T., Chen, D.F., Xu, G.J., 1992a. Anti-lipid peroxidative effect of an extract of the stems of kadsura heteroclita and its major constituent, kadsurin, in mice. Chem. Pharm. Bull. 40, 406–409.

Yang, X.W., Miyashiro, H., Hattori, M., Namba, T., Tezuka, Y., Kikuchi, T., Chen, D.F., Xu, G.J., Hori, T., Extine, M., Mizuno, H., 1992b. The Anti-HBsAg (Human Type B Hepatitis, Surface Antigen) and Anti-HBeAg (Human Type B Hepatitis, e Antigen) C18 DibenzocyclooctadieneLignans from Kadsura matsudai and Schizandra arisanen. Chem. Pharm. Bull. 40, 1510–1516.

Ying, C., Li, Y., Leung, C.H., Robek, M.D., Cheng, Y.C., 2007. Unique antiviral mechanism discovered in anti-hepatitis B virus research with a natural product analogue. Proc. Natl. Acad. Sci. U.S.A. 104, 8526–8531.

Zembower, D.E., Lin, Y.M., Flavin, M.T., Chen, F.C., Korba, B.E., 1996. Robust flavone, a potential non-nucleoside anti-hepatitis B agent. Antiviral Res. 39, 81–88.

Zhang, P., Zhang, N., Korba, B.E., Hosmane, R.S., 2005. Synthesis and in vitro anti-hepatitis B and C virus activities of ring-expanded ('fat') nucleobase analogues containing the imidizo[4,5-e][1,3]diazepine-4,8-dione] ring system. Bioorg. Med. Chem. Lett. 25, 5397–5401.

Zhang, Q., Jiang, Z.Y., Luo, J., Cheng, P., Ma, Y.B., Zhang, F.X., Zhou, J., Chen, J.J., 2008. Anti-HBV agents. Part 1: synthesis of alisol A derivatives: a new class of hepatitis B virus inhibitors. Bioorg. Med. Chem. 18, 4647–4650.

Zhao, C., Zhao, Y., Chai, H., Gong, P., 2006. Synthesis and in vitro anti-hepatitis B virus activities of some ethyl 5-hydroxy-1H-indole-3-carboxylates. Biorg. Med. Chem. 14, 2552–2558.

Zhu, Y.L., Pai, S.B., Liu, S.H., Grove, K.L., Jones, B.C., Simons, C., Zemlicka, J., Cheng, Y.C., 1997. Inhibtion of replication of hepatitis B virus by cytallene in vitro. Antimicrob. Agents Chemother. 41, 1755–1760.

Zhuang, M., Jiang, H., Suzuki, Y., Li, X.G., Xiao, P., Tanaka, T., Ling, H., Yang, B.F., Saitoh, H., Zhang, L.F., Qin, C.A., Sugamura, K., Hattori, T., 2009. Design, synthesis, and molecular hybrids of caudatin and cinnamic acids as novel anti-hepatitis B virus agents. Antivir. Res. 82, 73–81.

Zoulim, F., 2004. Antiviral therapy of chronic hepatitis B: can we clear the virus and prevent drug resistance. Antiviral Chem. Chemo. 15, 299–305.

Zoulim, F., Locarnini, S., 2009. Hepatitis B virus resistance to nucleos(t)ide analogues. Gastroenterology 137, 1593–1616.

CHAPTER 7

The Medicinal Chemistry of Antihepatitis Agents III: Anti-HCV Agents

1. INTRODUCTION

The infection of hepatitis C virus (HCV) has been more serious than that of any other hepatitis virus. It has been a leading cause of end-stage liver disease and hepatocellular carcinoma (HCC) (Lavanchy, 2009). Chronic hepatitis leads to approximately 25% of primary HCC and approximately to the same extent liver cirrhosis (Perz et al., 2006). Thus, HCC and HCV-related liver failure invite liver transplantation, seen more frequently in Western countries (Carbone et al., 2012). In liver transplant recipients with recurrent hepatitis C, the pegylated interferon (PEG-IFN) treatment was the standard of care, and after transplantation, sustained virologic response rates ranged from 26% to 50% (Berenguer et al., 2008; Bernuth et al., 2017). With the approval of new direct-acting antivirals (DAAs), the management of recurrent HCV infection has greatly improved (Bernuth et al., 2017). DAAs inhibit specific HCV nonstructural (NS) proteins that are vital for its replication. Drugs like boceprevir, telaprevir, simeprevir, asunaprevir, grazoprevir, and paritaprevir inhibit NS3/4A. Likewise, omvitasvir, ledipasvir, daclatasvir, elbasvir, and velpatasvir are NS5A inhibitors, and sofosbuvir and dasabuvir inhibit NS5B polymerase. A schematic representation of the targets of these drugs is given in Fig. 7.1 (Geddawy et al., 2017). These drugs in combination with each other have been fruitful, but the potential of drug–drug interactions and resistance variance have been a great challenge to the success of these drugs in combination or otherwise (Geddawy et al., 2017). A detail of the timeline of approval of these drugs by the FDA, as compiled by Geddawy et al., is given in Table 7.1. Further, side effects were a frequent problem in HCV

Studies on Hepatitis Viruses
ISBN 978-0-12-813330-9
https://doi.org/10.1016/B978-0-12-813330-9.00007-7

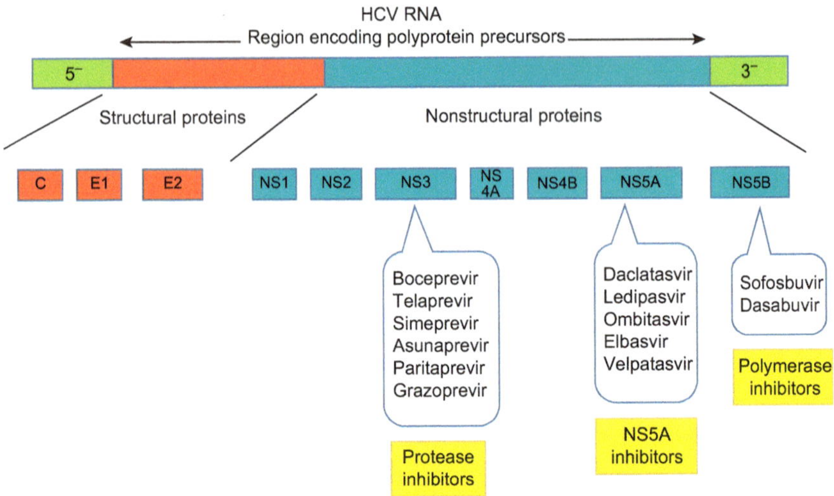

Figure 7.1 A schematic representation of the targets of direct-acting anti-hepatitis C virus (HCV) drugs. *(Reprinted from from Geddawy, A., Ibrahim, Y.F., Elbahie, N.M., Ibrahim, M.A., 2017. Direct acting anti-hepatitis C virus drugs: clinical pharmacology and future direction. J. Transl. Int. Med. 5, 8–17. Open access publication.)*

Table 7.1 A timeline detail of approval of direct-acting antihepatitis C virus (HCV) drugs by the FDA (Geddawy et al., 2017)

Date of approval	Drug/Drug combination
May 13, 2011	Boceprevir in combination with peginterferon alfa for adult patients
May 23, 2011	Telaprevir in combination with peginterferon alfa and ribavirin for adult patients
November 2013	Simeprevir in combination with peginterferon alfa and ribavirin or in combination with sofosbuvir
December 2013	Sofosbuvir in combination with ribavirin or pegylated interferon and ribavirin
October 2014	Ledipasvir/sofosbuvir in one tablet
December 2014	A combination of ombitasvir/paritaprevir/ritonavir/ dasabuvir for the treatment of patients with genotype 1
July 2015	Daclatasvir with sofosbuvir, a combination of omvitasvir/paritaprevir/ritonavir in one tablet to be used in combination with ribavirin for the treatment of HCV genotype 4 infections
January 28, 2016	A combination of elbasvir/grazoprevir, with or without ribavirin for the treatment of genotype 4 infections
July 2016	A combination of sofosbuvir/velpatasvir, with or without ribavirin for the treatment of adult patients with all genotype infections

treatment even with combination therapy. Therefore efforts are still continued to find more effective drugs against HCV infection.

As already discussed in Chapter 3, the most important targets that have been mostly exploited to find the potent anti-HCV drugs have been its NS proteins, NS3/4A, NS4B protease, NS5A, and NS5B. Additionally, attempts have also been made to find the drugs that inhibit the entry of the virus in human cells and the drugs that inhibit HCV replication. Both experimental and theoretical attempts have been made to find anti-HCV drugs of therapeutic importance.

2. EXPERIMENTAL STUDIES

2.1 NS3 Inhibitors

The NS protein of HCV, NS3, is a bifunctional protein with both serine protease and helicase/NTPase activities that is crucial for the delivery of essential components for viral replication (Bartenschlager and Lohmann, 2000). Inhibitors of this protease have been found to be effective in reducing the viral load in humans (Kolykhalov et al., 2000; Lamarre et al., 2003). The structural detail of this protease has been recently reviewed by Sharma et al. (2017). NS3 actually works with its cofactor NS4A, so its activated form is NS3-NS4A, which is simply written as NS3/4A (Sharma et al., 2017). Earlier studies on NS3/4A inhibitors were related to peptide-based protease inhibitors, where an N-terminal hexapeptide product of substrate cleavage (**1**, Fig. 7.2) was found at Boehringer-Ingelheim to inhibit NS3/4A (Llinas-Brunet et al., 1998). In succession to this, a hexamer (**2**, Fig. 7.2) was investigated with an IC_{50} value of 55 nM. Structures of these two compounds led to suggest that P1 carboxylic was acid was crucial to activity. P's refer to amino acid residues in the peptide substrates. As shown in Fig. 7.3, A "Pn" refers to an amino acid residue on the N-terminal side and a Pn′ to a residue on the carboxyl side. The corresponding binding sites in the enzyme are shown by Sn and Sn′.

NS3/4A protease inhibitors were divided into four major classes: (1) first-generation or breakthrough structures, (2) second-generation NS3/4A protease inhibitors, (3) novel structural features or classes, and (4) third-generation NS3/4A protease inhibitors (McCauley and Rudd, 2016). Structures given in Fig. 7.2 represent the first-generation or breakthrough structures, and their development is described subsequently. A large number of NS3/4A inhibitors belonging to other classes, along with other FDA approved anti-HIV compounds, are given in Table 4.2, and their

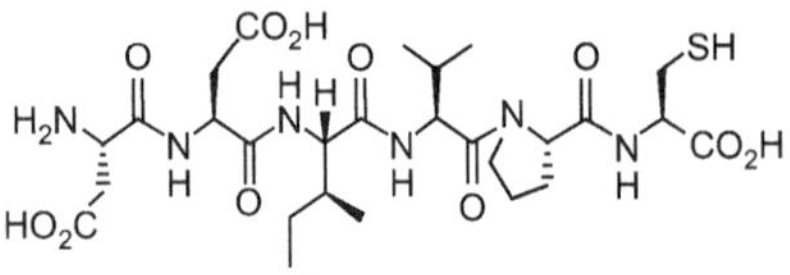

1

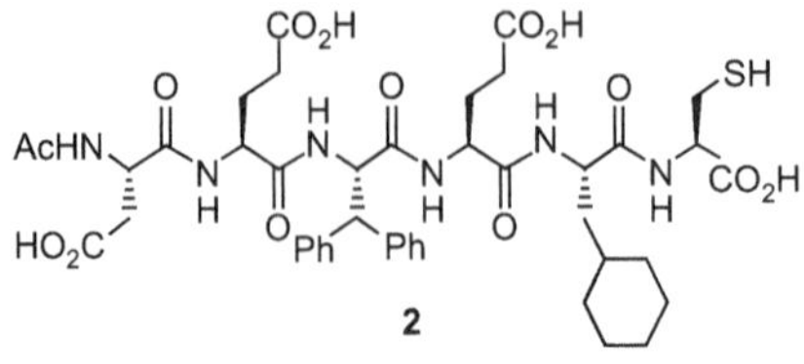

2

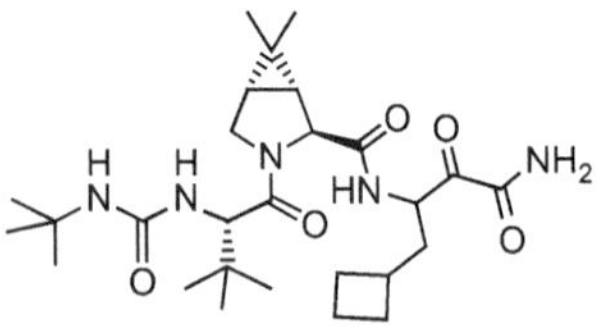

3, Boceprevir (SCH503034, VICTRELIS™)

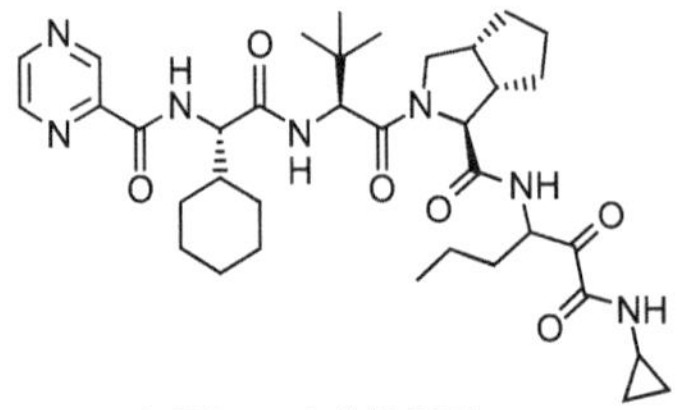

4, Telaprevir (VX-950)
INCIVEK™, INCIVIO™

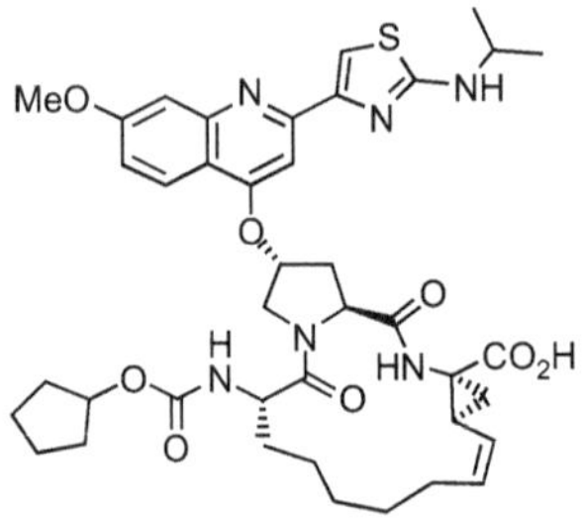

5, Ciluprevir (BILN-2061)

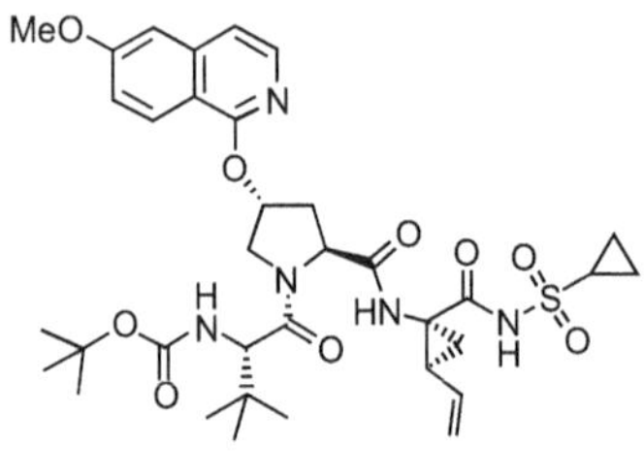

6, BMS-605339

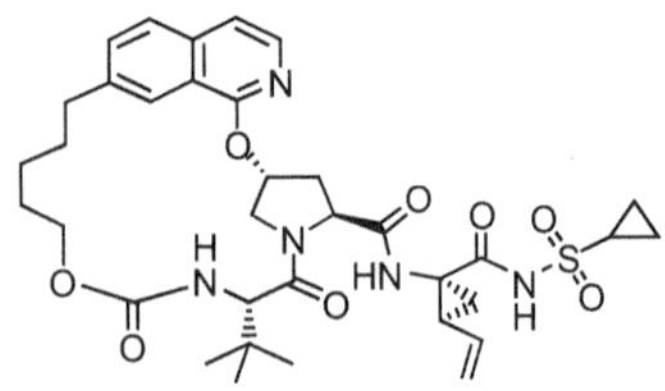

7, MK-4519

Figure 7.2 Structures of breakthrough HCV NS3/4A inhibitors. **1** and **2**, substrate-derived inhibitors, from Boehringer and IRBM; **3**, VICTRELIS, from Merk; **4**, INCIVEK, from Vertex; **5**, Ciluprevir, from Boehringer-Ingelheim; **6**, BMS-605,339, from Bristol-Myers Squibb; **7**, MK-4519, from Merck.

--------P4–P3–P2–P1–P1′–P2′–P3′–P4′--------

↑

Scissile bond

Figure 7.3 Nomenclature of amino acid residues in peptide substrate. A Pn refers to a residue on N-terminal direction and Pn′ to a residue on the carboxyl side.

structures in Fig. 4.2 (Chapter 4). Their description here may be superfluous; therefore we present only the details of first generation.

As already described, compounds 1 and 2 are substrate-based inhibitors that were developed by Boehringer-Ingelheim and IRBM. Compound 3, boceprevir, trade name Victrelis, was initially developed by Schering-Plough, then by Merck after it acquired Schering in 2009. It was approved by the FDA in 2011. In January 2015, Merck announced that it would be voluntarily withdrawing Victrelis from the market due to overwhelming superiority of development of newer DAA agents, such as ledipasvir and sofosbuvir.

Compound 4, telaprevir (VX-950) with the trade names Incivek and Incivo, was developed by Vertex Pharmaceuticals and Jonson and Johnson. It is recommended specifically to be used to treat HCV genotype 1 infection. Ciluprevir (5) was manufactured and developed by Boehringer-Ingelheim and prescribed to treat HCV genotype 1a and 1b infection under the trade name BILN 2061. It was found to be highly selective for NS3 protease over serine and cysteine proteases, human leukocyte elastase, and cathepsin B. However, due to toxicity in animals, its further development was discontinued in Phase Ib trial, but it was exploited to study new macrocyclic inhibitors, like simeprevir (TMC-435) and danoprevir, which have structures like 8 and 9, respectively, as given subsequently. Both simeprevir and danoprevir are NS3/4A protease inhibitors. However, while simeprevir with the trade name Olysio is in combination with other drugs for HCV genotype 1 and 4 infections, danoprevir has a favorable potency profile against multiple genotypes 1–6 and key mutants (genotype 1b, $IC_{50} = 0.2$ nM; replicon genotype 1b, $EC_{50} = 1.6$ nM) (Levin, 2015).

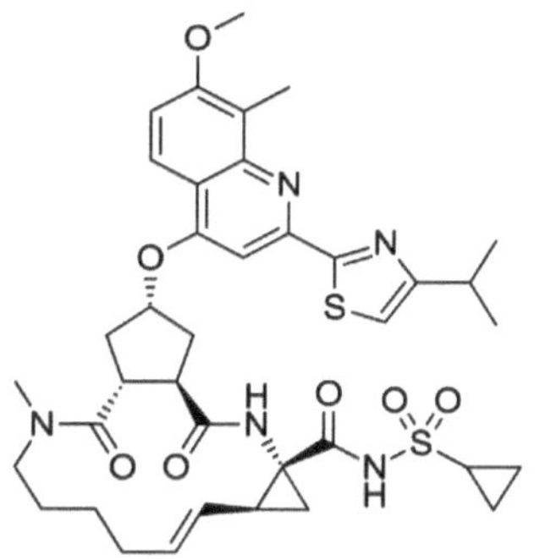

8, Simeprevir (TMC-435)

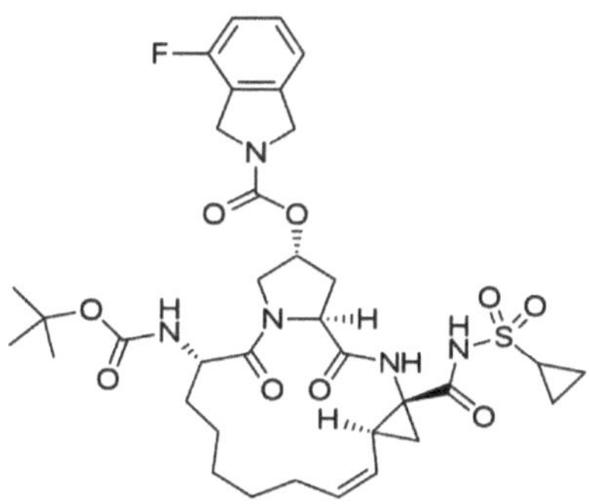

9, Danoprevir

BMS-605339 (6), developed by Bristol-Myers-Squibb, is a novel linear tetrapeptide α-ketamide inhibitor of NS3/4A protease. It has shown specificity for HCV genotype 1a and 1b, but mutation at some points in the protease has been found to induce resistance to this drug (Jenkins et al., 2014; Wang et al., 2017). A methoxy group at the C6-position of the isoquinoline ring of this compound was found to be optimal with respect to potency and pharmacokinetics of this compound. Compound 7, MK-4519, investigated by Merck, is a close analog of Ciluprevir, having a cyclic structure. Design of this compound was based on molecular modeling studies on P2—P4 macrocyclic inhibitors of NS3/4A protease (Liverton et al., 2008).

Some more studies were conducted on macrocyclic inhibitors of NS3/4A protease. Bäck et al. (2007) developed two such series containing either a P2 trisubstituted macrocyclic cyclopentane- or a P2 cyclopentene dicarboxylic acid moiety as surrogates for the widely used N-acyl-(4R)-hydroxyproline in the P2-position. These inhibitors were optimized for anti-HCV activity by examining different ring sizes in the macrocyclic ring systems and further by exploring the effect of P4 substituent removal on potency. A tetrapeptide inhibitor (**10**) developed by Llinás-Brunet et al. (2004b) was adopted as a motif to develop many potent inhibitors. Even the development of Ciluprevir (**5**, BILN-2061) was based on this motif. Optimization of the P2 quinoline substituent of this compound and macrocyclization had led to the discovery of Ciluprevir (Llinás-Brunet et al., 2004a). It was observed that a 14-membered ring system was the most potent and that compounds lacking P4-substituents were most favorable to NS3/4A inhibition. Further, the corresponding P1 acylsulfonamides were observed to be much superior to the corresponding P1 carboxylic acids in potency for NS3/4A protease inhibition. Thus, in the study of Bäck et al. (2007) the most potent compound reported was **11** that lacked P4-substituent and had an acyl sulfonamide group at P1 position.

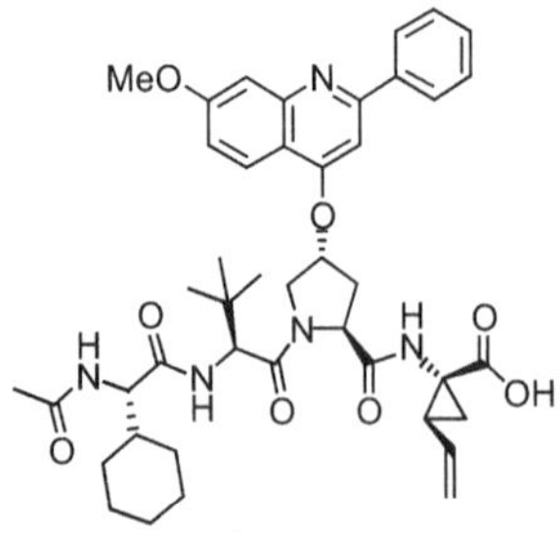

10

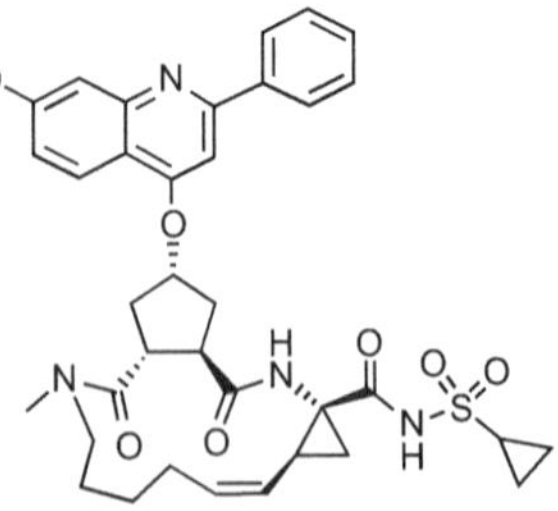

11

In NS3/4A protease inhibition, importance of P1 acylsulfonamide group in tri- and tetrapeptide inhibitors was also shown by Rönn et al. (2006).

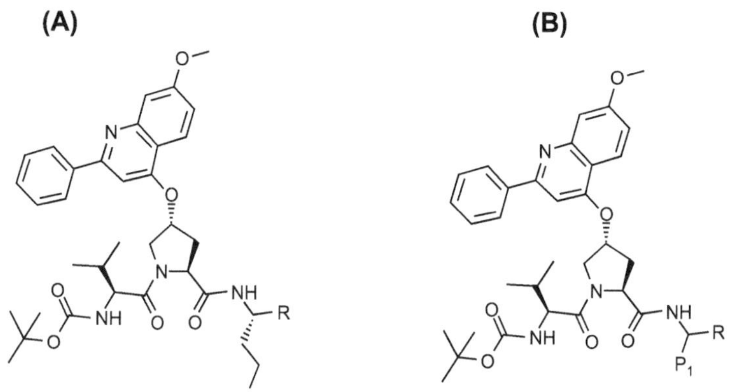

Figure 7.4 Tetrapeptide scaffolds. (A) variations in acylsulfonamide groups (R); (B) variations in sulfonamide groups (R) as well as in P1 substituents (P1).

Different series of tri- and tetrapeptide-based acyl sulfonamide inhibitors, with scaffolds as shown in Figs. 7.4 and 7.5, and their structure–activity relationships (SARs) from both enzymatic and cell-based in vitro were studied by these authors. The acylsulfonamide inhibitors were also observed by some authors (Poliakov et al., 2004) to exhibit an exceptional specificity for the NS3 protease. All the compounds studied by Rönn et al. were found to exhibit inhibitory potencies higher than those of corresponding carboxylic acids in both enzymatic and cell-based assays. In sulfonamide functionality, the sulfonyl oxygen was found to play an important role in interaction with the enzyme. It was also observed that the positive outcome of the sulfonamide substituents was highly dependent on the character of

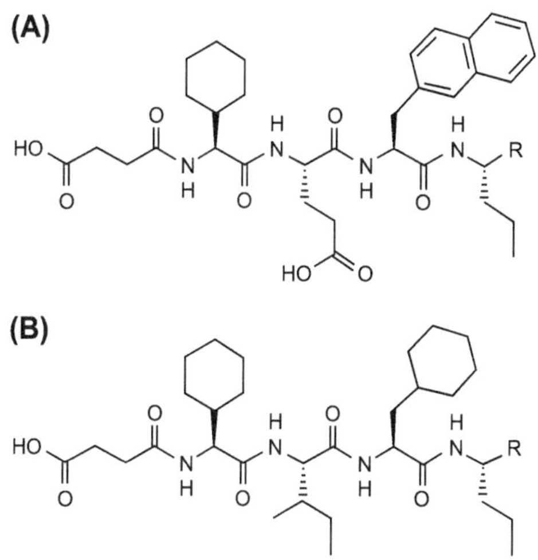

Figure 7.5 Two different tetrapeptide scaffolds having different P2-substituents.

the P1 side chain. The study finally emphasized the potential of acylsulfonamides as carboxylic replacement in product-based NS3 protease inhibitors.

A comparison between compounds **10** and **11** shows that while the former has a proline ring in its structure, the latter has a cyclopentane ring instead. Johansson et al. (2006) had identified some potent NS3/4A protease inhibitors from a series of compounds that had a trisubstituted cyclopentane moiety (**12**) in the P2 position in place of N-acyl-(4R)-hydroxyproline (**13**), widely used to develop inhibitors against several enzymes. Thus, with proper choice of substituents, these authors obtained some very potent and promising NS3/4A inhibitors, such as **14** and **15**, with K_i values of 22 and 16 nM, respectively.

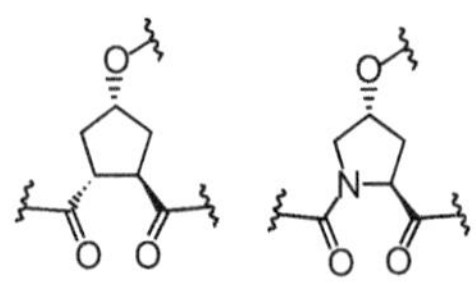

12 **13**

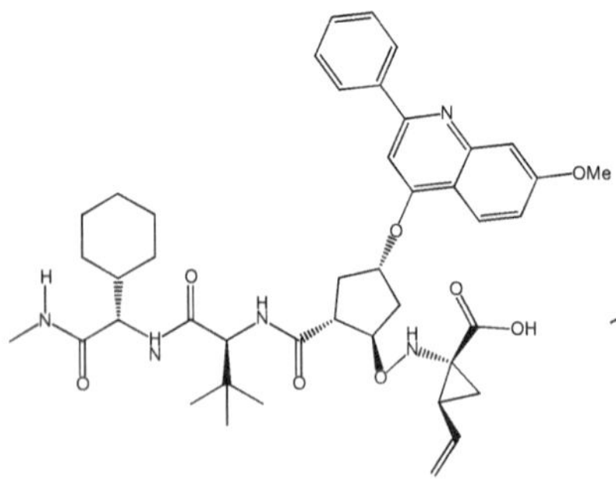

14

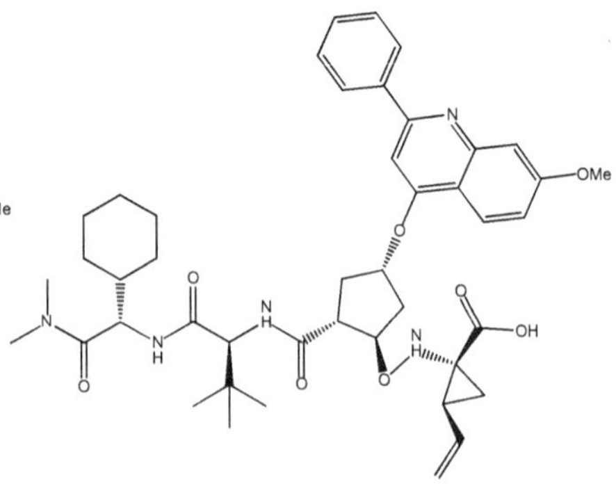

15

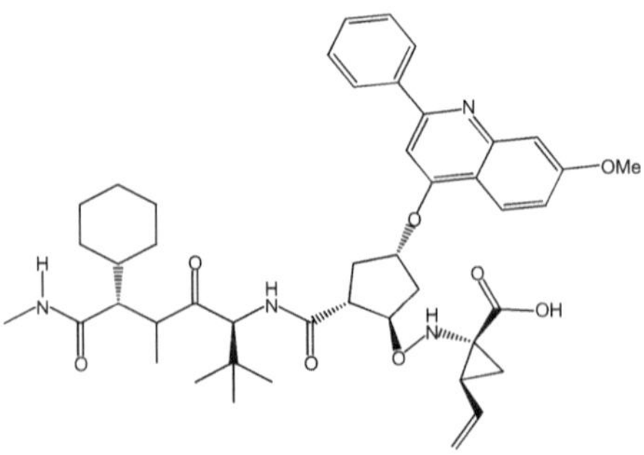

16

Compounds **15** differs structurally from **14** in that it has a methyl substituent on the nitrogen of the capping group, but it hardly differs from the latter in respect of potency, so it was suggested that the hydrogen on the capping methyl amide was not participating in any hydrogen-bond interaction. However, when the amide nitrogen of cyclohexyl glycine moiety was methylated, e.g., in compound **16**, a total loss of activity was observed ($K_i > 10^4$ nM), highlighting the importance of hydrogen of this position. The study of Johansson et al. had, however, suggested that P2 hydroxyproline scaffold could successfully be displaced by a properly substituted cyclopentane group.

Some authors studied a series of hydroxyanthraquinones (**17**) as NS3 helicase inhibitors (Furuta et al., 2015). Since some anthracyclines, containing hydroxyanthraquinone moiety, were known to be NS3 helicase inhibitors, these authors assumed that hydroxyanthraquinone moiety itself might be able to inhibit this enzyme. Therefore, these authors performed an SAR study on a series of hydroxyanthraquinone analogs and found that they can be potent NS3 helicase inhibitors and that their potency was a function of the number and positions of the phenolic hydroxyl groups. Thus, 1,4,5,8-tetrahydroxyanthraquinone (**17**, $R^1 = R^4 = R^5 = R^8$) was found to strongly inhibit NS3 helicase with an IC_{50} value of 6 μM. This result showed that a hydroxyanthraquinone moiety alone could be able to inhibit NS3 and that keto-phenols systems positioned on the same benzene ring as well as increasing the number of pairs of such systems positioned on the same ring were crucial to the activity.

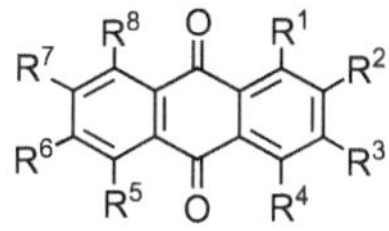

17

HCV NS3 protease has several unique features. Most important of them are a shallow, solvent exposed site, a preference for cysteine as the P1 amino acid in peptide substrate, and inhibition by the C-terminal carboxylic acid cleavage product (De Francesco et al., 1998). With the observation that 1-aminocyclopropanecarboxylic acid is a potent, nonsulfur-containing replacement for the P1 cysteine in peptide carboxylate inhibitors and that peptide boronic acids are substantially more potent inhibitors of NS3 protease than the peptide carboxylates (Kettner and Shenvi, 1984), Priestley and Decicco (2000) thought to attempt a combination of these two and prepared a concise synthesis of 1-aminocyclopropane boronic acid (**18**) and several related α,α-disubstituted amino boronic acids and then to incorporate them into peptides, e.g., **19**, and determine their potency to inhibit NS3 protease. In their study, however, these authors found only one compound, **19a**, to exhibit some potency against NS3 ($K_i = 1.6\ \mu M$). Two more compounds were tested (**19b** and **19c**) but found to have K_i values greater than 100 μM. Thus the whole results may not be called very satisfactory, but this study gave the hope that such chemicals may be developed as potent NS3 inhibitors. However, notwithstanding to such studies, a natural compound, Vitisin B (**20**), was found to be a potent NS3 helicase inhibitor by Lee et al. (2016). Vitisin B, a resveratrol, is a grapevine root extract. It was found to be highly potent against HCV replication ($EC_{50} = 6$ nM) with relatively low toxicity ($EC_{50} > 10\ \mu M$). Analysis of a number of Vitisin B–resistant HCV variants suggested that it produces its effect through NS3 helicase inhibition. A combined treatment of Vitisin B with sofosbuvir, an NS5B polymerase inhibitor, was found to exhibit a synergistic or at least additive anti-HCV effect. Lee et al. confirmed a direct binding between Vitisin B and purified NS3 helicase in vitro and found that it could inhibit NS3 with an IC_{50} value of 3 nM. These authors claimed that this compound was one of the most potent NS3 helicase inhibitors identified so far and was thus a prime candidate to be developed as the first drug derived from natural products.

18

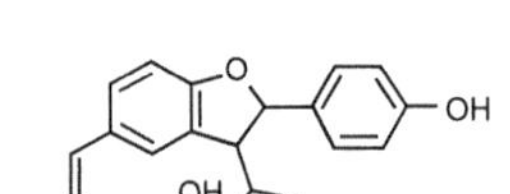

19

19a, R-R' = $-CH_2CH_2-$; **19b**, R=R' = CH_3;
19c, R-R' = $-CH_2(CH_3)_2CH_2-$

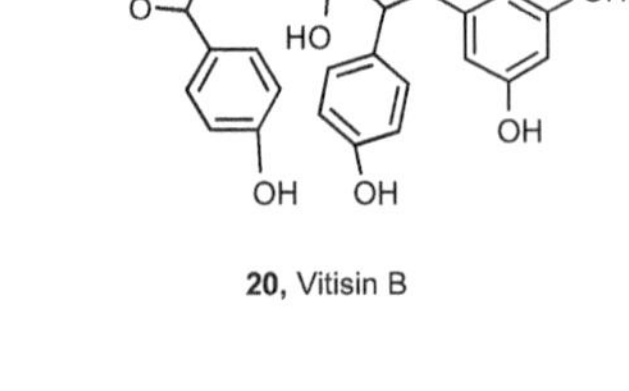

20, Vitisin B

2.2 NS4B Inhibitors

NS4B protease of HCV is much less characterized than other proteases of HCV (Hugle et al., 2001). It is an integral endoplasmic reticulum (ER) membrane protein, which is predominantly hydrophobic in nature. The NS4B N-terminal consists of two amphipathic α-helices, AH1 (aa 6–29) and AH2 (aa 42–66). The AH2 is well conserved in all HCV genotypes and critical for HCV replication (Gouttenoire et al., 2009a, 2014; Elazar et al., 2004). During HCV genome replication, an ER membrane–derived membranous web (MW) is formed, which is one of the most documented functions of NS4B (Egger et al., 2002; Gosett et al., 2003), so it has been proposed to play a structural role in HCV replication, as it can recognize the intracellular membrane into new membranous (e.g., MW) structures hosting the HCV replication complex (Gouttenoire et al., 2009b; Elazar et al., 2004; Romero-Brey et al., 2012). However, NS4B could not draw much attention for a long time as a drug target to be exploited for the discovery of anti-HCV agents, but in the last few years, impressive efforts have been made to exploit it. It could be found that small molecules targeting NS4B could suppress HCV replication by impairing AH2-mediated MW formation or RNA binding property (Cannalire et al., 2016). Several clemizole (**21**) analogs were reported as HCV replication inhibitors with EC_{50} values ≤ 5 μM against HCV genotype (gt) 1b or 2b (Choong et al., 2010, 2012). Among all the derivatives, however, only clemizole has entered Phase 1B trials in HCV-infected patients (gt 1 and 2). Another compound, anguizole (**22**), had resulted as the first NS4B ligand having submicromolar activity ($EC_{50} = 0.56$ and 0.31 μM against genotypes 1a and 1b, respectively). Lead optimization studies based on derivatives structurally related to anguizole led to an imidazo[1,2-a]pyridine series starting from **23** as a hit (Shotwell et al., 2012). Optimization of this series had led to a series of **24** in which compound 25 had exhibited the best antiviral potency with EC_{50} values of 7.4 and 18 nM for gt 1a and 1b, respectively. The high potency of this compound was attributed to the presence of pyrrolidinone ring (R-substituent), which has H-bond acceptors, able to lead to improved interactions with the receptor vis-à-vis a 1-pyrrolidnyl ring (in **26**) that had led to complete loss in activity (Shotwell et al., 2012).

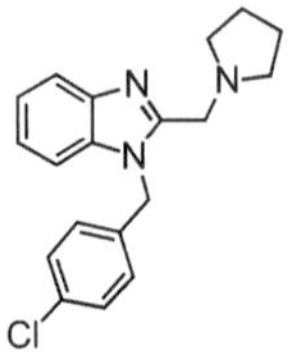

21, Clemizole

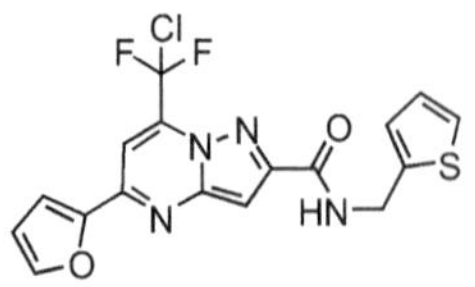

22, Anguizole

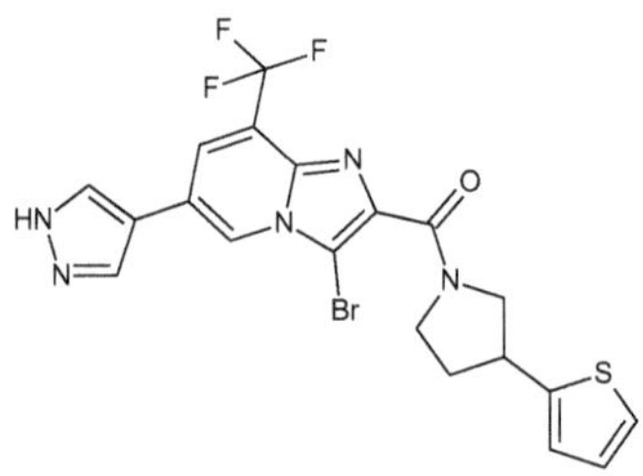

23

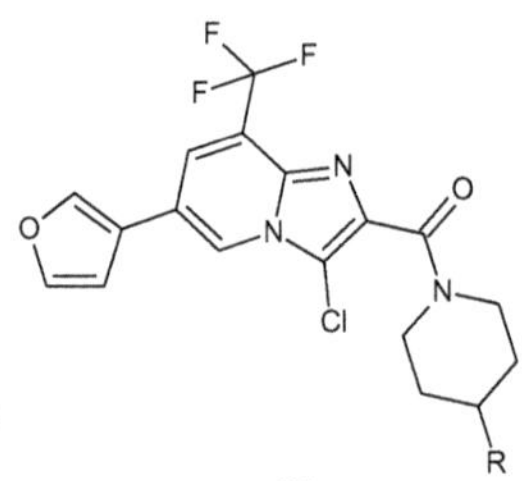

24

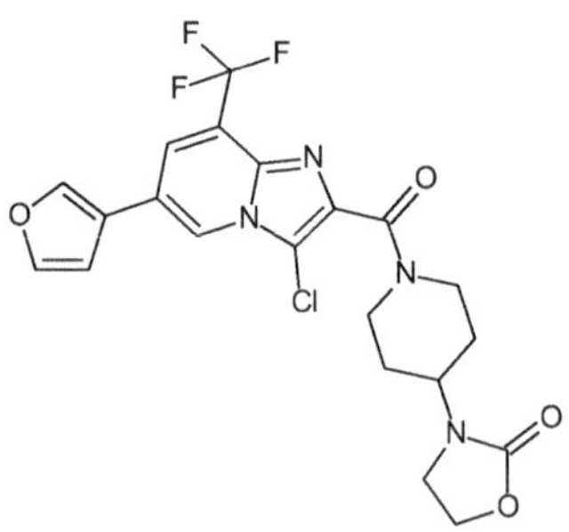

25

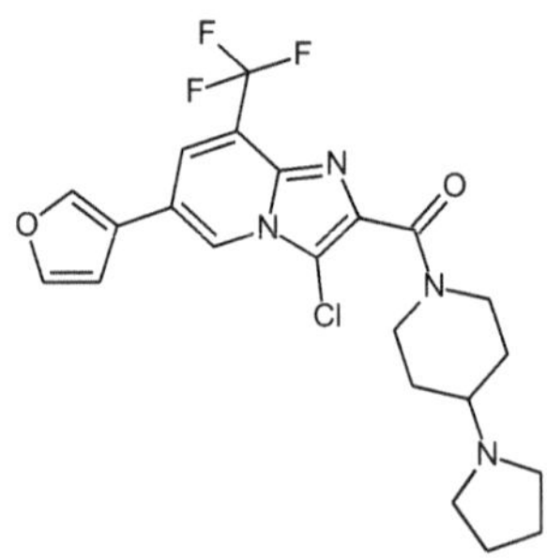

26

Like the compounds discussed earlier, Cannalire et al. (2016) in their review have presented a number of different series of chemicals studied by various authors for NS4B inhibition. Overall analysis of this review was that all known NS4B inhibitors are heterocyclic compounds with suitable physicochemical features and chemical characteristics in agreement with Lipinski's rules. However, so far the most advanced NS4B inhibitor is the antihistamine drug clemizole (**21**) hydrochloride, which is the only one in Phase Ib clinical trials for its safety and tolerability. Other promising preclinical compounds reported have been a pyrazolopyridine (**27**), an imidazopyridine (**28**), an indole (**29**), and a 2-oxadiazoloquinoline (**30**).

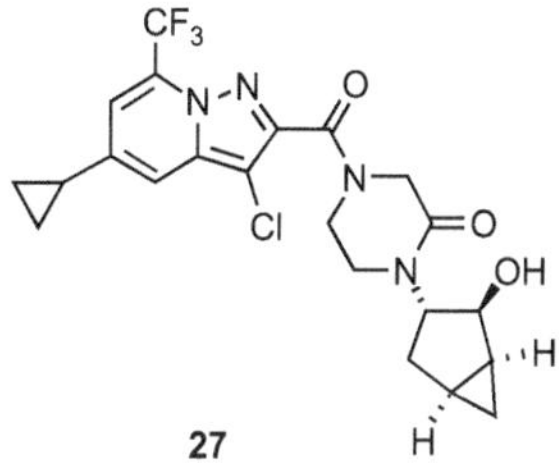

27

K_D = 8 nM
EC_{50} (gt 1a) = 0.8 nM
EC_{50} (gt 1b) = 5.4 nM
EC_{50} (gt 1b-H94N) = 200 nM
EC_{50} (gt 1b-V105M) = 800 nM
EC_{50} (gt 2a) ≥ 10 μM
EC_{50} (gt 2b) = 2.4 μM
EC_{50} (gt 3a) = 42.6 nM
EC_{50} (gt 5a) =6.98 nM
EC_{50} (gt 6a) = 4.47 μM
EC_{50} (gt 6o) ≥ 10 μM

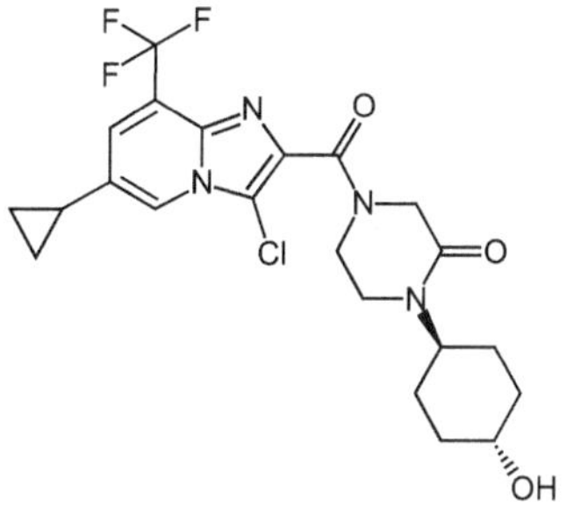

28

K_D = 60 nM
EC_{50} (gt 1a) = 0.6 nM
EC_{50} (gt 1b) = 0.3 nM
EC_{50} (gt 1b-H94N) = 5.0 nM
EC_{50} (gt 1b-V105M) = 38 nM
EC_{50} (gt 2a) ≥ 10 μM
EC_{50} (gt 2b) = 2.4 μM
EC_{50} (gt 3a) = 42.6 nM
EC_{50} (gt 5a) =6.98 nM
EC_{50} (gt 6a) = 4.47 μM
EC_{50} (gt 6o) ≥ 10 μM

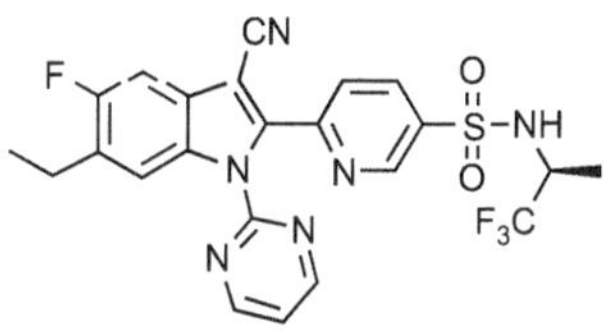

29

EC_{50} (gt 1a) = 7 nM
EC_{50} (gt 1b) = 2 nM
EC_{50} (gt 2a) = 2.2 μM

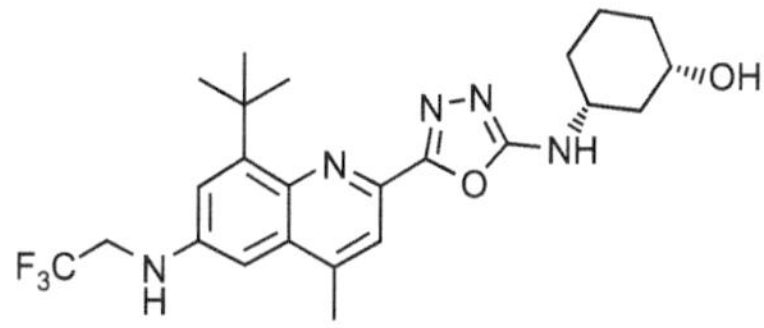

30

EC_{50} (gt 1a) = 0.08 nM
EC_{50} (gt 1b) = 0.09 nM
EC_{50} (gt 2a) = 3 nM
EC_{50} (gt 1b-H94N) = 0.6 nM
EC_{50} (gt 1b-H94S) = 0.3 nM
EC_{50} (gt 2b) = 0.6 nM; EC_{50} (gt 3a) = 3.7 nM
EC_{50} (gt 4a) = 0.9 nM; EC_{50} (gt 6a) = 3.1 nM

Kakarla et al. (2014) had made a high-throughput screening study to find that a piperazine derivative (**31**) was a potent HCV genotype 1 inhibitor and had produced its effect through NS4B inhibition. Further, it was found that amide function and C3/C6 stereochemistry of this compound were crucial for its activity. A replacement of its chiral phenyl cyclopropyl amide side chain with p-fluorophenylisoxazole-carbonyl moiety had led to a compound (**32**) that had relatively higher activity. When a further modification was made in 32 by replacing C6 nonpolar hydrophobic moiety with a phenyl moiety, no change was found in activity, but when the same was replaced with a thiophene moiety to get compound **33** or with an isoxazole moiety to get compound **34**, significant improvements were found in gt-1b and gt-1a potency, respectively.

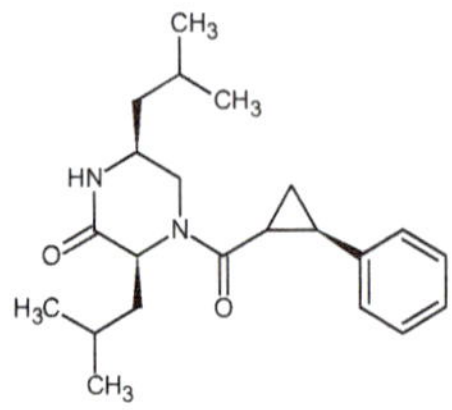

31
EC_{50}(gt 1a) = 1.0 μM
EC_{50}(gt 1b) = 0.8 μM

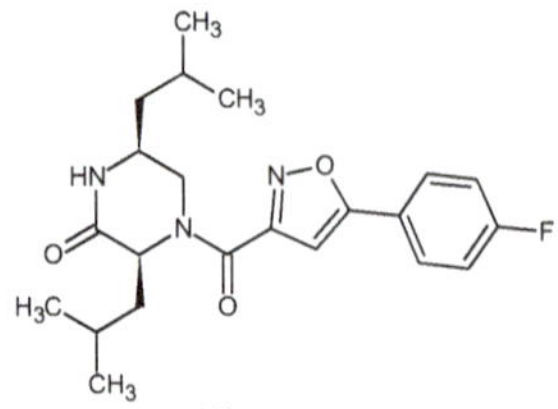

32
EC_{50}(gt 1b) = 0.13 μM

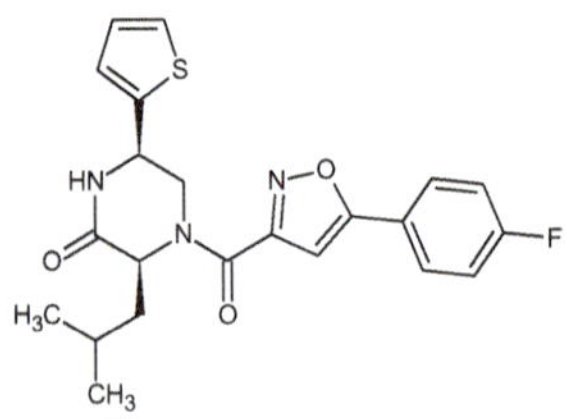

33
EC_{50}(gt 1b) = 0.006 μM

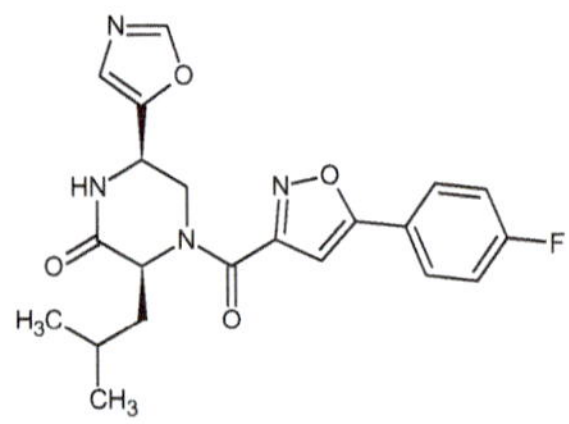

34
EC_{50}(gt 1a) = 0.016 μM

A series of 2-oxadizoloquinoline analogs were studied for their NS4B inhibition activity by Phillips et al. (2014), in which a compound **35** was found to exhibit potency against almost all genotypes of HCV with EC_{50} values as shown below the compound.

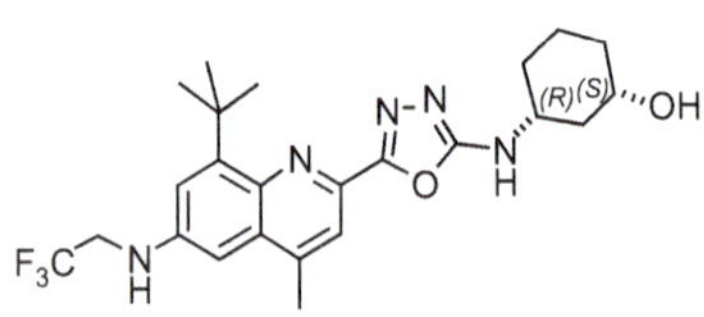

35
EC_{50} (gt 1a) = 0.08 nM; EC_{50} (gt 1b) = 0.10 nM
EC_{50} (gt 2a) = 3 nM; EC_{50} (gt 2b) = 0.6 nM
EC_{50} (gt 3a) = 3.7 nM; EC_{50} (gt 4a) = 0.9 nM
EC_{50} (gt 6a) = 3.1 nM

2.3 NS5A Inhibitors

A compound, BMS-790052 (Daclatasvir) with the structure as **36**, developed by Bristol-Myers-Squibb (BMS), was found to produce dramatic reductions in viral load and few side effects in a Phase I clinical trial (Gao et al., 2010; Belema et al., 2014). This compound was found to target NS5A protein of HCV. The researchers at BMS have shown that small molecule NS5A inhibitors are potent and specific inhibitors of HCV replication. In a small review, Asswlah (2011), based on some preliminary data, has concluded that HCV NS5A inhibitors offer considerable promise for HCV infection therapy.

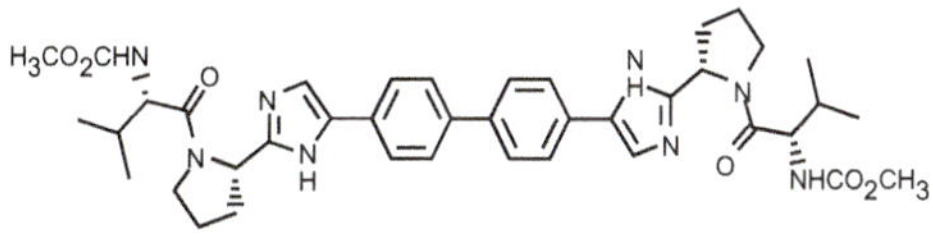

36, BMS-790052 (Daclatasvir)

In another review, Pawlotsky (2013) has pointed out that several novel NS5A inhibitors have exhibited high-potency, pan-genotypic activity, and a high barrier to resistance. A list of some potent NS5A inhibitors developed by various drug companies is given in Table 7.2. However, the exact mechanism of action of NS5A inhibitors is not yet fully known. A molecular modeling study on daclatasvir (**36**) showed that it has two modes of action and yields a shorter estimate of HCV half-life (Guedji et al., 2013). For all NS5A inhibitors, one putative mechanism is that they inhibit hyperphosphorylation, a requirement of NS5A in its pathway to RNA replication (Fridell et al., 2011). According to another thought, NS5A inhibitors alter the subcellular localization of NS5A, leading to faulty viral assembly (Targett-Adams et al., 2011; Lee et al., 2011).

Compound **36**, daclatasvir, was found to have an important feature that it had a C_2-symmetrical structure with two imidazole moieties linked by an aromatic linker. In some NS5A inhibitors, such as ledipasvir (Table 7.2), the one or both imidazole rings could be fused to a phenyl ring resulting into benzimidazole ring(s). This led Giroux et al. (2015a) to surmise if a [5,5]-bicyclic moiety, such as a thienoimidazole, could also serve the same purpose as benzimidazole and thus to attempt to find thienoimidazole-based NS5A inhibitors. In their efforts, they could find only one compound, **37**, that had satisfactory result, a picomolar potency ($EC_{50} = 0.006$ nM) against gt 1b. This encouraging result led Giroux et al. to explore if the breaking of C_2-symmetry may have any further improvement in the potency. Therefore, in the next communication, these authors (Giroux et al., 2015b) tried to find out the importance of non-symmetry as well as of linker aromaticity in the potency of the compounds. In this study, Giroux et al. found that thienoimidazole moiety was a suitable bioisostere for the benzimidazole moiety when used in conjunction with (1) a daclatasvir-like arrangement on one side of the molecule and (2) a biaryl linker. They concluded that nonsymmetrical thienoimidazole-based compounds may be well exploited to find potent NS5A inhibitors.

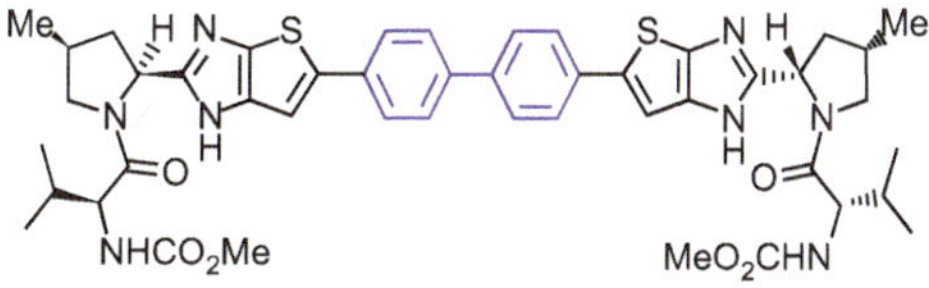

37

Table 7.2 Some NS5A inhibitors recently developed by various drug companies

Compound	Structure	GT inhb	Company
Omvitasvir (ABT-267)	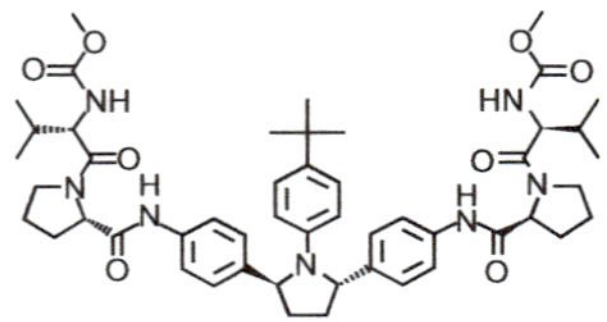	1	AbbVie
Ledispasvir (GS-5885)	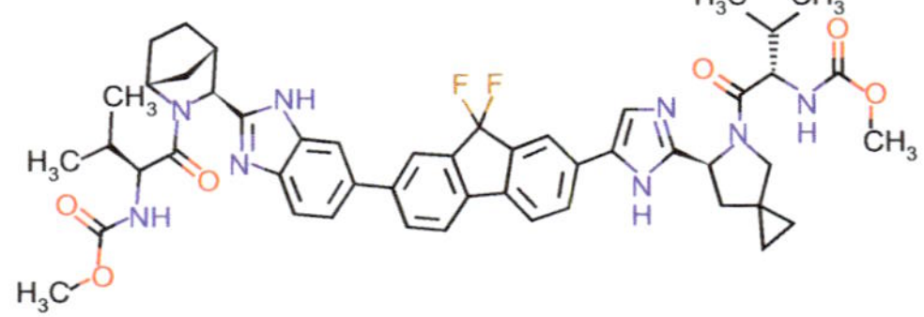	1	Gilead
GSK-2336805	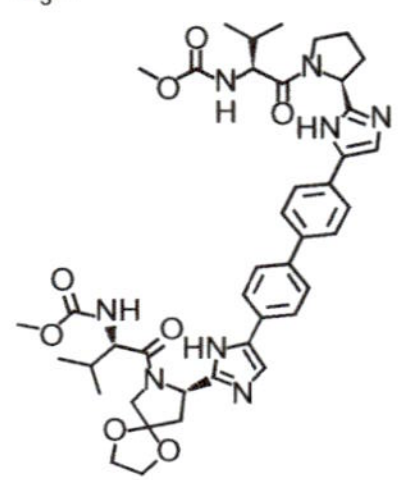	1b	GSK

Odalasvir (ACH-3102)	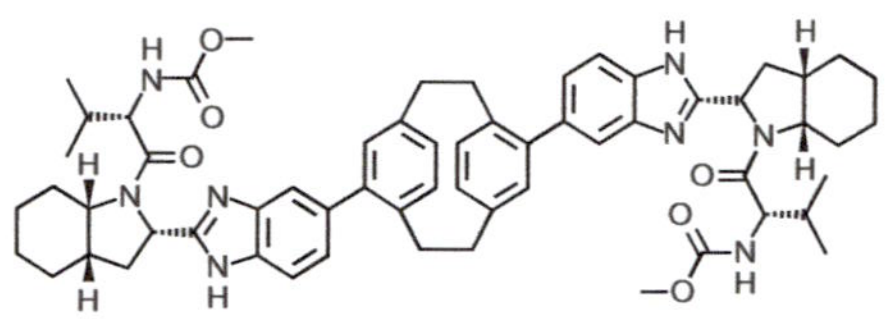	All GTs	ACH
Samatasvir (IDX-719)	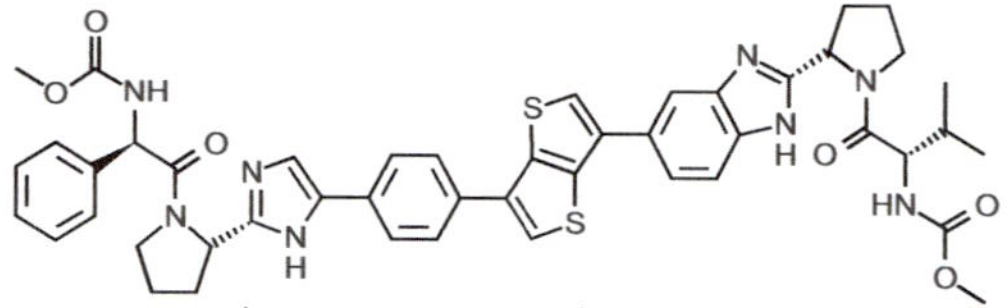	1a,1 b,2a, 3a, 4a, 5a	IDX
Ravidasvir (PPI-668)	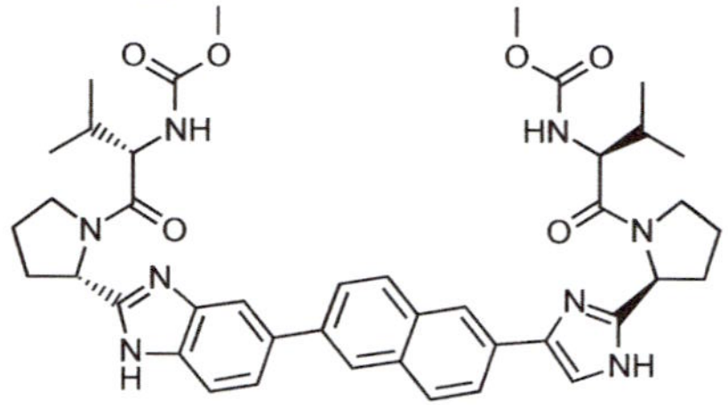	1	PPI
ACH-2928	Structure not disclosed	1a	ACH
BMS-824393	Structure not disclosed	1a	BMS
PPI-461	Structure not disclosed	1b	PPI

ACH, AChillion Pharmaceuticals; *BMS*, Bristol-Myers-Squibb; *GS*, Gilead Sciences; *GSK*, Glaxo-SmithKline; *IDX*, Idenix Pharmaceuticals; *PPI*, Presideo Pharmaceuticals, Inc.

The role of benzimidazole moiety in NS5A inhibitors was exclusively explored by Giroux group in another study (Henderson et al., 2015). Henderson et al. studied a series of bis-benzimidazole analogs with substituents at the 4-position of both the pyrrolidine rings (**38**). A (4S)-methyl substituent at pyrrolidine rings was found to possess very high potency in the picomolar range (38a, gt-1a $EC_{50} = 26$ p.m.; gt-1b $EC_{50} = 37$ p.m.). This methyl effect was further pronounced when scaffold symmetry was broken (**39**) and a compound **39a** was obtained (39a, gt-1a $EC_{50} = 4$ p.m.; gt-1b $EC_{50} = 5$ p.m.). This conducive effect of methyl substituent might be due to its involvement in hydrophobic interaction with the protease.

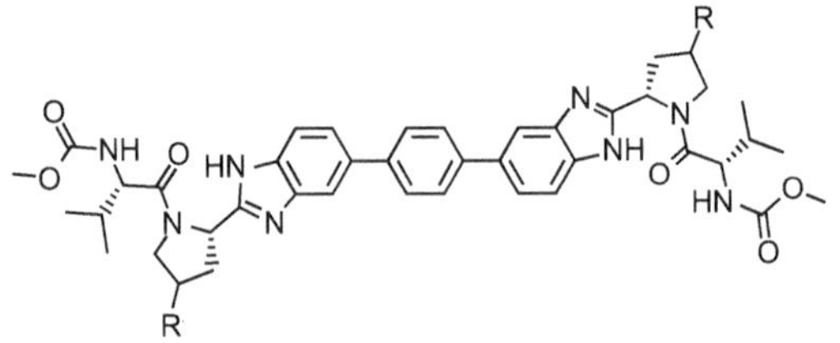

38

38a: R = (S)-Me

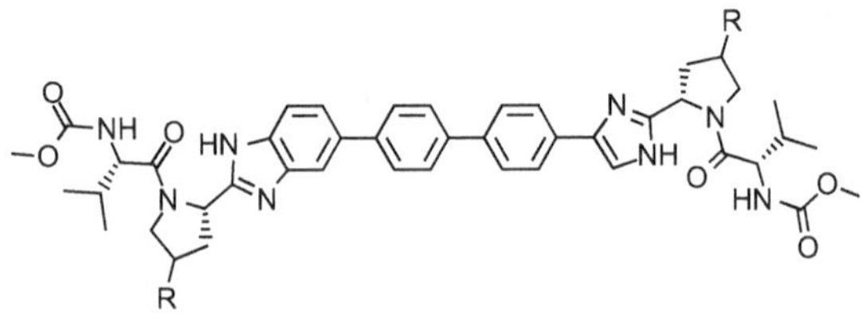

39

39a: R = (S)-Me

Coburn et al. (2013) had observed that changes in the core structure of chemicals studied so far to develop NS5A inhibitors, while holding the rest of the portion, could lead to better potency. Consequently, they had developed a compound **40** (MK-8742) that had a good activity profile against many HCV genotypes, as shown below the compound structure. Encouraged by this, Coburn group (Yu et al., 2016) studied the effect of a variety of fused tricyclic cores of different shape and length. This study led these authors to find a number of compounds with fused tricyclic core that had had pan-genotype activity (acting against numerous genotypes, not just one) and were orally bioavailable in the rat. These compounds are shown in Fig. 7.6. From this study, Yu et al. had concluded that a linear fused

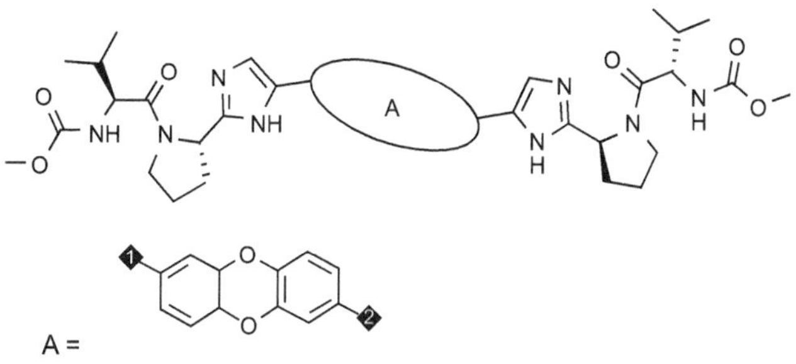

I

EC_{50} (gt 1a) = 10 pM; EC_{50} (gt 1b) = 9 pM
EC_{50} (gt 2a) = 10 pM

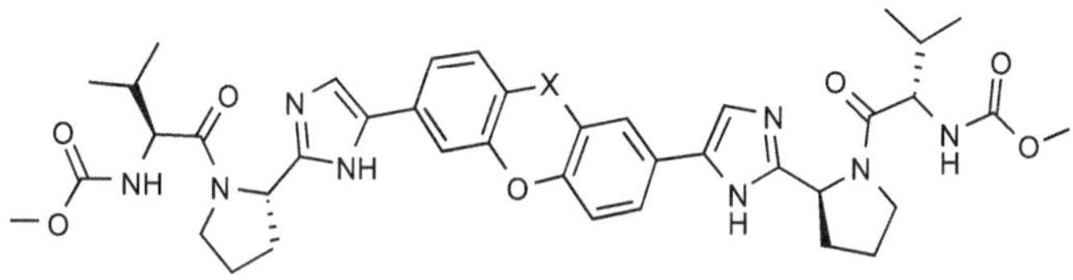

II

IIa, X= 

EC_{50} (gt 1a) = 0.04 nM; EC_{50} (gt 1b) = 0.002 nM
EC_{50} (gt 2a) = 0.06 nM; EC_{50} (gt 3a) = 0.30 nM;

IIb, X= 

EC_{50} (gt 1a) = 0.09 nM; EC_{50} (gt 1b) = 0.007 nM
EC_{50} (gt 2a) = 0.07 nM; EC_{50} (gt 3a) = 0.30 nM

IIc, X = 

EC_{50} (gt 1a) = 0.02 nM; EC_{50} (gt 1b) = 0.004 nM
EC_{50} (gt 2a) = 0.005 nM; EC_{50} (gt 3a) = 0.07 nM

IId, X = 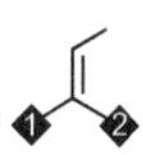

EC_{50} (gt 1a) = 0.06 nM; EC_{50} (gt 1b) = 0.007 nM
EC_{50} (gt 2a) = 0.03 nM; EC_{50} (gt 3a) = 0.4 nM

Figure 7.6 Structures of some NS5A inhibitors with fused tricyclic core structures and their potency profiles.

tricyclic core was beneficial to NS5A inhibition potency of the compounds and that a linear structure arrangement on the core was necessary, but a basic nitrogen, a single carbonyl, or a sulfonyl group in the core was not tolerable. In continuation to such study, these authors attempted to find the effects of fused tetracyclic cores in the structure (Tong et al., 2016a). The group of these authors had already identified a potent compound with a fused tetracyclic core (**40**, MK-8742) (Coburn et al., 2013), but this tetracyclic core was indole-based. Now they tried to explore the impact of changing the indole-based core and thus studied the role of a variety of alternative fused tetracyclic cores (Tong et al., 2016b) just to find that both the overall geometric makeup of the core and the substitution on the core were crucial for the potency of the compounds. In a recent communication, Tong et al. (2017), however, identified a compound with a unique tetracyclic indole core (**41**) that could currently enter clinical trials and is under evaluation as part of an all-oral DAA regimen for the treatment of chronic HCV inhibition. This compound has been named Ruzasvir (MK-8408). Its activity profile is given below its structure. Ruzasvir represents an example of an HCV NS5A inhibitor that has a flat gt 1, pan-genotypic profile.

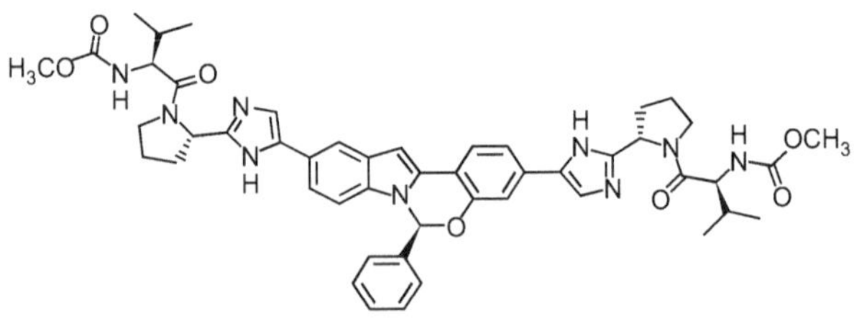

40, MK-8742

EC_{50} (gt 1a) = 4 pM; EC_{50} (gt 1b) = 3 pM
EC_{50} (gt 2a) = 3 pM; EC_{50} (gt 3a) = 20 pM
EC_{50} (gt 1a, Y93H) = 200 pM

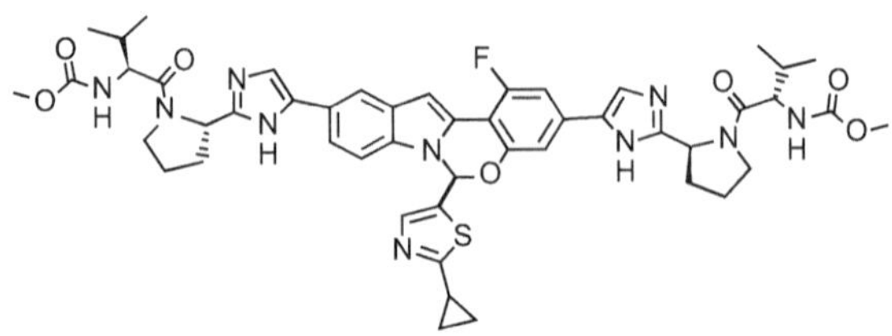

41, Ruzasvir (MK-8408)
EC_{50} (gt 1a) = 0.003 nM; EC_{50} (gt 1a, Y93H) = 0.067 nM
EC_{50} (gt 1a, L31V) = 0.016 nM; EC_{50} (gt 2b) = 0.036 nM

Coburn et al. (2013) had also mentioned the development of another potent NS5A inhibitor, MK-4882 (**42**), with the activity profile as given below the compound. Further efforts of Nair et al. (2016) led to develop an improved analog of this compound, where the left-hand-side proline ring had a silicon atom substituted with two methyl groups and the right-hand-side proline group had two fluorine atoms at its C3 carbon (**43**). This compound was named MK-8325. It had a better activity profile than **42**. To find a still better compound, Yu et al. (2016) attempted to insert a macrocyclic structure in the compounds. Consequently, they tried to constrain the conformations of amino acid side moieties attached to proline rings. These amino acid moieties, which may be called "caps," were found to be necessary components for the potency of NS5A inhibitors. Thus with the scaffold as **44a** or **44b**, Yu et al. (2016) attempted different macrocyclic R-groups, as the one shown, for example, in Fig. 7.7, and synthesized several compounds but could not find any compound better than MK-4882 or MK-8325.

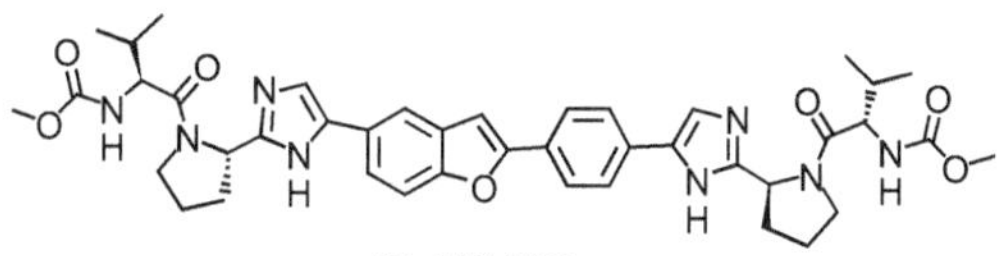

42, MK-4882

EC_{50} (gt 1a) = 0.01nM; EC_{50} (gt 1b) = 0.001 nM
EC_{50} (gt 2a) = 0.4 nM; EC_{50} (gt 3a) = 4 nM
EC_{50} (gt 1a, Y93H) = 24 nM

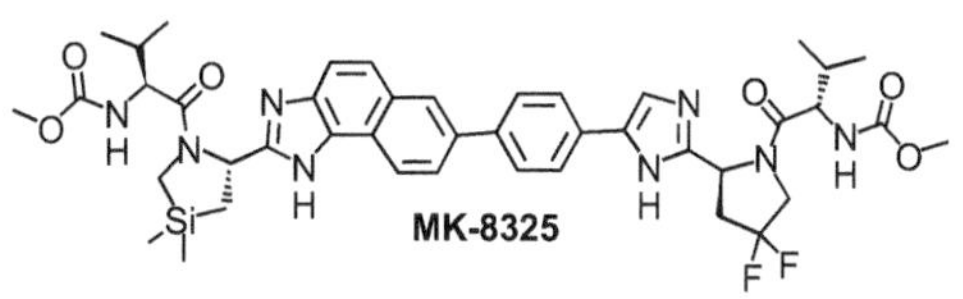

MK-8325

43

EC_{50} (gt 1a) = 0.01nM; EC_{50} (gt 1b) = 0.003 nM
EC_{50} (gt 2a) = 0.002 nM; EC_{50} (gt 3a) = 0.4 nM
EC_{50} (gt 1a, Y93H) = 47 nM

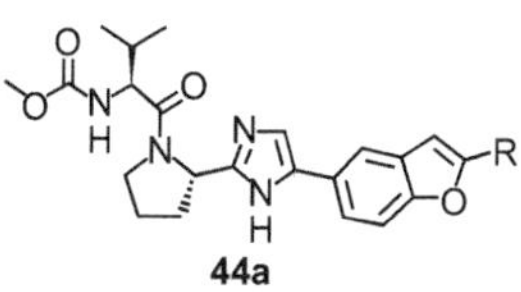

44a

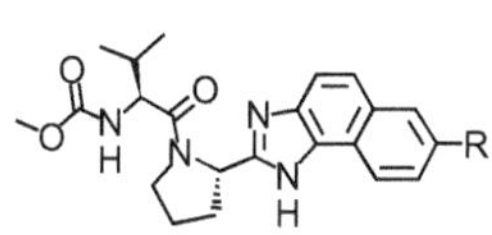

44b

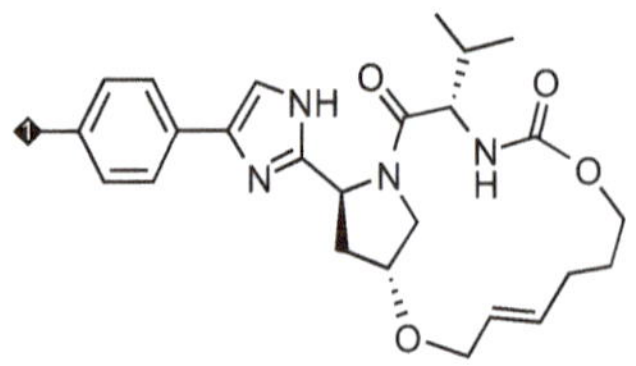

Figure 7.7 An example of an R-group used to find macrocyclic analogs of MK-4882 and MK-8325.

2.4 NS5B Inhibitors

Of all the HCV targets, NS5B polymerase has been most widely exploited to develop anti-HCV drugs. The reason is that the crystal structure of this polymerase is extensively studied. Thus the availability of its several crystal structures, together with the conservative nature of its nucleotide binding, has drawn great attention for the development of anti-HCV agents both theoretically and experimentally. Consequently, a large number of NS5B inhibitors have been reported, many of which are currently in use and many in clinical trials, and several reviews have been written on the study of NS5B polymerase inhibitors (Patil et al., 2011a; Mayhoub, 2012; Varshney et al., 2012a; Das, 2017). Here, we present some recent developments in design and discovery of NS5B polymerase inhibitors. There have been several classes of chemicals that have been studied for NS5B inhibition, but most widely studied are nucleosides on nucleotides. Other chemicals developed can be put into the nonnucleoside/nucleotide class of NS5B inhibitors.

2.4.1 Nucleosides/Nucleotides

As already discussed in Chapter 3, NS5B polymerase has three domains, palm, fingers, and thumb domains. The catalytic site of the polymerase lies in the palm domain. It is where the conserved aspartic acid residue–containing Gly317-Asp318-Asp319 (GDD) motif common to viral polymerases is located (Sofia et al., 2012). This is the site where a ribonucleoside 5′-triphosphate binds through the coordination of a divalent Mg^{2+} or Mn^{2+} and is subsequently added to the 3′-end of the growing RNA chain through the formation of 3′,5′-phosphodiesterase linkage (Fig. 7.8). Nucleosides or nucleotides function as chain terminators and thus block the further replication of the virus. They are thus known as "chain terminator inhibitors" (Perrone et al., 2007a,b). Nucleoside/nucleotide inhibitors (NIs) can also interfere with the cellular proliferative machine, causing undesirable side effects. They can be classified into two subclasses: (1) sugar-modified NIs

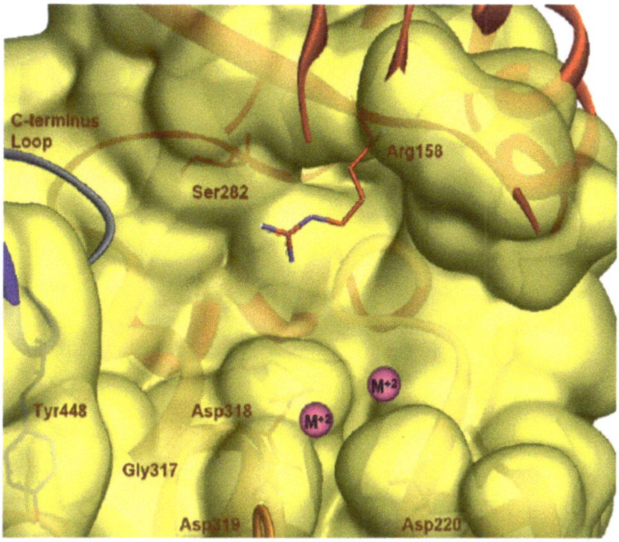

Figure 7.8 An exclusive representation of NS5B polymerase exhibiting the GDD motif and divalent cations, Mg^{2+} and Mn^{2+}, in the palm domain. *(Reprinted with permission from Sofia, M.J., Chang, W., Furman, P.A., Mosley, R.T., Ross, B.S., 2012. Nucleoside, nucleotide, and non-nucleoside inhibitors of hepatitis C virus NS5B RNA-dependent RNA-polymerase. J. Med. Chem. 55 (6), 2481–2531. Copyright 2012 American Chemical Society.)*

and (2) base-modified NIs, where the former in general are prodrugs that are activated into corresponding triphosphate forms by the host cell kinases (Perrone et al., 2007a,b). In the sugar-modified NIs, the sugar portion is modified, and in base-modified NIs, the base portion (pyrimidine or purine bases) is modified. Simple ester prodrugs have better pharmacokinetic properties of nucleosides.

Among the most promising sugar-modified NIs are those shown in Fig. 7.9. In this list, the first two compounds, namely 2′-C-methyladenosine and 2′-O-methylcytidine, were studied by Merck Research Laboratories, which had in vitro NS5B inhibition potency in the low micromolar range. Other compounds were also equally potent. The 2′-deoxy-2′-fluoro-C-methylcytidine was found to exhibit submicromolar activity in a biochemical assay (Clark et al., 2005). Like these compounds, many other compounds having high potency and belonging to different series of sugar-modified nucleosides studied by various authors have been reviewed by Mayhoub (2012). However, due to development of viral resistance, very few of them have entered the clinical trials. So far, only

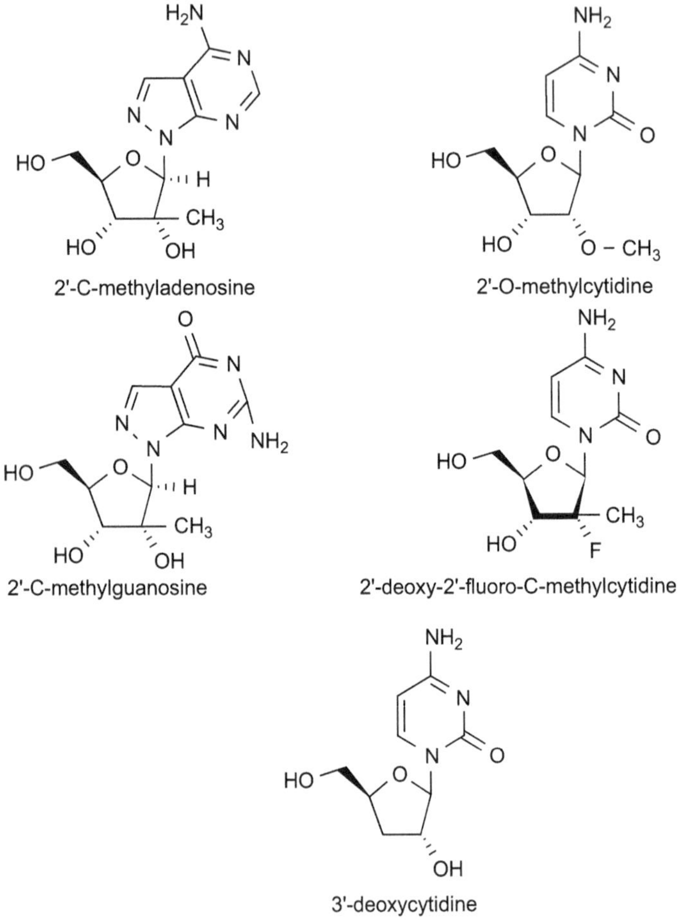

Figure 7.9 Structures of some important sugar-modified nucleosides acting as NS5B inhibitors.

2′-deoxy-2′-fluoro-C-methylcytidine (Fig. 7.9) and a 4′-azid nucleoside analog (**45**, R-1479) have entered the clinical trials in their prodrug forms, **46** (R-7128) and **47** (R-1626), respectively. These compounds were found to compete with the substrate and act as chain terminators. The NIs that passed the clinical trials and were approved by FDA can be found in Table 4.1 in Chapter 4. Some authors had studied a series of pyrrolo[2,3-d] pyrimidine nucleoside analogs (**48**) acting as NS5B inhibitors to find that larger alkyl groups at R_1-and R_2-positions were found to be conducive to the activity but also had cytotoxic effects (Varaprasad et al., 2007).

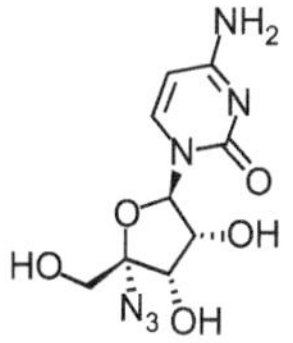

45, R-1479

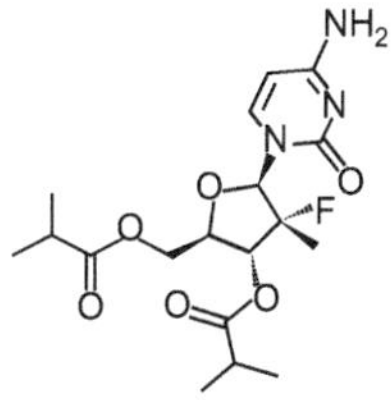

46, R-7128

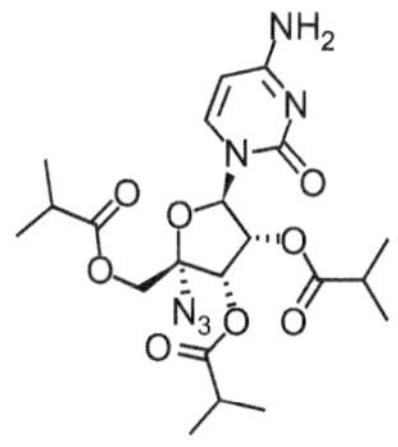

47, R-1626

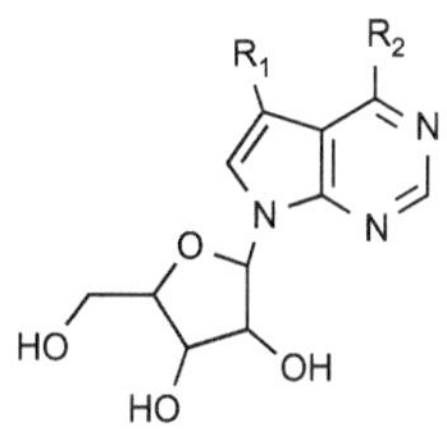

48

Compound 45 and its prodrug **47** are heterobase modified nucleosides. The advantage with them is that they are inert, or at least more resistant to phosphorolysis by purine nucleoside phosphorylase and deamination by adenosine deaminase, which deactivate other unmodified NIs (Eldrup et al., 2004). Because of this, these compounds possess higher oral bioavailability and better pharmacokinetic profiles in general (Olsen et al., 2004).

Search for safe NS5B inhibitors led to attempt a variety of non-nucleoside inhibitors (NNIs). A description of them is presented next.

2.4.2 Nonnucleoside Inhibitors

Nonnucleoside NS5B inhibitors may belong to different classes of chemicals, such as benzimidazole derivatives, aryl diketo acids, acyl pyrrolidines, proline sulfonamides, indole derivatives, thiouracil derivatives, pteridine derivatives, thiazolone derivatives, imidazopyridine derivatives, isatin derivatives, sesquiterpene lactones, and miscellaneous agents (Patil et al., 2011a). Mayhoub (2012) described the development of some NNIs belonging to different classes and their clinical stage, as mentioned subsequently.

NNIs have been found to exert their effects through allosteric inhibition of the polymerase, i.e., they bind to the different allosteric pockets of the enzyme leading to the conformational changes of enzyme-catalytic active site, while NIs directly bind to the polymerase active site. At least four allosteric sites have been found from the NS5B cocrystal structures

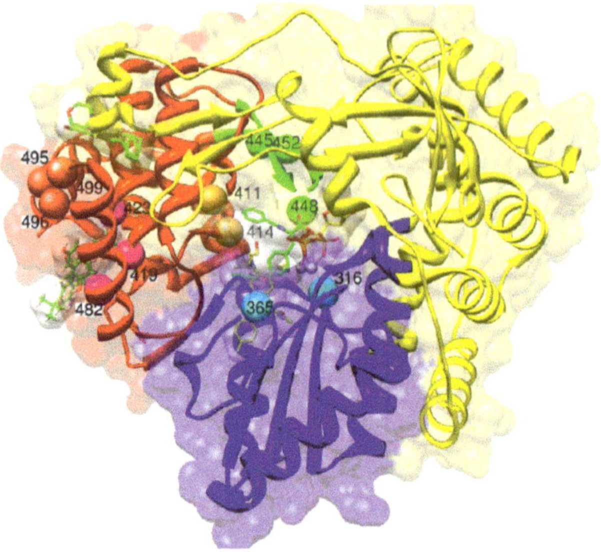

Figure 7.10 A diagram showing four allosteric sites in NS5B polymerase (two thumb sites and two palm sites). Thumb I site is shown in *salmon color* with residues P495, P496, and V499, thumb II site in *magenta* with residues L419, M423, and I482, palm I site in *gold color* with residues N411 and M414, and palm II site in *cyan* with residues C316 and S365. *(Reprinted from Delang, L., Coelmont, L., Neyts, J., 2010. Antiviral therapy for hepatitis C virus: beyond the standard of care. Viruses 2, 826–866. Open access publication.)*

(Fig. 7.10) (Delang et al., 2010; Tan and He, 2011). They are known as Thumb I site, Thumb II site, Palm I site, and Palm II site. Thumb I site lies between the thumb and fingertips region close to GTP noncatalytic binding site, Thumb II site lies at the border between thumb and palm domains 35 Å away from the active site, Palm I site is found at the top border of palm domain site close to Thr418, and Palm II site is also the part of the palm region close to Asp316, and it is the closest one to the active site (Hirashima et al., 2006; De Francesco and Migliaccio, 2005).

2.4.2.1 Benzimidazole Derivatives

The first NNI that entered the clinical trial was a benzimidazole derivative, 2-(4-((4′-chloro-4-(2-oxopyrrolidin-1-yl)-[1,1′-biphenyl]-2-yl)methoxy)-2-fluorophenyl)-1-cyclohexyl-1H-benzo[d]imidazole-5-carboxylic acid. It was active in its hydrochloride form (**49**) with the name JTK-109 (Hirashima et al., 2006). It inhibits NS5B polymerase in the nanomolar range (IC50 = 17 nM) affecting HCV gt 1b. It possesses favorable pharmacokinetics and high selectivity index and is highly selective to HCV polymerase. Like all benzimidazole derivatives, it binds to allosteric thumb site I. It had been in a Phase I clinical trial by Japan Tobacco, but now, its study has been discontinued.

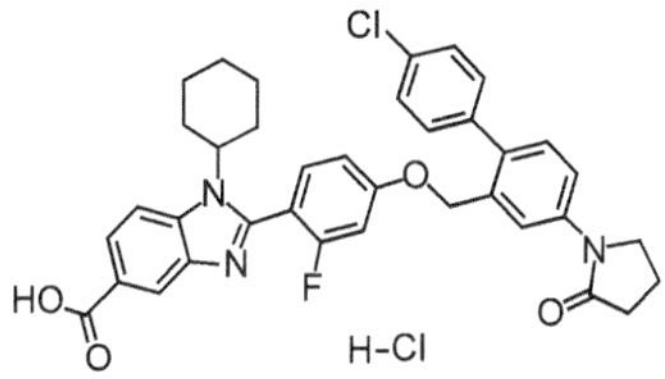

49, JTK-109

2.4.2.2 Benzofuran Derivatives

HCV-796 (**50**), a benzofuran derivative, is an investigational nonnucleoside polymerase inhibitor for the treatment of hepatitis C that is being evaluated in ongoing clinical trials in combination with PEG-IFN by ViroPharma and Wyeth. According to ViroPharma, Phase 2 program for this compound was started with Wyeth in October 2006, but both decided to discontinue the dosing due to some side effects and in 2008 decided to discontinue its development, but this decision was not finalized.

In biochemical assay, HCV-796 was found to affect the genotypes 1a and 1b with EC_{50} values of 5 and 9 nM, respectively. In combination with interferon, additive effect was noted in naïve patients with several chronic genotypes (Kneteman et al., 2009). In fact, the triple therapy of HCV-796/PEG-INF/ribavirin has been found to be highly effective. A structural combination of HCV-796 with ribavirin had given a new class of compound with improved potency (Zhu et al., 2010). Otherwise, several mutations that included Leu314, Cys316, Ile363, Ser365, and Met414 were found to be associated with HCV-796 that directly interact with it, leading to a 10−1000-fold decrease in its potency (Howe et al., 2008).

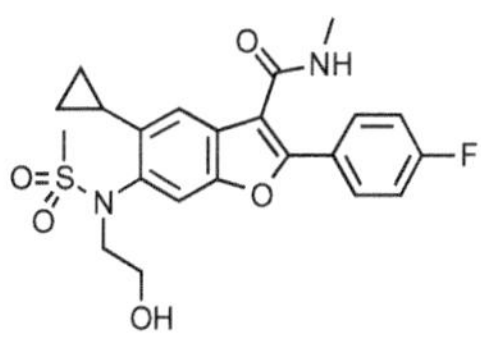

50, HCV-796

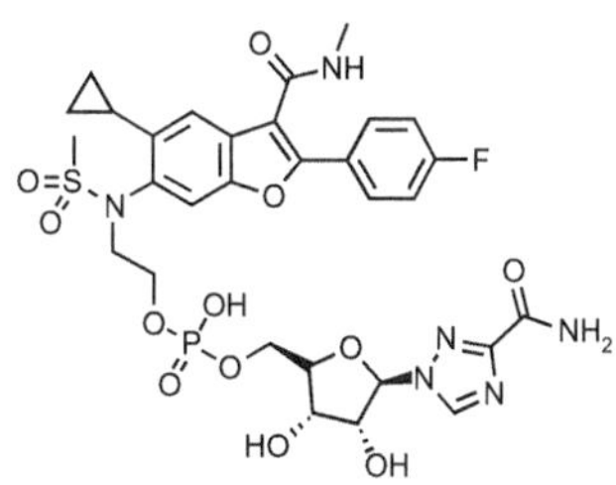

51, HCV-796-ribavirin-prodrug

2.4.2.3 Pyrimidine Derivatives

A pyrimidine derivative, N-[6-[3-tert-butyl-5-(2,4-dioxopyrimidin-1-yl)-2-methoxyphenyl]- naphthalen-2-yl]methanesulfonamide (**52**), known as ABT-333 or Dasabuvir, was studied by Kowdley et al. (2014) for its NS5B inhibition. ABT-333, in combination with ribavirin and ABT-450/ritonavir (where ABT-450, also known as paritaprevir, is an inhibitor of NS3/4A and ritonavir is an anti-HIV agent), had shown the rates of sustained virologic response of 93%–95% among previously untreated patients and 47% among patients who had not had a response or who had had only a partial response to prior therapy with peginterferon and ribavirin. ABT-333 was found to have IC_{50} values against HCV gt 1a and 1b polymerase of 2.2 and 10.7 nM, respectively. The reported mutations associated with this are C316Y, S368T, M414T/I/V, Y448H/C, A443V, and S556G (Kati et al., 2014).

Another analog of ABT-333, ABT-072 (**53**), was found to be equally potent and an HCV gt 1 selective NS5B inhibitor. It had even 5 to 9 times greater potency than ABT-333 in the presence of 40% plasma (Howe et al., 2008). A 12-week regimen of ABT-072 with ABT-450/ribavirin was found to be well tolerated with 9 out of 11 patients achieving sustained virologic response through 36 weeks of posttreatment observation (Lawitz et al., 2013). These findings suggested that a peginterferon-free regimen may have the potential to cure a high proportion of HCV genotype 1–infected patients.

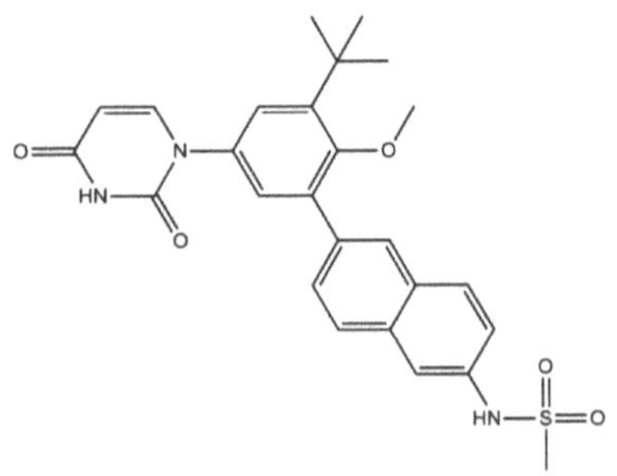

52, ABT-333 (Dasabuvir)

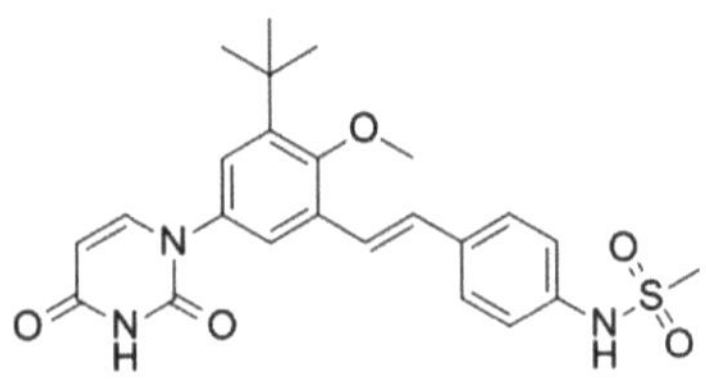

53, ABT-072

2.4.2.4 Indole-Based Compounds

Some authors also studied the indole-based compounds for the inhibition of NS5B polymerase of HCV. Recently, Chen et al. in their two consecutive studies made the fruitful efforts to find a few very potent indole analogs (Chen et al., 2012a,b). In their first study, these authors analyzed several indole-based series, in which the optimization of a series of **54** led to a highly potent compound, **54a** (Chen et al., 2012b). This compound had RdRp inhibition $IC_{50} = 8$ nM and replicon inhibition $EC_{50} = 20$ nM and an excellent pharmacokinetic profile in rats and dogs. An X-ray structure of this compound bound to NS5B polymerase of HCV was also reported (Fig. 7.11A). The compound is shown to bind to the palm site of the enzyme within the active site cavity, where it is involved in the formation of number of hydrogen bonds, in which both the NH and oxygen of the pyridone ring and all three oxygen atoms of C-2 acyl sulfonamide chain have taken part. Additionally, the indole core is shown to have π-stacking interactions with the benzene ring of protein backbone Tyr-448 and the N-1 benzyl moiety of the compound stacks atop the thiol side chain of Cys-366. Further, the indole C-5 methyl and C-6 fluoro groups penetrate into a constricted tunnel that extends to a protein surface on the other side of the thumb domain. Thus indole core, C-3 pyridone, C-2 acyl sulfonamide, C-5 methyl, and C-6 F all contributed to the high potency of the compound.

In their next study, Chen et al. (2012a) attempted a few more series of indole analogs in which a compound, **55a**, belonging to the series of **55** was found to be the most potent. This compound was found to possess excellent pharmacokinetic profile and an EC_{50} value of only 1 nM, which made it an interesting candidate for further investigation. This compound was an irreversible inhibitor that had covalent interaction with the protein. An X-ray crystal structure of this compound bound to HCV polymerase (PDB code: 3TYQ) was also obtained (Fig. 7.11). This compound was also found to bind with the palm site of apoprotein within the active site cavity. Like compound **54a**, this compound is also shown to have several hydrogen bondings. Additionally, it is also involved in the formation of a covalent bond between its phenyl ring and the side chain of the residue Cys-366 in the protein with a 1.8 to 1.9 Å S—C distance. This covalent bond is a result of the side chain thiol attacking the electron-deficient benzene ring at the position para to the nitro group with departure of the fluoro group via a presumed S_NAr reaction.

Although compound **55a** had shown very good activity ($EC_{50} = 1$ nM) and excellent oral exposure in rats, dogs, and monkeys, its NO_2 group could be considered a potential toxic liability. This led Zeng et al. (2013) to make certain modifications in substituents at indole ring. Thus they attempted a series of **56** in which an analog **56a** was found to have an

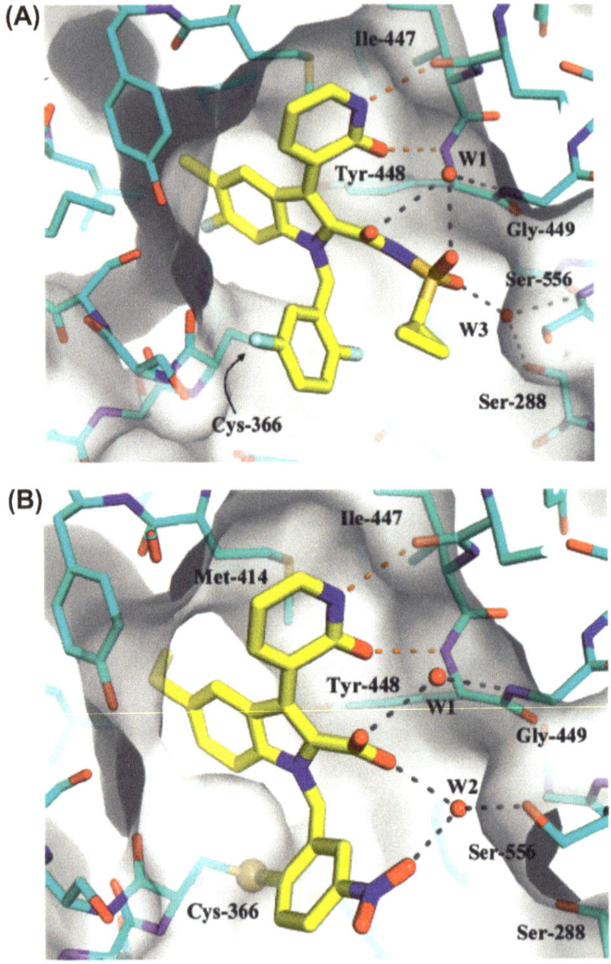

Figure 7.11 (A) X-ray crystal structure of compound **54a** bound to NS5B polymerase. (B) X-ray crystal structure of compound **55a** bound to NS5B polymerase. *(A) Reprinted with permission from Chen, K.X., Vibulbhan, B., Yang, W., Sannigrahi, M., Velazquez, F., Chan, T.-Y., Venkatraman, S., Anilkumar, G.N., Zeng, Q., Bennet, F., Jiang, Y., Lesburg, C.A., Duca, J., Pinto, P., Gavalas, S., Huang, Y., Wu, W., Selyutin, O., Agrawal, S., Feld, B., Huang, H.-C., Li, C., Cheng, K.-C., Shih, N.-Y., Kozlowski, J.A., Rosenblum, S.B., Njoroge, F.G., 2012b. Structure-activity relationship (SAR) development and discovery of potent indole-based inhibitors of the hepatitis C virus (HCV) NS5B polymerase. J. Med. Chem. 55, 754–765. Copyright 2012 American Chemical Society. (B) Reprinted with permission from Chen, K.X., Lesburg, C.A., Vibulbhan, B., Yang, W., Tin-Yau Chan, T.-Y., Venkatraman, S., Velazquez, F., Zeng, Q., Bennett, F., Anilkumar, G.N., Duca, J., Jiang, Y., Pinto, P., Wang, L., Yuhua Huang, Y., Selyutin, O., Gavalas, S., Pu, H., Agrawal, S., Feld, B., Huang, H.-C., Li, C., Cheng, K.-C., Shih, N.-Y., Kozlowski, J.A., Rosenblum, S.B., Njoroge, F.G., 2012a. A novel class of highly potent irreversible hepatitis C virus NS5B polymerase inhibitors. J. Med. Chem. 55, 2089–2101. Copyright 2012 American Chemical Society.)*

improved toxicity acceptability profile, though its EC_{50} value was slightly higher (3 nM). However, in both in vitro and in vivo assays, **56a** was found to possess good oral exposure and safety profile, so it was expected to be a promising candidate to be developed as a potent anti-HCV agent.

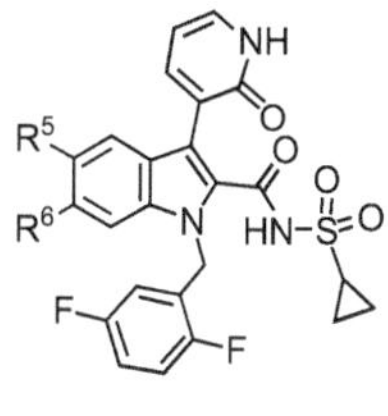

54

54a: R_5 = Me, R_6= F

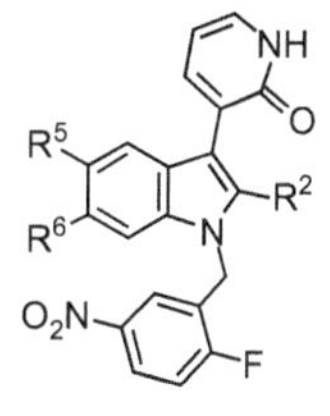

55

55a: R^2 = COOH, R^5 = Et, R^6 = H

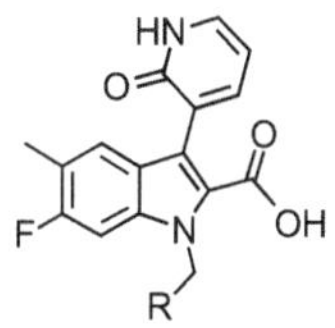

56

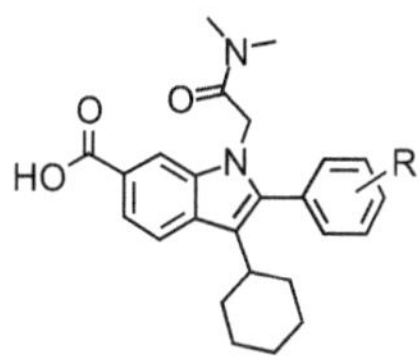

57

56a: R= 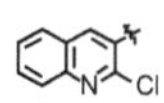

Earlier, Harper et al. (2005) had analyzed several series of indole-N-acetamides, in which a series as represented by **57** was found to be highly promising. In this series, the N-substituent was optimized with a $-CH_2CON(Me)_2$ group. Now the variation at the C2-phenyl ring had given several promising compounds as given in Table 7.3. These compounds show submolecular blockade of subgenomic RNA replication in HUH-7 cells (a type of human liver cells). These compounds have an encouraging activity profile and thus potential for further development.

A structure-based drug design had led Anilkumar et al. (2011) to study the effects of incorporation of novel heterocyclic moieties at the indole C3-position, which had a bidentate interaction with the protein backbone. This study gave a highly potent compound (**58**) with impressive enzyme and replicon inhibition activities (NS5B $IC_{50} = 0.017$ μM and Replicon $EC_{50} = 0.3$ μM). In their next communication, these authors attempted the modifications at the C2-position of the indole that led them to find an

Table 7.3 Intrinsic potency (IC_{50}) and cell-based potency (EC_{50}) of a series of indole-based NS5B polymerase inhibitors (Harper et al., 2005)

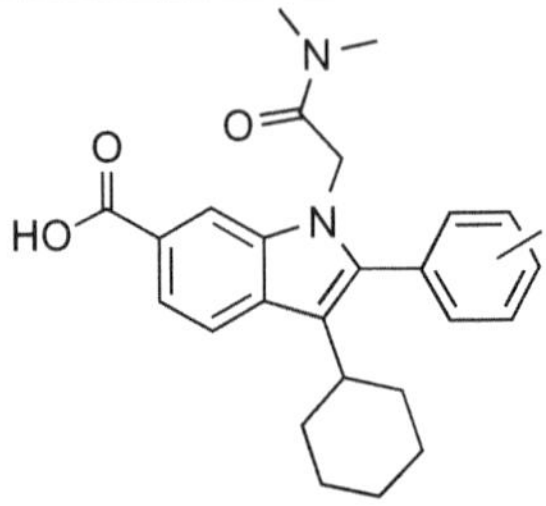

56

Compound	R	IC_{50} (nM)	EC_{50} (μM)
56a	H	59 ± 6	1.5 ± 0.6
56b	4-Cl	15 ± 9	0.9 ± 0.3
56d	3-F	18 ± 15	0.6 ± 0.1
56e	4-Me	17 ± 2	0.5 ± 0.2
56f	4-OMe	18 ± 5	0.5 ± 0.1
56g	4-OBn	6 ± 2	2.6 ± 0.6

equally potent compound as **59** (NS5B IC_{50} = 0.039 μM and Replicon EC_{50} = 0.011 μM) (Anilkumar et al., 2012). Both the compounds, **58** and **59**, were palm site inhibitors.

A genetics-based attempt by Jin et al. (2014) to discover novel indole derivatives as NS5B inhibitors had given a highly potent compound (**60**) that had minimal toxicity (EC_{50} = 1.1 μM and CC_{50} = 61.8 μM). That its effect was through NS5B RNA polymerase inhibition was verified by in vitro experiment, where its IC_{50} value was found to be 0.292 μM.

More recently, Han et al. (2016) screened a library of in-house small molecules using a cell-based assay to identify a compound containing an N-protected indole scaffold (**61**) as a novel anti-HCV agent. Although compound **61** had good inhibition activity, to investigate a better compound, several analogs of it were studied, among which a compound belonging to series of **62** was found to be highly potent against NS5B polymerase. This compound was **62a.** Its (R)-enantiomeric form had EC_{50} = 0.72 μM and SI > 69.44. Its (S)-isomer was found to be comparatively very inferior (EC_{50} = 7.12 μM and SI = 4.86). Its racemic mixture was thus preferred (EC_{50} = 1.02 μM and SI = 45.56). (*EC_{50}*, concentration of a drug that gives half-maximal response; *IC_{50}*, concentration of an inhibitor that reduces the activity of an enzyme by half; *ST*, selectivity index of the drug).

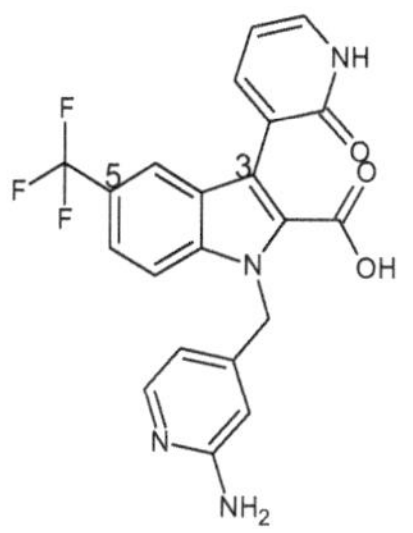

58

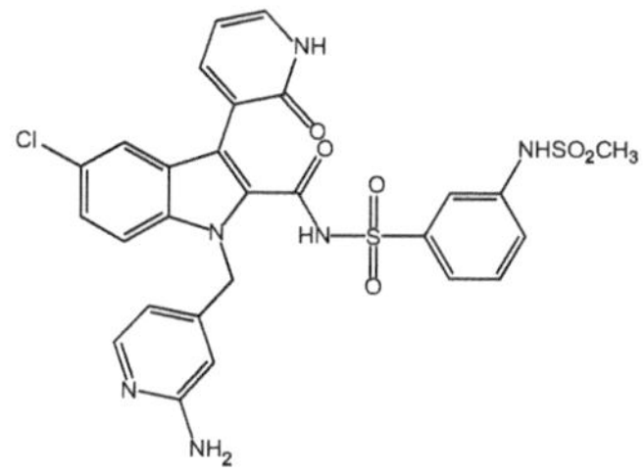

59

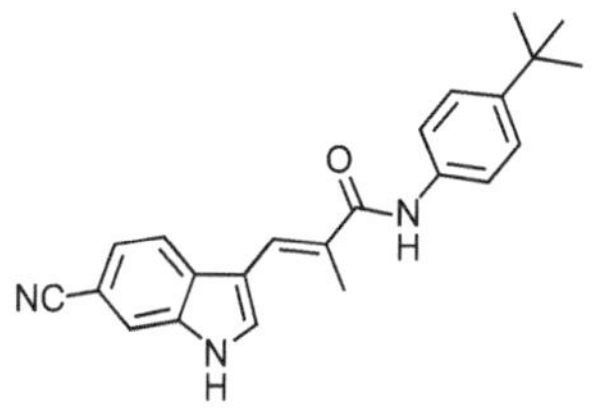

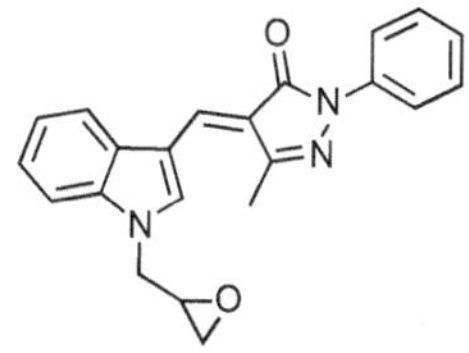

60 (EC_{50} = 1.1 μM; CC_{50} = 61.8 μM) **61** (EC_{50} = 5.78 μM; CC_{50} = 46 μM)

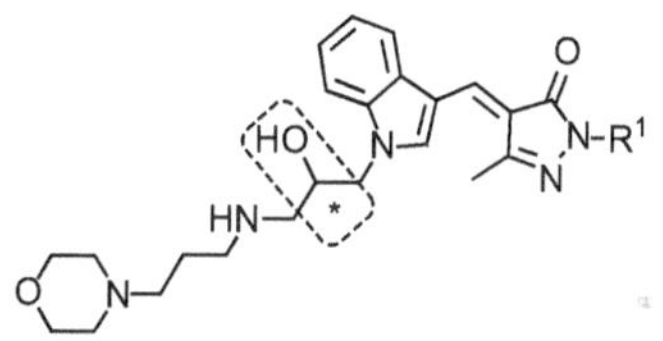

62

62a: R^1 = p-F-P

2.4.2.5 Aurones

Aurone (**63**) is a heterocyclic chemical, containing a benzofuran element associated with a benzylidene linked in position 2 (i.e., a 2-benzylidenebenzofuran-3-one). It is a type of flavonoid (Nakayama, 2002). However, in aurone, a chalcone-like group is closed into a 5-membered ring instead of the 6-membered ring, more typical of flavonoids. The aurone exists in two isomers, with (*E*)- and (*Z*)-configuration, and it contains a benzofuran element associated with a benzylidene linked in position 2. The pharmacological potential of aurones was reported several years ago (Boumendjel, 2003). An analog of aurone, aureusidin (**64**), a naturally occurring aurone, was found to inhibit HCV RdRp with an IC_{50} value of 5.2 μM (Nakayama et al., 2000). Since aurone was identified to interact with thumb I site, Haudecoeur et al. (2011) tried to exploit this information to find aurone derivatives that could inhibit NS5B polymerase. They synthesized a large number of aurone analogs and screened their potency, but the most active compound found was **65**, having an IC_{50} value of 2.2 μM for NS5B polymerase. A docking study was performed for the compounds studied with an X-ray structure of RdRp (PDB code: 2DXS) in which the interactions of **64** as well as that of aureusidin (**65**) were as shown in Fig. 7.12. This docking study identified the important pharmacophoric features crucial in the inhibitory activity of aurones against RdRp of HCV. These were as (1) a hydroxy group at position 4 or at positions 4 and 6 and (2) a 2′,4′- or 3′,4′-dihydroxylated B-ring or a hydrophobic and bulky substituent or alternative core at the B-ring. This highlighted the crucial importance of the A-ring hydroxyl substitution

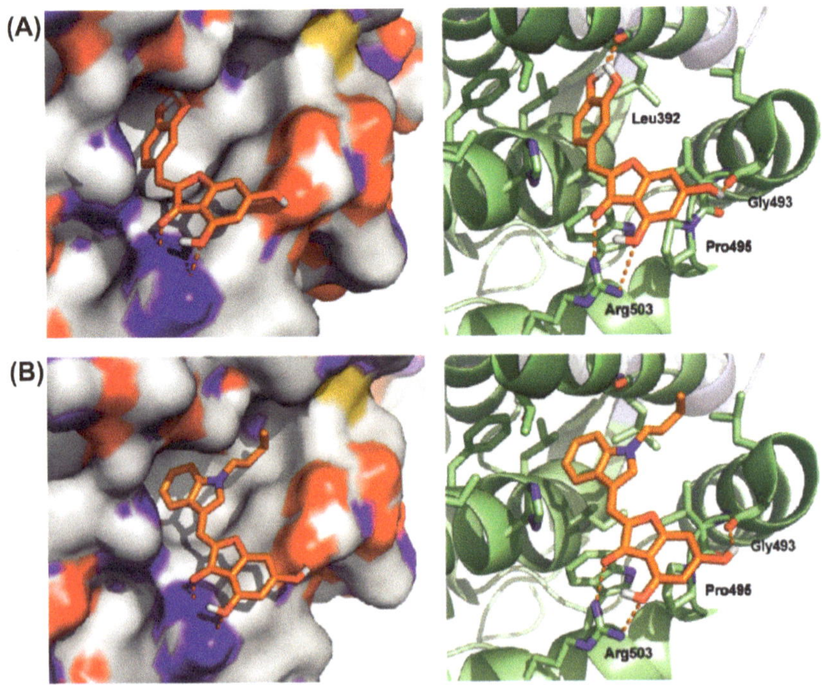

Figure 7.12 A representation of structures of compounds **64** and **65** docked in reported X-ray structure of RdRp of hepatitis C virus (PDB code: 2DXS). Part **A** represents the binding of **64** and Part **B** the binding of **65**, where in each the left-hand portion shows the surface view and the right-hand portion the ribbon view. In Part **A**, compound **64** is shown to have five hydrogen bonds, two with Arg503, two with Leu392, and one with Gly493, while part **B** compound **65** is shown to have only three hydrogen bonds, two with Arg503 and one with Gly493. *(Reprinted with permission from Haudecoeur et al. (2011). Copyright 2011 American Chemical Society.)*

pattern as a hydrophilic and hydrogen-bond donor moiety and as well as the anchor role of this part of the inhibitors.

Following this observation, this group of authors targeted the B-ring moiety of aurones and found that the replacement of the B-ring with an *N*-substituted indole moiety leads to improved inhibitory effect, and some compounds with this modification were found to be highly active, possessing IC_{50} values 2.3–2.4 μM RdRp inhibition (Meguellati et al., 2014). The most active compound had the structure as **66**. The docking experiment performed with the most active compounds again confirmed that the binding pocket for aurone analogs was only the allosteric thumb pocket I of RdRp.

Based on their two previous studies as described earlier, the same group of authors (Meguellati et al., 2016) focused their attention on pseudodimeric aurones that had never been investigated. So, they tried a series of compounds having the scaffold as **67** that had a spacer between the benzofuranone moieties. In this class of aurones, the most active compound was found to be **67a** with an IC_{50} value of 1.3 μM. However, the mutagenic and molecular modeling studies on this series of compounds had revealed that the binding site for the most active compounds was probably the palm pocket I instead of thumb pocket I.

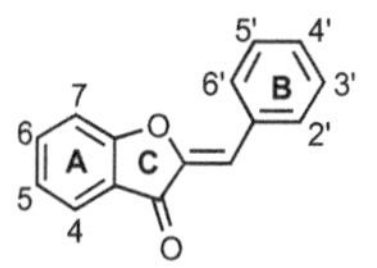

63

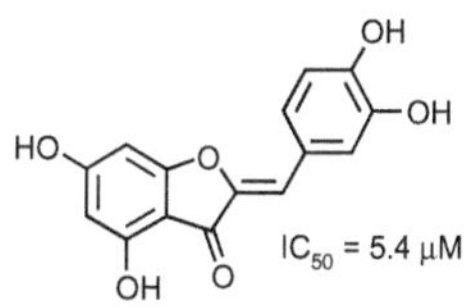

64, Aureusidin

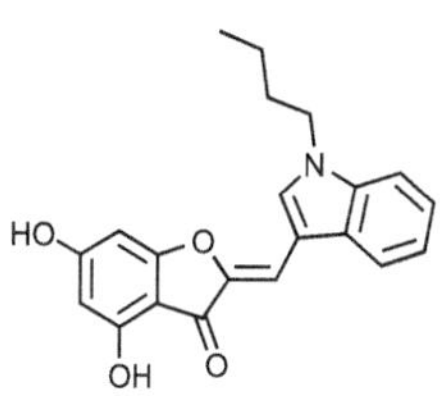

65, IC_{50} = 2.2 μM

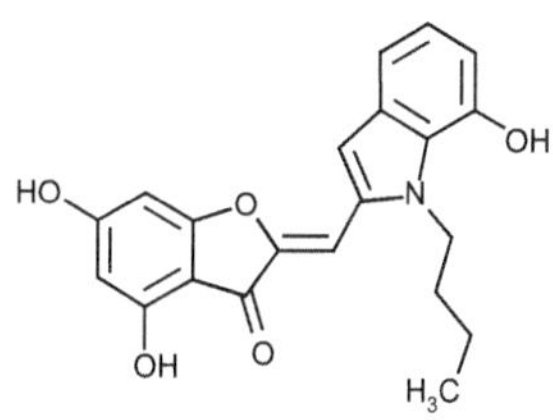

66

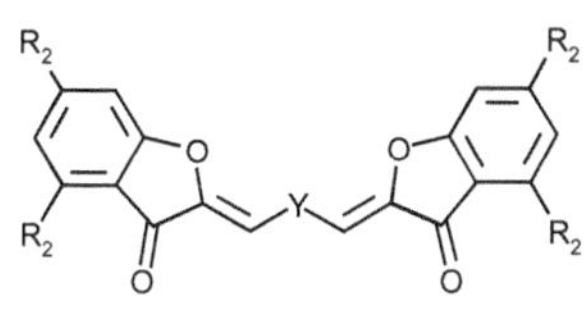

67

67a, Y = Ph (*1,6* position), R_2 = OH

The docking study of **67a** had shown that one of the benzofuranone moieties was mostly stabilized by hydrophobic interactions in an area surrounded by apolar residues, normally occupied by palm I site, such as Pro197, Met414, Tyr415, Ille447, and Tyr448. On the other hand, the second benzofuranone moiety occupied a more polar zone in which hydroxyl substituents may be involved in hydrogen bondings with residues such as Asn291 or with possible water molecules acting as bridges between the ligand and the protein. However, the aromatic linker was stabilized by van der Waals contacts with Phe193 and Tyr448.

2.4.2.6 4-Thiazolidinones

Thiazolidinone-4-one scaffold (**68**) has been exploited to develop a variety of drugs including anti-HCV agents. To explore potent anti-HCV agents, Rawal et al. (2008) had synthesized a series of **68** and screened their potency against NS5B polymerase. This had led them to find two most potent compounds, **69** and **70**, with IC_{50} values of 31.9 and 32.2 μM, respectively. Later, Küçükgüzel et al. (2013) performed another study to find potent NS5B inhibitors of this class of compounds. They attempted a series of **71**, resulting into a most potent compound, **72**, with IC_{50} value of 5.6 μM. In a docking study, the arylidene moiety (the two rings on the left side of imine nitrogen) was found to have the maximum interaction with the protein: the chlorine of the aryl ring and both the nitrogens of the 5-membered heterocyclic ring were found to interact with different residues of the enzyme. Encouraged by this, these authors made another attempt on another series of arylidene-4-thiazolidinones, but did not find any compound better than **72**. The most active compound was found to have an IC_{50} value of 25.3 μM. However, a docking study had led to confirm that these compounds bind in thumb II site (Cakir et al., 2015).

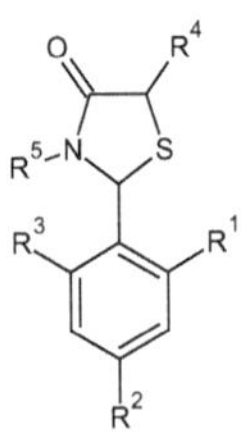
68

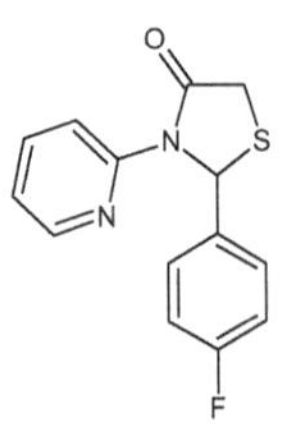
69

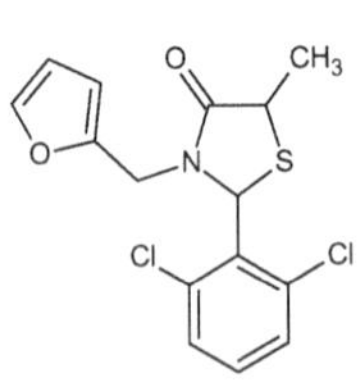
70

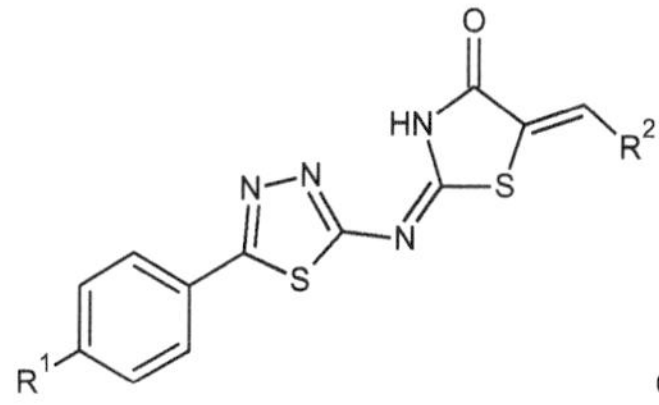

71

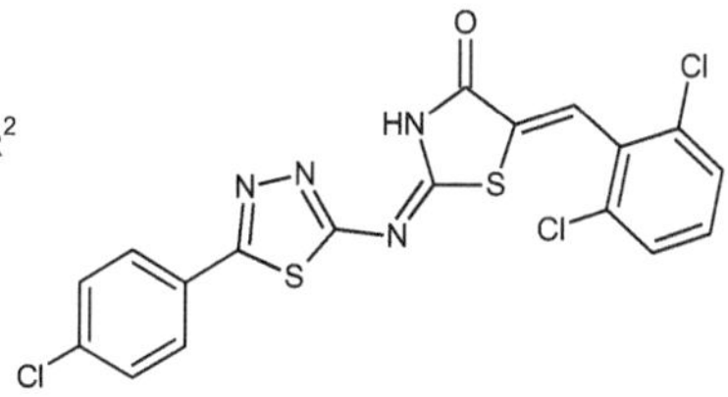

72

2.4.2.7 Benzothiadiazines

Like 4-thiazolidinones, benzothiadiazine have also exhibited multiple biologic activities. They were thus also tried to explore anti-HCV agents. In a review, Das et al. (2011) discussed the status of benzothiadiazine derivatives developed as NS5B polymerase inhibitors. Many compounds derived were based on scaffold **73**. A mechanistic study of this class of HCV polymerase inhibitors had revealed that they can be potent anti-HCV agents (Liu et al., 2006). A derivative of this (**74**) was identified as a more potent inhibitor. This compound was found to interfere with RNA synthesis and to be able to form a ternary complex with the protease and RNA template (Liu et al., 2006; Gu et al., 2003). After the discovery of **74** as a potent inhibitor of NS5B, the general core **73** was studied extensively by several groups to find compounds with improved potency and pharmacokinetic properties. In the past, several benzothiadiazine derivatives were discovered as potent NS5B polymerase inhibitors with pharmacokinetic properties by companies like GSK, Abbott, InterMune, and Roche. Abbot's benzothiadiazine analog A-782,759 (**75**) (Mo et al., 2005) was a potent NS5B inhibitor. Another compound, ANA-598 (setrobuvir) (**76**), belonging to benzothiadiazine class and containing a tricyclic dihydropyridinone structural motif was identified as a potent HCV NS5B polymerase inhibitor by Ruebsam et al. (2009) at Anadys Pharmaceuticals. It binds to palm I site close to Tyr448. It was found orally bioavailable and a potent inhibitor of HCV genotype 1a and 1b, having in vitro potency with subnanomolar IC_{50} values. However, when Anadys Pharmaceuticals was acquired by Roche in 2011, Roche terminated the development of ANA-598 in July 2015 when it was in Phase IIb clinical trials.

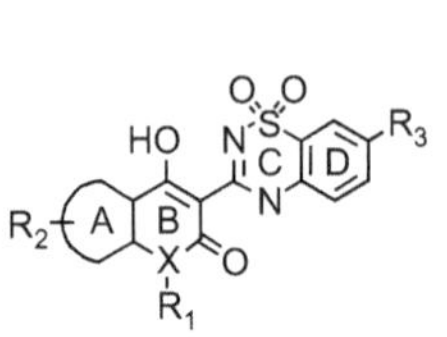

73

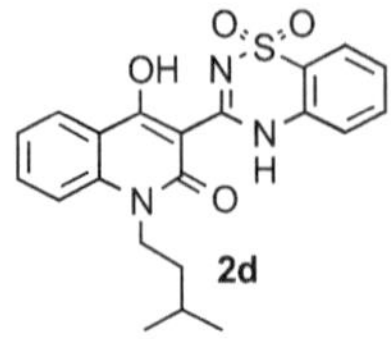

74

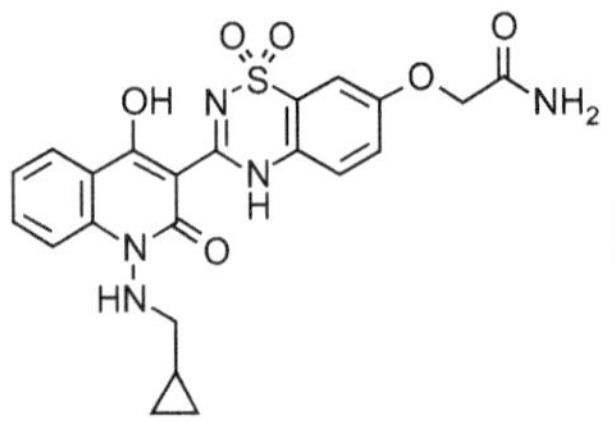

75 (A-782759)

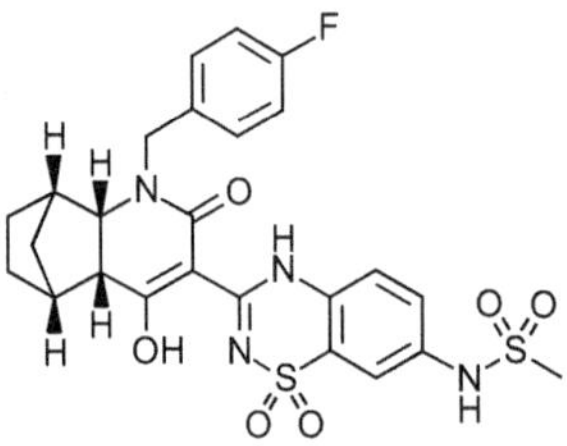

76 (ANA-598, Setrobuvir)

2.4.2.8 Rhodanines

For the development of potent NS5B polymerase inhibitors, Talele et al. (2010) had performed structure-based virtual screening that had led them to find two rhodamine leads. Rhodanine (**77**) is a 5-membered heterocyclic organic compound possessing a thiazolidine core. The two lead compounds were as **1** and **2** shown in Fig. 7.13. They were found to possess IC_{50} values of 7.7 and 10.6 μM, respectively. Patel et al. (2013) attempted to synthesize a molecular hybrid of these two leads and find its activity. The molecular hybrid was a derivative of **78**, i.e., **78a**, which had an IC_{50} value of 6.7 μM. These authors then explored the optimal substituents on the terminal phenyl ring of this hybrid structure (**78a**). This resulted in compound **78b** having $IC_{50} = 2.6$ μM. Both the compounds **78a** and **78b** were more active in their L-isomeric form than D-isomeric form. Their D-isomeric forms had IC_{50} values of 19.3 and 5.4 μM, respectively. Earlier than this, however, Powers et al. (2006) had found an active rhodamine analog (**79**) binding to palm II site. The halogen substituents at the benzylidine ring were found to be favorable to the activity: particularly, 3,4-di-Cl, 3,4-di-Br and 2,4,5-tri-F had given the best IC_{50} values.

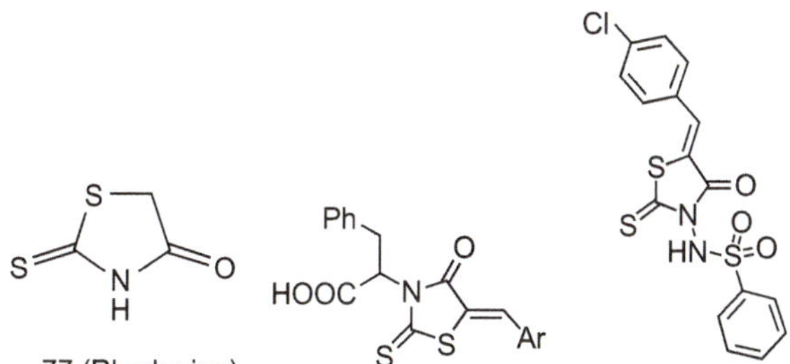

77 (Rhodanine)

78

79

78a: Ar = 3-OPh-Ph
78b: Ar = 3-(3,4-diCl)OPh-Ph

2.4.2.9 Thiophenes

Thiophenes are another class of compounds containing sulfur. However, they are not so well studied for anti-HCV activity. However, Mayhoub (2012) has reported three thiophene analogs acting against NS5B polymerase. These are VCH-759 (**80**), VCH-916 (**81**), and VCH-222 (**82**).

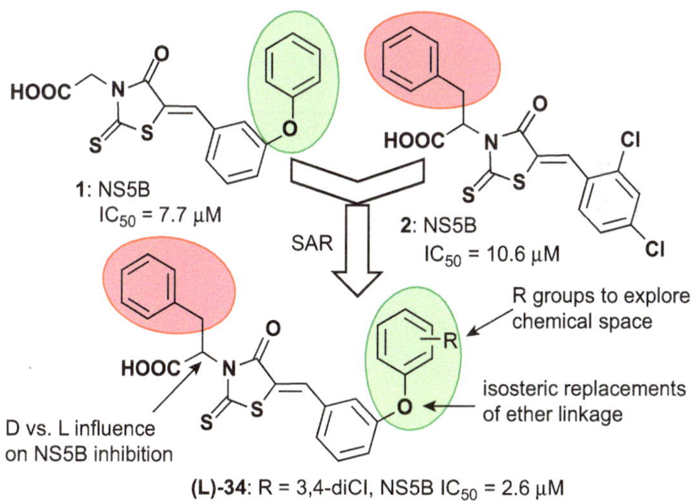

Figure 7.13 A pictorial representation of some active rhodamines and their further development. *(Reprinted with permission from Patel, B.A., Krishnan, R., Khadtare, N., Gurukumar, K.R., Basu, A., Arora, P., Bhatt, A., Patel, M.K., Dana, D., Kumar, S., Kaushik-Basu, N., Talele, T.T., 2013. Design and synthesis of L- and D-phenylalanine derived rhodanines with novel C5-arylidenes as inhibitors of HCV NS5B polymerase. Bioorg. Med. Chem. 21, 3262–3271. Copyright 2013 Elsevier.)*

They all bind to thumb II site and are well tolerated with submicromolar IC_{50} values versus gt 1a and 1b replicons.

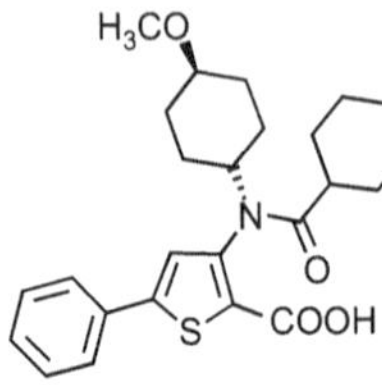

80 (VCH-759)

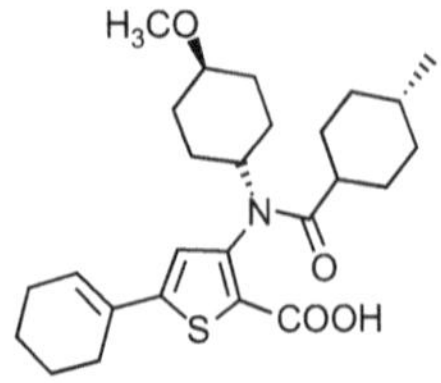

81 (VCH-916)

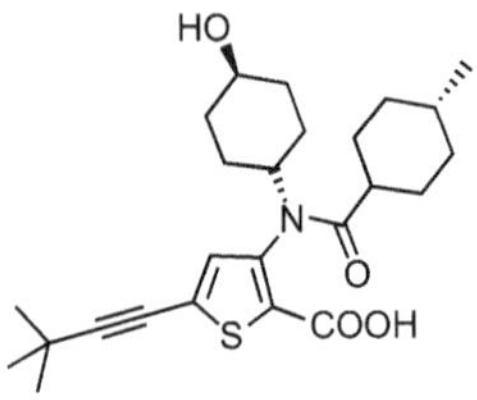

82 (VCH-222)

2.4.2.10 Quinolones

Some quinolones were also studied for NS5B polymerase inhibition activity. In a study, Kumar et al. (2011) had reported a 4-quinolone analog (**83**) having very high enzyme inhibition and cell-based antireplicon activities (NS5B $IC_{50} = 1.1\ \mu M$, Replicon $EC_{50} = 12\ \mu M$). An SAR study led these authors to move from this keto-quinolone to a more potent benzyl ester analog (**84**) having much higher potencies (NS5B $IC_{50} = 0.008\ \mu M$, Replicon $EC_{50} = 0.23\ \mu M$).

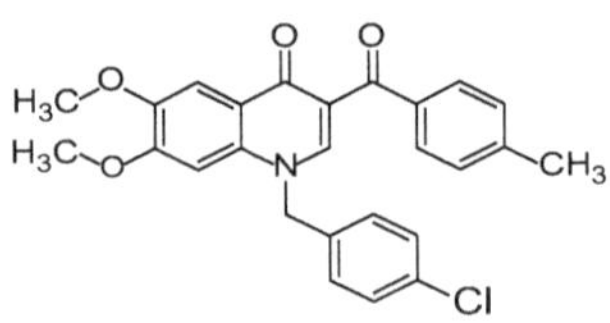

83

NS5B $IC_{50} = 1.1\ \mu M$
Replicon $EC_{50} = 12\ \mu M$

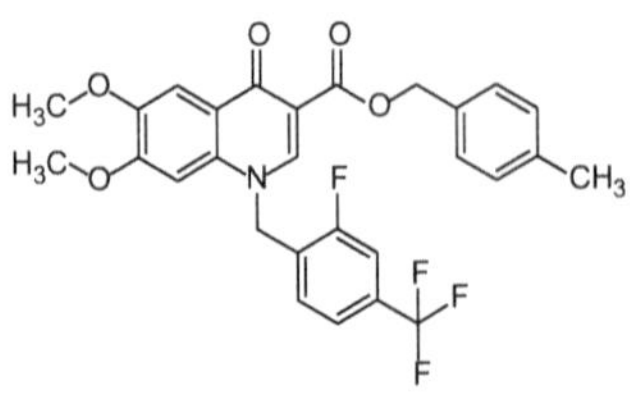

84

NS5B $IC_{50} = 0.008\ \mu M$
Replicon $EC_{50} = 0.23\ \mu M$

2.4.2.11 Aryl Diketoacids

Kim et al. (2008) had studied a series of aryl diketoacids (**85**) in which they found a very potent compound (**85a**) having very low enzyme inhibition, as well as cell-based replicon inhibition concentrations (NS5B $IC_{50} = 0.96$ μM, Replicon $EC_{50} = 0.82$ μM). A pharmacophore-based docking study performed by these authors had suggested that potency of aryl diketoacids depends upon aryl linkers as well as the size and position of the aromatic substituents.

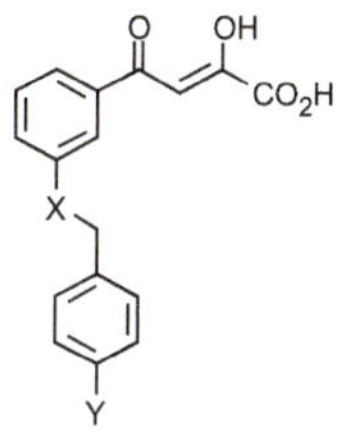

85

85a: X = NH, Y = Cl

NS5B IC_{50} = 0.96 μM, Replicon EC_{50} = 0.82 μM

In addition to previously discussed classes of NIs and NNIs, there were a few more classes of compounds studied for anti-HCV agents, but they were not found so important. The NIs and NNIs that have emerged so far as the most important and useful were recently summarized by Polamreddy et al. (2016), given in Tables 7.4 and 7.5, respectively. Mayhoub (2012) has given in Fig. 7.14 a pictorial detail of binding sites of different classes of NNIs of HCV NS5B polymerase.

Table 7.4 Nucleoside inhibitors in the market and under investigation

Molecule	Clinical phase[a]	Company
GS-7977 [sofosbuvir]	Launched	Gilead
VX-135	Phase 2	Vertex Pharmaceuticals
ACH-3422	Phase 1	Achillion
PSI-938	Discontinued	Gilead
RG7128 [mericitabine]	Discontinued	Roche
IDX21437 [MK 3682]	Phase 2	Idenix/Merck
IDX184	Discontinued	Idenix
PSI-7851[isomer of sofosbuvir]	Discontinued	Pharmasset
TMC649128	Discontinued	Medivir and Tibotec
GS-6620	Phase 1	Gilead
BCX5191	Discontinued	Biocryst
Valopicitabine	Discontinued	Idenix/Novarties

[a]Clinical phase details of all compounds are captured from Adisin site.

Table 7.5 Details of clinical phase, company, and allosteric sites of nonnucleoside inhibitors of therapeutic importance

Molecule	Clinical phase[a]	Company	Allosteric site
Beclabuvir [BMS-791325]	Phase 3	Bristol-Myers Squibb Company	Thumb I
MK-3281	Discontinued	Merck	Thumb I
BI 207127 [deleobuvir]	Discontinued	Boehringer-Ingelheim	Thumb I
TMC647055	Phase 2	Janssen	Thumb I
BILB 1941	Discontinued	Boehringer-Ingelheim	Thumb I
Filibuvir	Discontinued	Pfizer	Thumb II
VCH759	Phase 2	ViroChemPharma	Thumb II
VCH916	Phase 2	ViroChemPharma	Thumb II
VCH222 [lomibuvir]	Discontinued	ViroChemPharma	Thumb II
GS-9669 [radalbuvir]	Phase 2	Gilead	Thumb II
GSK625433	Phase 1	GlaxoSmithKline	Palm I
ANA598 [setrobuvir]	Discontinued	Anadys/Roche	Palm I
ABT-333 [Dasabuvir]	Launched	AbbVie	Palm I
ABT-072	Discontinued	AbbVie	Palm I
GS-9190 [tegobuvir]	Phase 2	Gilead	Palm II
IDX375	Discontinued	Idenix	Palm II
HCV-79	Discontinued	Viropharma/Wyeth	Palm II
GSK5852	Phase 2	GlaxoSmithKline	Palm II
PPI-383	Phase 1	Presidio Pharmaceuticals, Inc.	Palm II

[a]Clinical phase details of all compounds are captured from Adisin site.

3. THEORETICAL STUDIES

Theoretical attempts have been made to mainly find out the following:

1. structural properties of the molecules of a given class of drugs that govern their biologic activities, so proper structural modifications can be made in the existing scaffold to have analogs of increased potency (quantitative structure–activity relationship [QSAR] study)
2. the nature of drug–receptor interactions by docking the molecules in the well-known 3D structure of the receptors (proteins or enzymes), so proper structural modification can be made in drug molecules to have optimum interactions with receptors (molecular modeling)

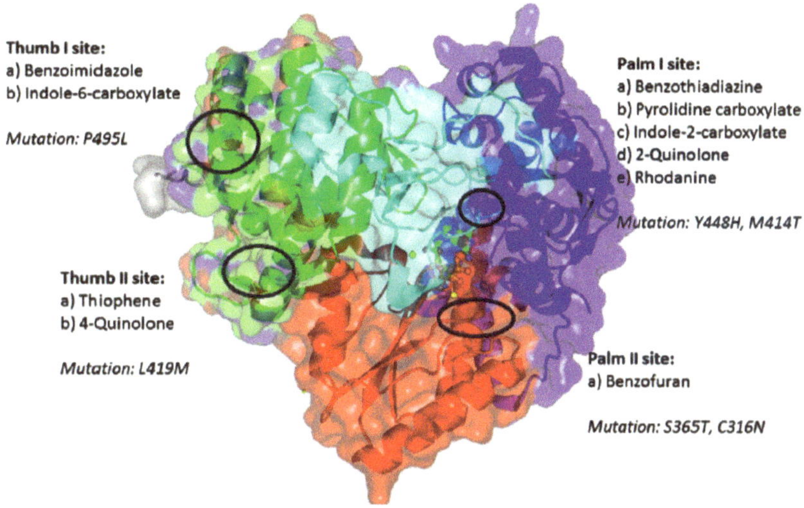

Figure 7.14 A pictorial detail of binding sites of different classes of nonnucleoside inhibitors of HCV NS5B polymerase. *(Reprinted with permission from Mayhoub, A.S., 2012. Hepatitis C RNA-dependent RNA polymerase inhibitors: a review of structure-activity and resistenace relationships; different scaffolds and mutations. Bioorg. Med. Chem. 20 (10), 3150–3161. Copyright 2012 Elsevier.)*

3. homology models of proteins/enzymes whose structures are not known, so drug–receptor interactions can be studied (homology modeling)
4. pharmacophoric models

For these four types of approaches, various software products are available. First, we present some studies on NS3/4A protease inhibitors of HCV.

3.1 NS3/4a Protease Inhibitors

For a series of tripeptide inhibitors such as **86**, incorporating a mix of α- and β-amino acids, Nurbo et al. (2008) found by both experimental and molecular modeling studies that P_3 position was least sensitive to β-amino acid substitution, but proper oxyanion hole interaction was found to be important for good inhibition potency.

Wei et al. (2008) performed several 3D-QSAR studies such as comparative molecular field analysis (CoMFA), comparative molecular similarity indices analysis (CoMSIA), and pharmacophore modeling, for a series of truncated ketoacid inhibitors with general structure as **87** designed for exploring the P2 and P3 binding pockets of HCV NS3 protease. In **87**, X refers to varying carboxybenzyl groups and P2 and P3 different amino acid moieties. In their overall studies, Wei et al. found that a hydrophobic group at P2 and a hydrophilic group at P3 could lead to higher potency of the compounds. By mapping the pharmacophoric features developed onto the molecular surface of NS3, they found that the binding pocket of position P3 was large and versatile because it was surrounded by Lys136, Val132, and Cys159, while that of position P2 appeared to be buried somewhat. This mapping led Wei et al. (2008) to believe that there was still room around P3 for designing even more potent inhibitors for HCV NS3 protease.

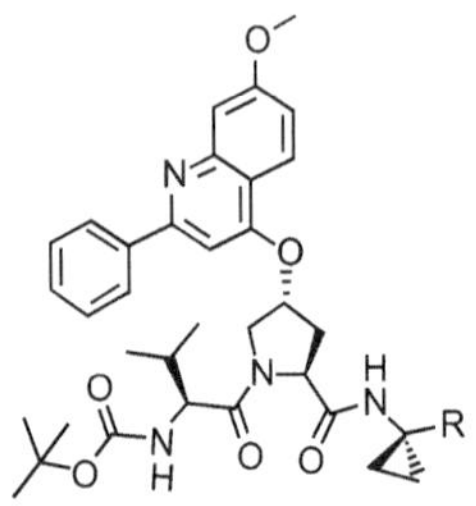

86

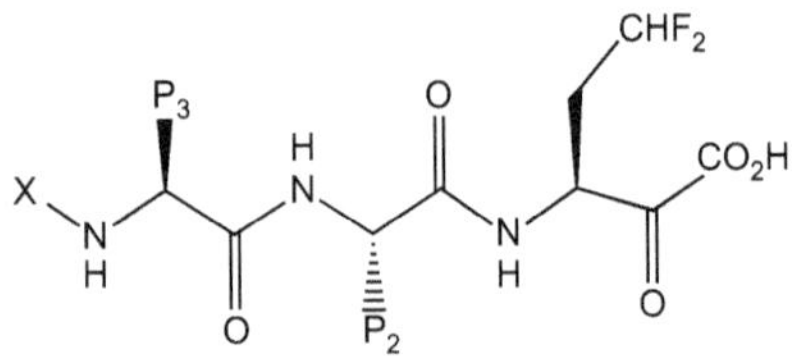

87

Through molecular modeling and CoMFA studies, attempts were made by da Cunha et al. (2006) to understand the interactions of BILN 2061 macrocyclic analogs (**88**) with the wild-type and D168A mutant NS3 serine proteases. BILN 2061 (**89**) itself was docked in both the receptors according to Tsantrizos et al. (2003), where the favorable orientation of the compound complexed with wild-type NS3 protease involved seven hydrogen bonds, two from quinolone ring, one from thiazole ring, and four from macrocyclic ring. However, in complexation with D168A, it formed only three hydrogen bonds, one involving sulfur atom and two involving −NH and −OH groups of macrocyclic ring. From CoMFA contour analysis, which tells the favorable and unfavorable regions for bulky and charged substituents, da Cunha et al. (2006) found that the region around the cyclopentyl carbonate group was sterically favorable, but the area around the thiazole ring was unfavorable for bulky substituents. Similarly, the region close to isopropyl group of the thiazole ring was favorable for positively charged substituents such as any alkyl group, and the region close to thiazole ring was favorable to negatively charged substituents.

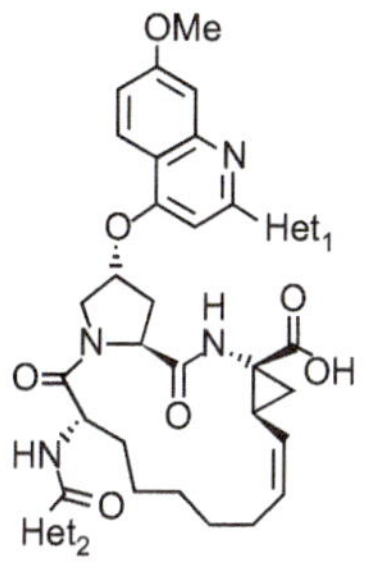

88, BILN 2061 analogues

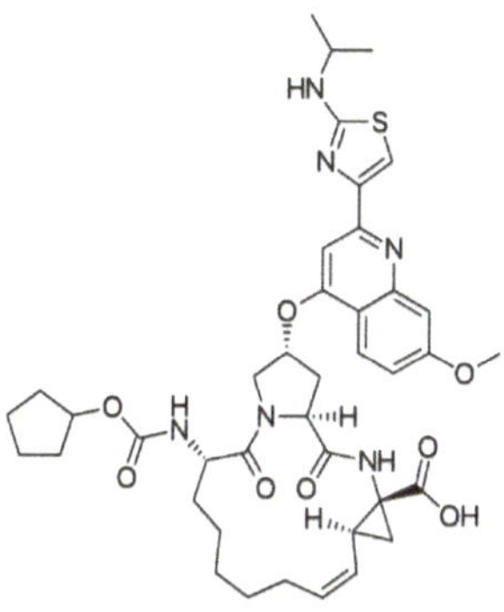

89, BILN 2061 (Ciluprevir)

3.2 NS5B Polymerase Inhibitors

3.2.1 Benzimidazoles

Benzimidazoles as NNIs have not only been experimentally well studied but also theoretically for NS5B polymerase inhibition. A series of benzimidazoles, as shown in Fig. 7.15, were studied by Patil et al. (2012a) for their NS5B polymerase inhibition activity and then subjected to some QSAR studies. By a molecular docking study of all the compounds in NS5B polymerase (PDB code: 2DXS) using FlexX software, these authors found that benzimidazole ring forms hydrophobic contacts with Val37, Ala395, Thr399, Leu492, Ala395, and Phe429. The 2-(4-Me-benzoxy) phenyl substituent attached to the C-2 position of the benzimidazole ring is located in a deep hydrophobic pocket formed by His428, Leu392, Leu425, Phe429, Pro495, Pro496, Val499, and Arg503. The 4-Me-benzoxy

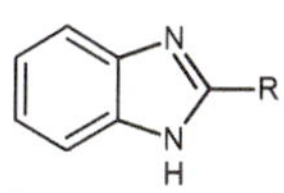

R = 3,4-diClPh, 3,4-diFPh, 4-(2-pyridyl)Ph, 4-pyridyl, N,N-diMePh, *i*-prPh, 2-benzoxyPh, 2-(4-Cl-benzoxy)Ph, 2-(4-Me-benzoxy)Ph

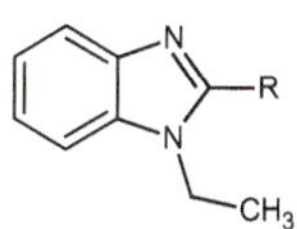

R= *i*-prPh, 2-benzoxyPh, 2-(4-Cl-benzoxy)Ph, 2-(4-Br-benzoxy)Ph

Figure 7.15 Benzimidazole derivatives studied by Patil et al. (2012).

substituent was found to nicely fit into a narrow hydrophobic pocket formed by His428, Pro495, Leu497, Val499, and Arg503. The ethyl group at the N-3 position of the benzimidazole ring was observed to interact with Leu489, Gly493, and Val494. Additionally, one side of the phenyl ring still had the scope to accommodate a large hydrophobic substituent. Such findings were parallel to what Patel et al. (2008) had found for a series of benzimidazole analogs as represented by **90**. They had, however, also found from CoMFA and CoMSIA analysis that a sterically bulky group around the cyclohexyl moiety at the N-1 position of benzimidazole ring as well as bulky substituents around the indole ring will be conducive to the activity. Further, a bulky group around R-substituent at the chiral carbon would be detrimental to the activity; rather a negatively charged group here would be conducive. An electropositive group around the R_1-position of the indole ring would be favorable to NS5B inhibition activity. From the CoMSIA model, it was indicated that the presence of a hydrophilic group around R-substituent at the chiral carbon would be favorable to the activity. Near the R1-position of the indole ring, hydrogen-bond donor groups have been indicated to be favorable, while around the R-substituent at the chiral carbon atom, hydrogen bond acceptor groups would be favorable to the potency of the compounds. In another study on a series of benzimidazole analogs, Gupta et al. (2010) had found that only steric factors are important for their anti-HCV activity.

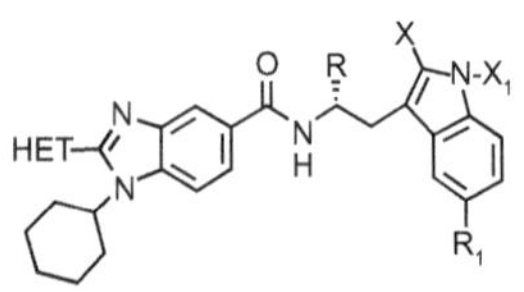

90

Consistent to what Patel et al. (2008) and Patil et al. (2012a) found regarding the interactions of benzimidazoles with NS5B polymerase of HCV, almost the same was found in the next study by Patil et al. (2013) regarding a new set of benzimidazoles identified by virtual screening by FlexX docking. The top score screened belonged to the general structure as **91**. The binding modes of two compounds having high scores, **91a** and **91b**, were described. Compound **91a** had two hydrophilic interactions with amino acid residues, Arg503 and His428. The benzyl substituent at the 6-position of the benzimidazole ring showed hydrophobic interaction with Val37, Leu 392, Thr399, His128, and Val494. Compound **91b** had hydrogen bond interaction with Leu 392. Its 4-pyridine moiety was located in the hydrophobic pocket formed by Ala396, Leu 492, and Val494. The CONH-phenyl moiety interacted with the hydrophobic pocket formed by Pro495, Tryp500, and Arg503. The basic benzimidazole nucleus had ionic interactions with Leu392, Ala395, Ile424, Leu425, His 428, and Phe429.

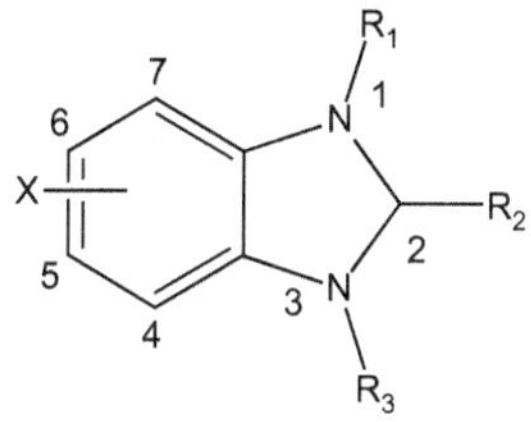

91

91a: R_1 = 1-naphthylmethyl, R_2 = 3,4-difluorophenyl, R_3 = Nil, X = 5-(N-(benzylamino)carboimidoyl

92a: R_1 = Nil, R2 = 4-pyridyl, R_3 = H, X = 7-phenylcarbonyl

However, when a few series of pyrimidine nucleoside inhibitors (Fig. 7.16) were subjected to 2D-QSAR analysis by Patil et al. (2010), in most of the cases the NS5B inhibition activity was found to be basically correlated with hydrophobic parameter (ClogP) or some parameters that defined the bulkiness of the molecule, such as molar refractivity or Kier's first-order molecular connectivity index ($^1\chi^v$) (Kier and Hall, 1976), suggesting that hydrophobic or bulky substituents would be favorable to the activity.

A FlexX docking study on a series of 2,5-disubstituted imidozo[4,5-c] pyridines (**92**) by us (Gupta et al., 2013a) had exhibited, with reference to the template molecule **92a**, that 2-fluorophenyl (R_2-substituent) had crucial interaction with the deep hydrophobic pocket made of Ala395, Leu425, Trp500, Phe429, Leu392, Ile424, and His428, and similarly, 4-i-PrPh group (R_1-substituent) was buried in a deep pocket where it interacted with Val137 and Ala396 residues (Fig. 7.17). The parent moiety, imidozo[4,5-c]pyridine, might have some hydrogen-bond interactions with residues Ala393, Val494, Trp397, and Leu492.

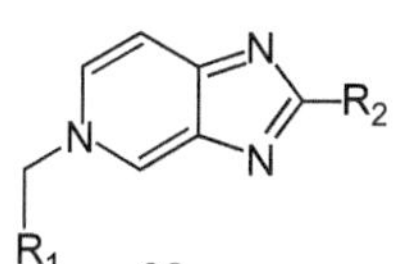

92

92a: R_1 = 4-i-Prph, R_2 = 2-FPh

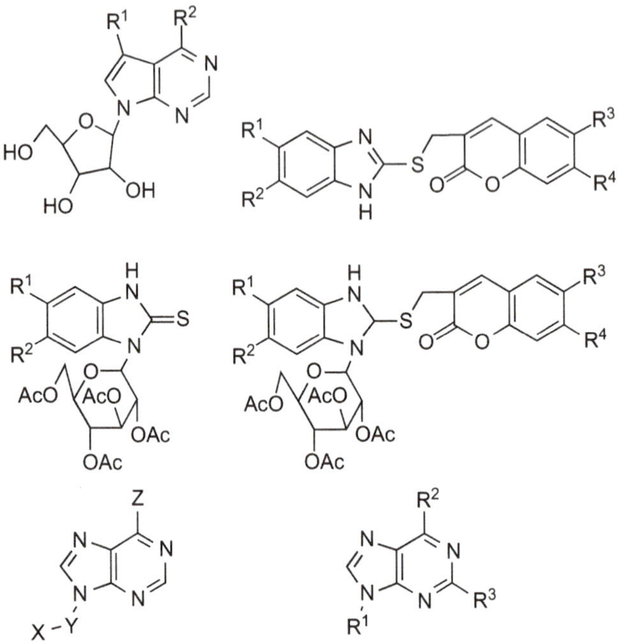

X = a nonphosphate group
Y = not necessarily 2-methyl ribose group
Z = not necessarily a methylsulfonyl group but small substituents like Cl, OH, NH_2, NMe_2 or SMe

Figure 7.16 A few different groups of pyrrolo[2,3-d]pyrimidine nucleoside derivatives.

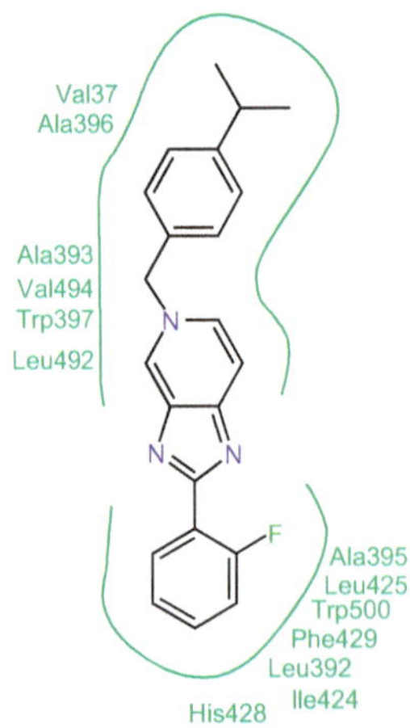

Figure 7.17 The binding of a 2,5-disubstituted imidozo[4,5-c]pyridine analog (**92a**) in hepatitis C virus NS5B polymerase (PDB code: 2DXS). *(Reprinted with permission from Gupta et al. (2012). Copyright 2012 Springer Nature.)*

3.2.2 Benzothiadiazines

Benzothiadiazines also have been theoretically well studied. A detailed theoretical study was made on a series of benzothiadiazines (**93**) by Ismail et al. (2012) acting against NS5B polymerase of various HCV genotypes (1a, 1b, and 4). The most important study was to find pharmacophore features that were extracted using molecular dynamics and an energy-based pharmacophore. As Fig. 7.18 shows, in the pharmacophore, there were two hydrophobic features (H7 and H8 in green), three aromatic rings (R11, R12, and R13 shown by orange circles), and one hydrogen-bond acceptor (A5 shown in red). The hydrophobic tail (R_1) and enolic oxygen of ring B (A5) were the most essential features. Benzothiadiazines have been found to bind to the palm site of the receptor, and these authors highlighted, as shown in Table 7.6, the characteristics of the main pockets of the binding site and the amino acids that constitute them. There are three pockets, where pocket I is constituted of Phe415, Met414, Asn411, and Gly410, pocket II is constituted of Pro197 and Leu384, and finally pocket III is

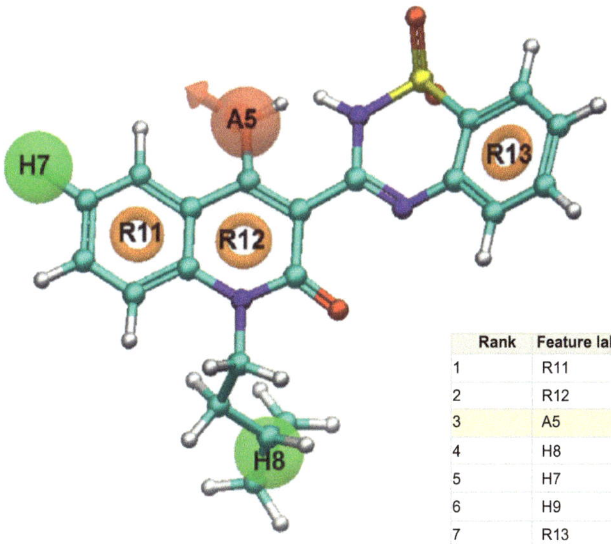

Rank	Feature label	
1	R11	−0.83
2	R12	−0.42
3	A5	−0.27
4	H8	−0.17
5	H7	0.00
6	H9	0.00
7	R13	−0.57

Figure 7.18 Energy optimized pharmacophore developed by Ismail et al. (2012). The essential features are two hydrophobic regions (H7 and H8), one hydrogen-bond acceptor (A5), and three aromatic moieties (R11, R12, and R13). *(Reprinted with permission from Ismail, M.A.H., Ella, D.A.A.E., Abouzid, K.A.M., Mahmoud, A.H., 2012. Integrated structure-based activity prediction model of benzothiadiazines on various genotypes of HCV NS5b polymerase (1a, 1b and 4) and its application in the discovery of new derivatives. Bioorg. Med. Chem. 20, 2455–2478. Copyright 2012 Elsevier.)*

Table 7.6 The descriptions of palm I pockets available to interact with inhibitors

Features of Palm I site	Components	Description
Pocket I	Phe415, Met414, Asn411 and Gly410	A hydrophobic shallow and wide subpocket
Pocket II	Pro197 and Leu384	Deep and narrow hydrophobic subpocket
Pocket III	Ser288, Ser556, Asn291 and Asp318	Relatively narrow hydrophilic subpocket
Polar amino acid residues	Tyr448, Gly449 and Ser556	Involved in polar interactions with inhibitor
Structural water molecules	2	Forms water bridge between ligand and the protein

shown to be constituted of Ser288, Ser556, Asn291, and Asp318. There were two structural water molecules also in the binding sites, which were found to be essential in the binding. The first water molecule was found to interact, acting as a bridge, with a sulfonyl oxygen of the ligand and Ser556, and the second one to interact with NH of Gly449 residue as well as the ligand enolic oxygen. Additionally, there were three polar amino acid residues, Tyr448, Gly449, and Ser556. In the drug-receptor interactions, Met414 was important in the σ-π interaction with ring A, and similarly, Phe193 was important in the σ-π interaction with ring D. The enolic oxygen proved to be an essential feature, as it forms a hydrogen-bond directly with Tyr448 and Gly449 via a water bridge (WB). By virtual screening a hit compound (**94**) was selected and docked in palm I site of a homological model of genotype 4 NS5B polymerase, which showed a mutation of Met414 and Val414. The interaction of this compound with the polymerase is shown in Fig. 7.19.

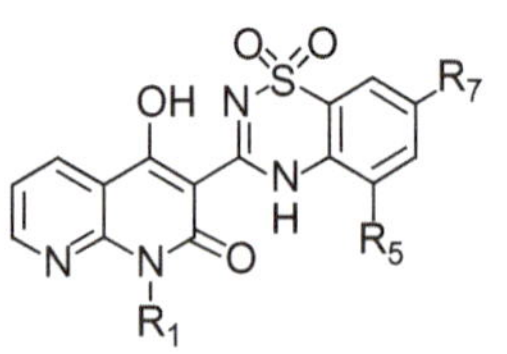

93

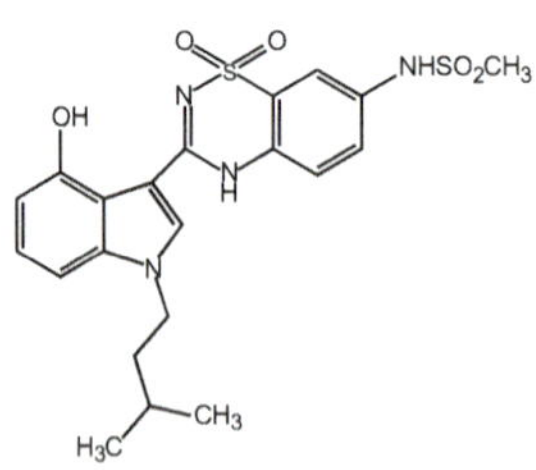

94

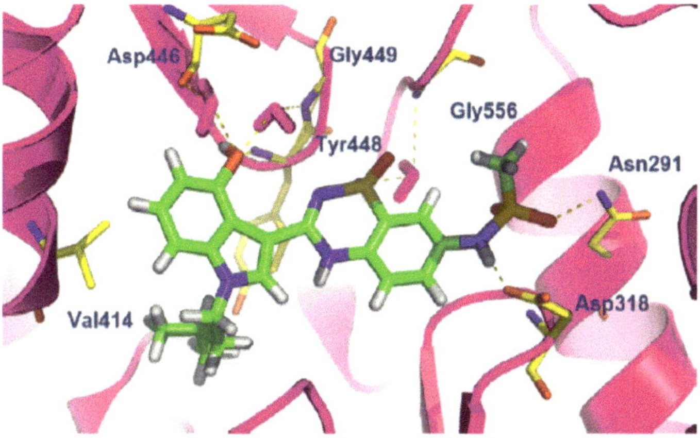

Figure 7.19 Docking of hit compound (**94**) in palm I site of homological model of genotype 4 NS5B polymerase. The methanesulfonamide group is shown to form H-bond with Asn291 and Asp318. *(Reprinted with permission from Ismail, M.A.H., Ella, D.A.A.E., Abouzid, K.A.M., Mahmoud, A.H., 2012. Integrated structure-based activity prediction model of benzothiadiazines on various genotypes of HCV NS5b polymerase (1a, 1b and 4) and its application in the discovery of new derivatives. Bioorg. Med. Chem. 20, 2455–2478. Copyright 2012 Elsevier.)*

A molecular docking and molecular dynamics study on a series of benzothiadiazines had led Wang et al. (2010) to conclude almost the same as Ismail et al. did. Their conclusions were as follows.

1. Hydrophobic substituents with a similar size of benzo group-like isosteres are preferential at positions 1 and 2 (ring B of benzothiadiazines).
2. Substituents at position-3 containing a linear alkyl chain (four or five carbon atoms) or a branched alkyl chain (five to eight carbons) can increase the inhibitory activity by one to two orders of magnitude.
3. A polar substituent like methanesulfonamide group at position-14 can enhance the activity of the drug by providing a hydrogen bonding interaction with the protein target.

When Li et al. (2010) investigated binding models of five different scaffold inhibitors of NS5B polymerase by using molecular dynamics simulations, free energy calculation, and decomposition, it was found that Tyr448 plays the most critical role in the binding of most inhibitors. In addition, favorable contributions of residues Pro197, Arg200, Cys366, Met414, and Tyr448 in a deep hydrophobic pocket proved to be important for the selectivity of inhibitors.

On some similar series of benzothiadiazines, our group (Patil et al., 2012b) also performed FlexX docking, and when the most active compound of series (**95**) was docked in genotype 1 NS5B polymerase (PDB code: 2DXS) the oxygen of 4-OH group was found to act as a hydrogen-bond acceptor to form H-bond with Leu492 and one of the oxygens of methylsulfonamide group to from the H-bond with Trp500 and the methyl group to interact with His428. The alkyl group at N1-position was found to be present in a hydrophobic pocket and interact with Trp397.

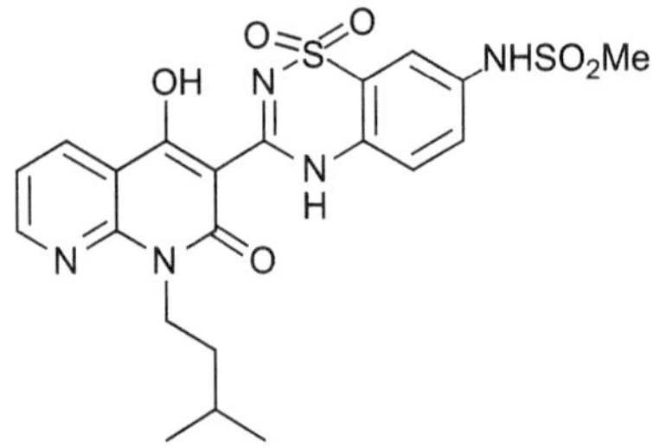

95

That methylsulfonamide group plays a very important role in NS5B inhibition was shown by Barreca et al. (2013). These authors identified a pyrazolobenzothiazine scaffold as a suitable template for getting novel anti-HCV agents. The best compound of this series was a derivative having a methylsulfonamide tail (**96a**). This derivative was able to have the desired π-π and three-point hydrogen-bond interactions through its para-methanesulfonamide phenyl group. The phenyl group had π-π interaction with Phe193 residue, and one oxygen of sulfonamide group was involved in H-bond interaction with Ser288 residue through a water molecule, and the other oxygen had formed two H-bonds, one with Asn291 and one with Asp318.

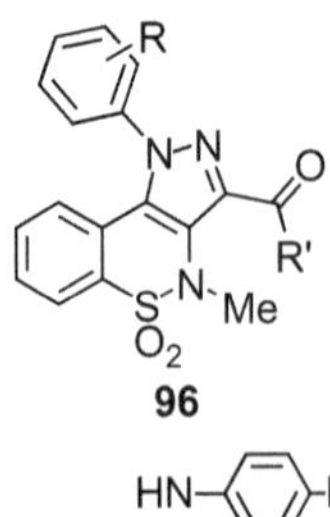

96

96a: R = 3-F, R_1 =

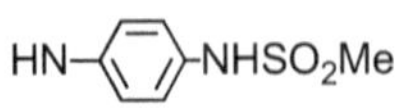

QSAR studies were made on many other classes of sulfur-containing compounds acting as NS5B polymerase inhibitors like thiophenes, thiouracils, thioureas, thiozolodines, and thiozolines, as described next.

3.2.3 Other Sulfur-Containing Compounds

A series of thiophene derivatives such as **97**, where different substituents were as given below the structure, was analyzed by our group (Varshney et al., 2012a). For this class of compounds, a 2D-QSAR equation obtained was as follows:

$$\begin{aligned}\log(1/IC_{50}) &= 0.638(\pm 0.182)\,{}^{1}\chi^{v}{}_{R1} + 0.793(\pm 0.326)\sigma_p \\ &\quad + 0.327(\pm 0.312)I_1 - 0.598(\pm 0.389)I_2 \\ &\quad - 1.104(\pm 1.777) \qquad (1)\\ &\quad n = 22,\ r = 0.89,\ r^2_{cv} = 0.69,\ r^2_{pred} = 0.76, \\ &\quad s = 0.25,\ F_{4,17} = 16.75(4.67)\end{aligned}$$

In this equation, n is the number of data points used, r is the correlation coefficient, r^2_{cv} is the square of correlation coefficient obtained by leave-one-out jackknife procedure, r^2_{pred} is for the test set, s is the standard deviation, and F is the F-ratio between the variances of calculated and observed activities. The data within parentheses with ± sign are 95% confidence intervals, and that given within parenthesis for F is the standard F-value at 99% level. Among the independent variables, ${}^{1}\chi^{v}{}_{R1}$ is Kier's first-order valence molecular connectivity index for R_1-substituent, σ_p is Hammet's electronic constant for para-substituent at the aryl ring of R_1-substituent, and I_1 and I_2 are two indicator variables used with a value of 1 each for a methyl group present at the ortho-position of the aryl ring of R_1-substituent and an additional methyl group, if any, present at the same aryl ring, respectively. Now, this equation showed that activity is primarily related to R1-substituent, where its molecular connectivity index ${}^{1}\chi^{v}{}_{R1}$, which refers to molecular size, has a positive effect, and I_1 parameter indicated that a methyl group at aryl ring of phenylsulfonamide group will be advantageous, but I_2 indicated the methyl group at any other position will be detrimental to the activity, which may be due to some steric problem. The positive coefficient of σ_p indicated that any electron-withdrawing substituent at the para-position of the phenyl ring will be conducive to the activity, and that may be due to the participation of the para-substituent in some electronic interaction with the enzyme. Thus this study showed the importance of phenylsulfonamide group, and this is consistent to what has been discussed earlier regarding the methylsulfonamide group in the case of benzothiadiazines.

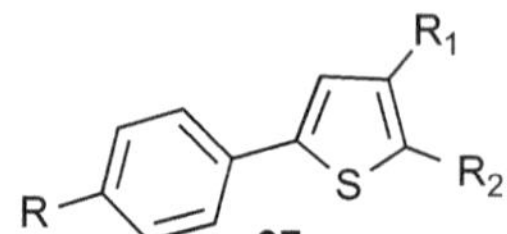

97

R = H, COMe, or SO_2Me;
R_1 = $NHSO_2Ph$, R_2 = $CONH_2$ or COOH

Similarly, our group also analyzed the SAR of a series of thiouracils (**98**) to conclude that an aromatic or large bulky R_1-substituent would be important for NS5B inhibition activity (Patil et al., 2011b). For a series of thioureas (**99**), Sharma et al. (2013) had derived a 2D QSAR equation, as shown by Eq. (2), in which ClogP refers to calculated hydrophobicity parameter, and out of the three indicator variables, I_1 and I_2 were used with a value of 1 each for R-substituent being an alkyl group or an aromatic group, respectively. The variable I_3 was used for the Ar-substituent being heterocyclic ring containing nitrogen. The negative coefficients of these three parameters showed that such substituents will not be beneficial to the activity. However, the positive coefficient of "m" that was used for the number of CH_2 moiety indicated that flexibility of the side chain may be important for the activity.

$$\begin{aligned}\log(1/EC_{50}) = {} & 0.179(\pm 0.061)\text{ClogP} - 0.758(\pm 0.306)I_1 \\ & - 2.267(\pm 0.322)I_2 - 0.331(\pm 0.143)I_3 \\ & + 0.086(\pm 0.055)m + 5.562(\pm 0.270) \qquad (2) \\ & n = 58,\ r = 0.926,\ r^2_{cv} = 0.83,\ r^2_{pred} = 0.62, \\ & s = 0.25,\ F_{5,52} = 62.12(3.39)\end{aligned}$$

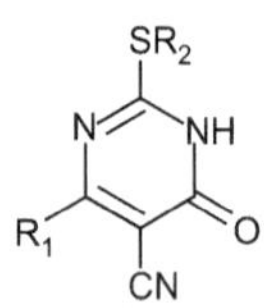

98

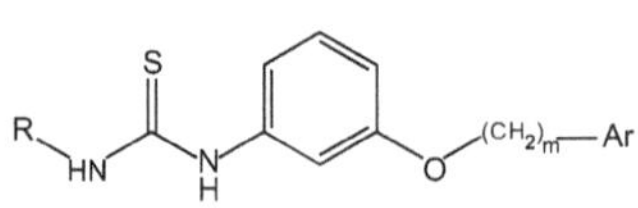

99

We had also performed a QSAR analysis for isothiozoles (**100**) and thiozolones (**101**) to find that their activity would be mainly governed by their molecular density, so essentially the bulk of the molecule was found to be important, which might influence the interaction of the molecules in some deep pocket of the receptor (Varshney et al., 2012b).

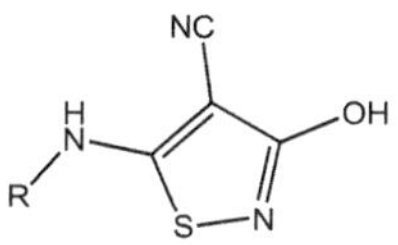

100

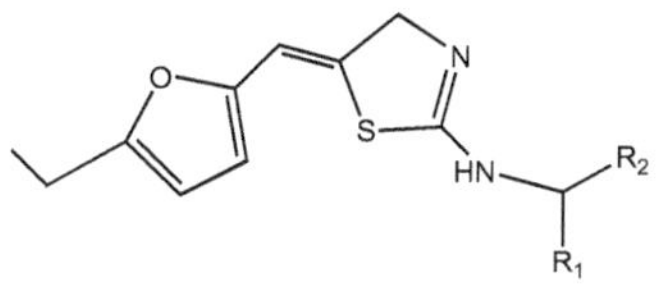

101

3.2.4 Pyrimidines

When pyrimidines were subjected to QSAR studies, the hydrophobic and hydrogen bond interactions were found to be crucial in them also. For a combination of several series of pyrimidine analogs studied by various authors (Arasappan et al., 2012; Bennett et al., 2012; Kwong et al., 2012; Girijavallabhan et al., 2012), which could be collectively represented as **102**, our group made a 2D-QSAR and docking study (Gupta et al., 2013b). The QSAR study had correlated the NS5B polymerase inhibition activity of these compounds with different physicochemical properties of the molecules, as shown by Eq. (3). In this equation, ClogP refers to hydrophobicity of the molecule, ${}^{1}\chi^{v}{}_{R2}$ to Kier's first-order valence molecular connectivity index for R_2-substituent, and Xt to total structure connectivity index of the whole molecule. An indicator parameter "I" was used with a value of 1 for the presence of a benzothiazole ring such as **103** at R_2-position. From this correlation, it was suggested that moderately hydrophobic compounds (ClopP $\leq$ 1.84) will have better activity, and also, since Xt has a negative coefficient, a very bulky molecule would not be tolerated. Also, at R_2-position a benzothiazole ring will not be preferred. However, a positive coefficient of ${}^{1}\chi^{v}{}_{R2}$ had indicated that overall size of this substituent will be favorable to the activity.

$$\begin{aligned}\log(1/EC_{50}) = {} & 0.514(\pm 0.165)\mathrm{ClogP} - 0.140(\pm 0.028)\,(\mathrm{ClogP})^2 \\ & + 0.332(\pm 0.112)\,{}^{1}\chi^{v}_{R2} - 31.554(\pm 5.835)\mathrm{Xt} \\ & - 0.426(\pm 0.132)\mathrm{I} + 12.933(\pm 1.487) \qquad (3) \\ & n = 90,\ r = 0.871,\ r^2_{cv} = 0.768,\ r^2_{pred} = 0.648, \\ & s = 0.24,\ F_{5,84} = 52.63(3.24),\ (\mathrm{ClogP})_o = 1.84\end{aligned}$$

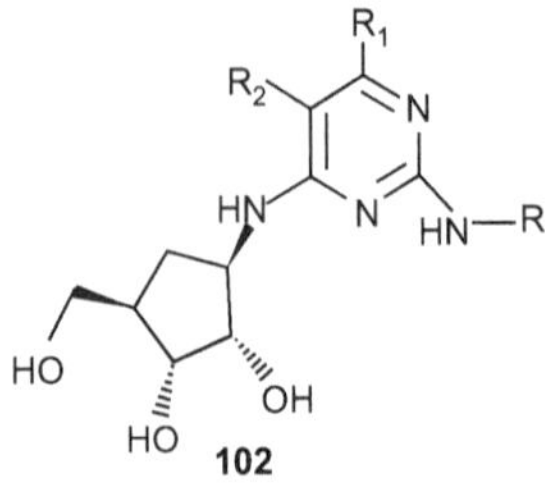

102

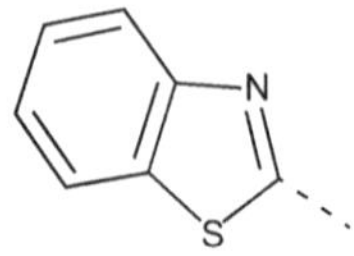

103

Gupta et al. (2013a,b) had also performed a docking study of some of these compounds using Molegro Virtual Docker. When a representative compound was docked in the thumb site of NS5B polymerase (PDB code: 4GMC) complexed with a thumb inhibitor, only −NH− and two −OH groups attached to cyclopentyl ring were found to form the H-bonds with few residues of the polymerase, while the pyrimidine ring was found to be engulfed in a deep hydrophobic pocket of the enzyme. Also for a series of pyrimidine nucleosides, Samanta et al. (2012) had found that steric and hydrophobic interactions were important for their activity.

3.2.5 Miscellaneous Classes

Worachartcheewan et al. (2015) classified around 1000 compounds into active and nonactive NS5B inhibitors based on their physicochemical properties, including quantum chemical descriptors, constitutional descriptors, functional groups, and molecular properties, by machine learning classifiers such as artificial neural networks, support vector machine, random forest, and decision tree. From this study, these authors could select the top 10 active and 10 inactive fragments, as shown in Figs. 7.20 and 7.21, respectively, with their molecular score and concluded that active inhibitors tend to be larger, more complex and flexible, more lipophilic, more capable for H-bonding, more reactive, and more charged than inactive inhibitors.

In another study, Weidlich et al. (2013) developed an SAR model with advanced machine learning classifiers random forest and k-nearest neighbor simulated annealing for a very large number of compounds possessing genotype 1b NS5B inhibition activity. This model led these authors to identify novel chemical scaffolds for NS5B inhibitors and suggest that Candesartan cilexetil (**104**), which is currently used to

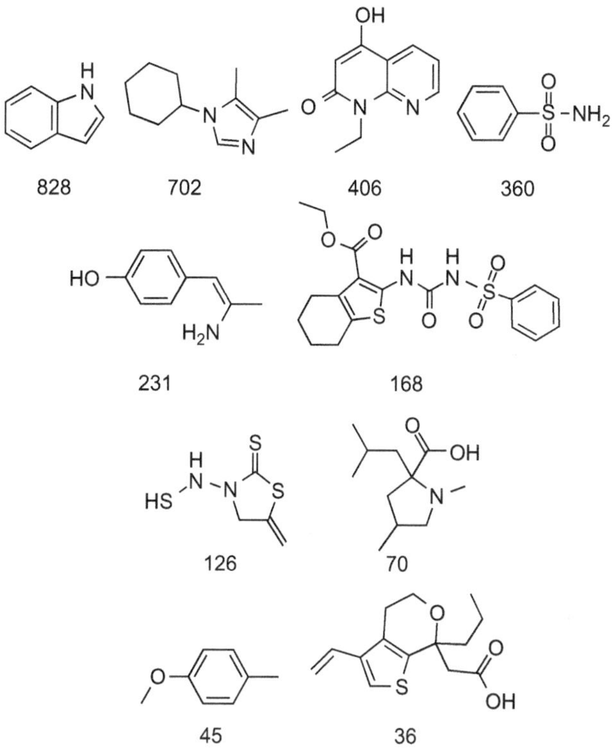

Figure 7.20 Top 10 active fragments of hepatitis C virus NS5B inhibitors ranked by their molecular scores (given below each structure) (Worachartcheewan et al., 2015).

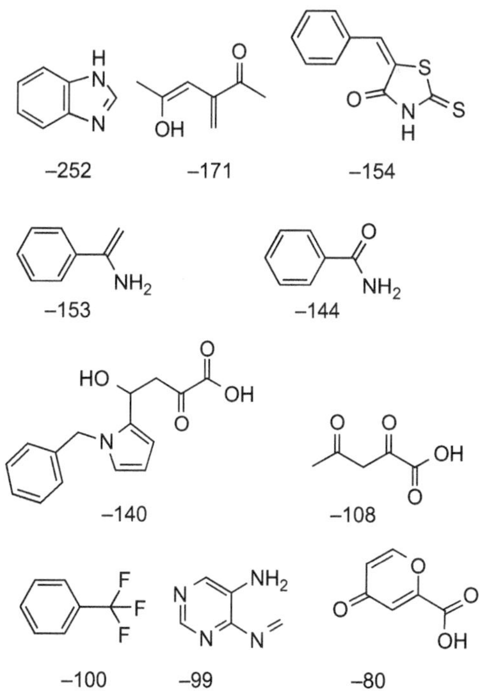

Figure 7.21 Top 10 inactive fragments of hepatitis C virus NS5B inhibitors ranked by their molecular scores (Worachartcheewan et al., 2015).

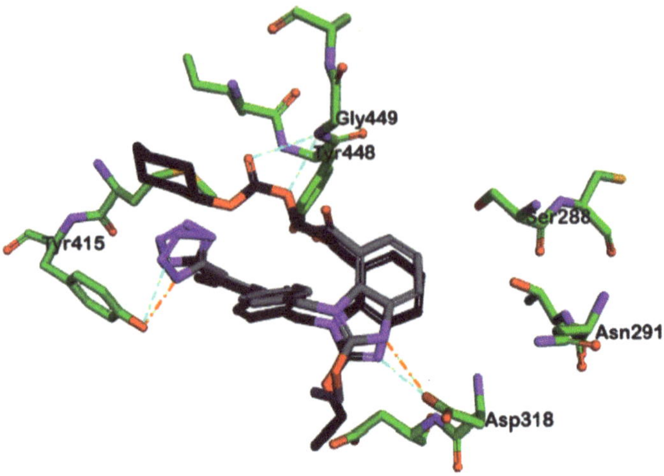

Figure 7.22 The docked structures of Candesartan Cilexetil (**104**) and Candesartan active moiety (**105**) shown in *black* and *grey*, respectively. Their hydrogen bondings are shown with *dashed lines in orange and black*, respectively. The active moiety forms H-bonds with Tyr415 and Asp318. *(Reprinted with permission from Wiedlich et al. (2013). Copyright 2013 Elsevier.)*

treat hypertension and heart failure, may be a scaffold for NS5B inhibitor. In the docking study, a pose of Candesartan cilexetil was found to have a similar binding mode as a small molecule found in the crystal structure with PDB code 3E51. The conformational analysis of docking poses of Candesartan cilexetil indicated that the palm site is the best fit for its active moiety, Candesartan (**105**). The binding of Candesartan cilexetil with NS5B is shown in Fig. 7.22. From the docking study, however, these authors suggested that replacement of benzyl ring in benzimidazole in Candesartan cilexetil with thiaine-1,1-dioxide heterocycle can improve the activity. The cilexetil ester is metabolized in the human body. N-phenyl methanesulfonamide derivatives could prevent metabolism and improve activity (Kalgutkar et al., 2010). From their analysis, Wiedlich et al. (2013) were able to find two additional NS5B inhibitors (**106** and **107**) with novel nonnucleoside chemical motifs.

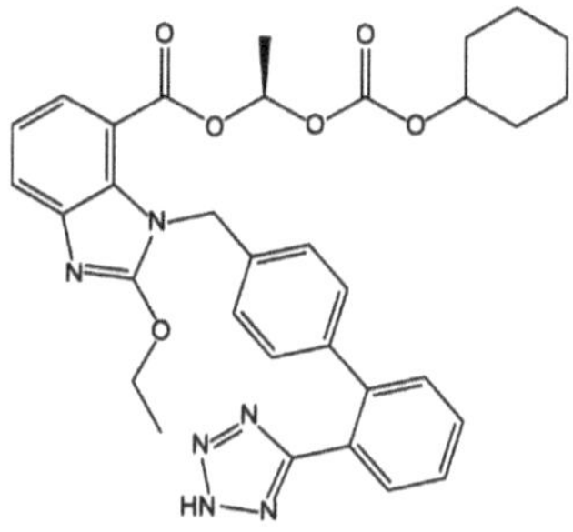

104, Candesartan Cilexetil

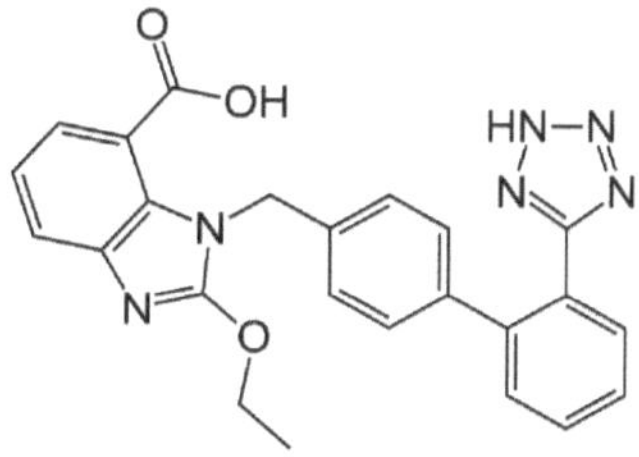

105, Candesartan active moiety

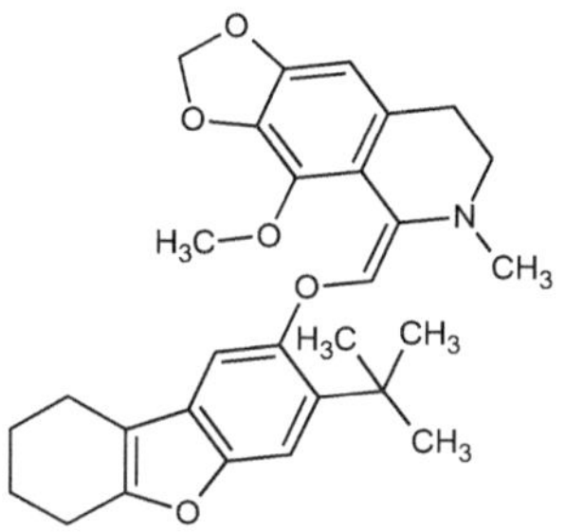

106

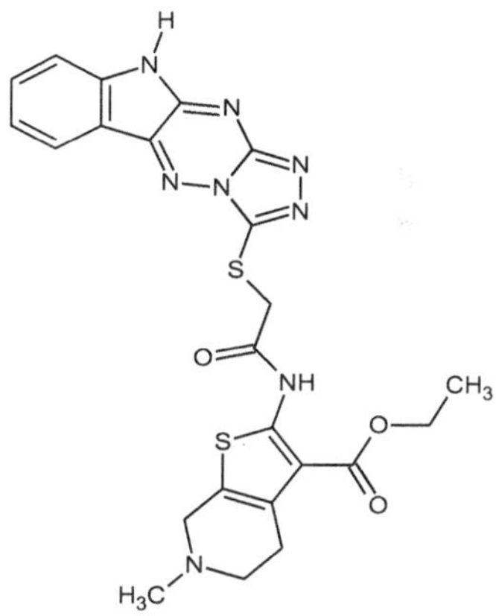

107

Through virtual screening of a diverse set of National Cancer Institute (NCI) data, Musmuca et al. (2010) had found 20 top-scoring molecules, among which four derivatives were found to possess IC_{50} values between 45 and 75 μM (Fig. 7.23). Among these compounds, NSC 123526 had exhibited a dock conformation that was in good agreement with thumb training set most active compound **108**. As Fig. 7.24 exhibits, the tricyclic ring of NSC 123526 points toward the interior of the thumb domain, the primary binding cavity of the domain, and makes van der Waals interactions with Leu419, Met423, His475, and Tyr528. This is similar to the interaction of 4-methyl-phenyl group of **108**. Similarly, its para-chloro-phenyl ring is located in the region occupied by fluoro-phenyl thiophene moiety of 107 and is characterized by van der Waals interactions with Leu419, Met423, Ile482, Ala486, Leu489, and Leu497. Additionally, one of the oxygen atoms of its carboxylic group was found to form a hydrogen bond with the guanidinium group of Arg501.

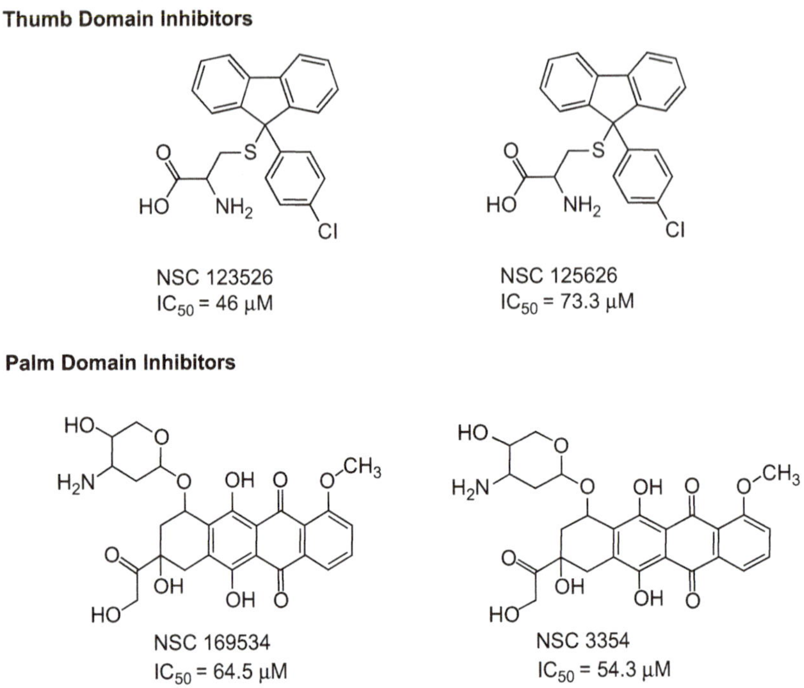

Figure 7.23 Four hit compounds as NS5B inhibitors screened by Mumuca et al. (2010).

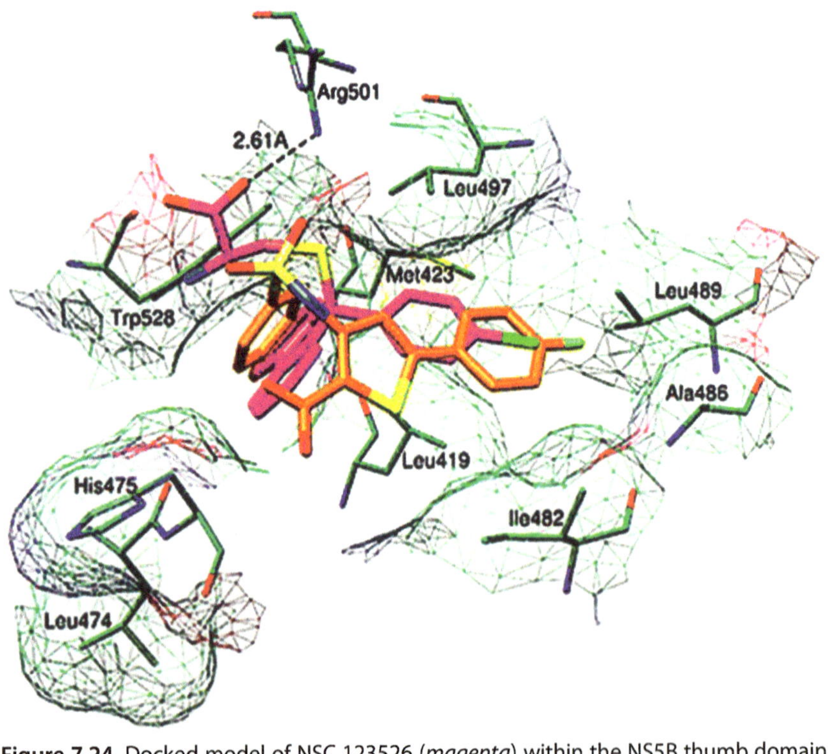

Figure 7.24 Docked model of NSC 123526 (*magenta*) within the NS5B thumb domain. Superimposed is compound **107.** Only a few residues of binding site are shown. H-bond interactions are shown by *black dotted lines. (Reprinted with permission from Musmuca et al. (2012). Copyright 2012 American Chemical Society.)*

The virtual screening of the compounds was based on the application of two receptor-based 3D-QSAR models on two external test sets, containing 81 and 223 molecules for thumb and palm models, respectively.

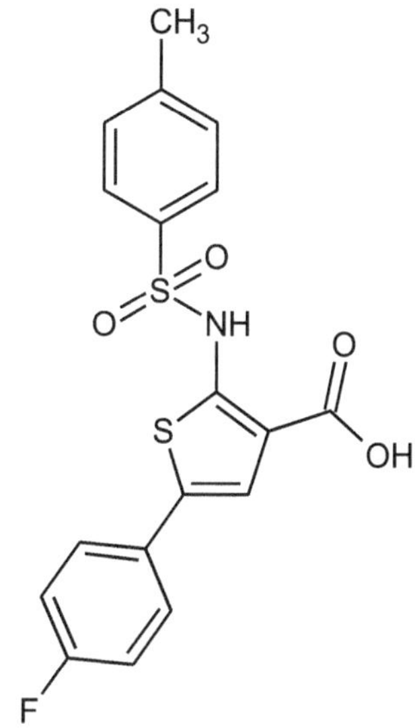

108

Bhatt et al. (2011) had conducted an SAR study on a series of α,γ-diketoacids to find a better pharmacophore. They synthesized a series of analogs of **109** and evaluated their NS5B inhibition activity. They chose an analog (**109a**) as a hit compound and sought to explore SAR around the aryl/heteroaryl specificity domain while keeping the diketo acid metal chelating pharmacophore constant. The 2,4-difluorophenyl ring as specificity domain was found to be the most potent inhibitor of NS5B polymerase. A Glide docking model of this compound in the polymerase exhibited the chelation with the two Mn^{2+} ions through the oxygens of both carboxyl and hydroxyl groups. Both the metal ions were also coordinated with conserved active site residues, Asp319, Asp320, and Asp318, and the backbone carbonyl oxygen atom of Thr221. The 2,4-difluorophenyl moiety makes hydrophobic interactions with the methylene groups of Arg158 and the side chain of Ile160. A hydrogen bond is formed between 4-fluoro group of the compound and the hydroxyl group of Ser282. An intramolecular hydrogen bond between the γ-keto and α-hydroxyl groups leads to the formation of a pseudo-six membered ring, which favors the metal chelation due to increase in electron density at the oxygen atom of α-hydroxyl group. The γ-keto group also forms a hydrogen bond, mediated by a water molecule, with strictly conserved Asp225. This may not permit the virus to easily acquire the mutations to escape inhibition by such inhibitors. Indeed, mutations of Asp225 have shown to completely inactivate the enzyme (Lohmann et al., 1997).

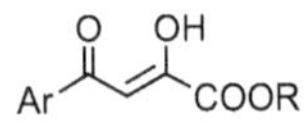

109

109a: Ar = furan-2-yl, R = H
109b: Ar = 2,4-difluorophenyl, R =H

Nichols et al. (2012) performed a molecular docking study on some S-trityl-L-cystein (STLC) derivatives (**110**) and identified three derivatives as potential NS5B inhibitor leads. These compounds were F-3205 (**110a**), F-3065 (**110b**), and F-3070 (**110c**). Of these, the last two exhibited much better NS5B RdRp inhibition potency (≥84%) than the first one (≈ 44%) at 100 μM. In docked structures of these compounds in NS5B (PDB code: 2D3U), the R-enantiomer of F-3065 and the S-enantiomer of F-3070 were found to have the lowest interaction energy, and hydrogen bondings and hydrophobic interactions were the dominant interactions. In hydrogen-bond interactions, both carboxylic and amino groups of F-3065 were found to be involved, whereas only the carboxylic group of F-3070 was involved. In hydrophobic interactions, the trityl moieties in both the compounds were found to be buried in the thumb allosteric binding site, and similarly in both, one phenyl moiety was found to be placed in a pocket formed by Leu419, Arg422, Met423, and Try528, while the other two phenyl rings occupied two depressions on the enzyme surface. Thus, Nichols et al. assumed that STLC can be used as good scaffold to find still better NS5B inhibitors.

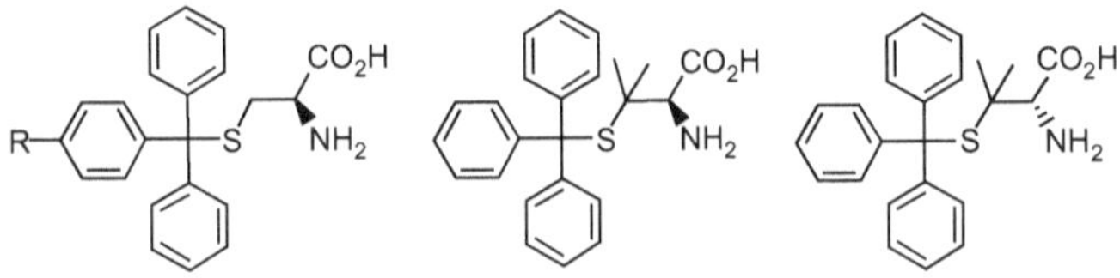

110
110a, R= MeO

110b, F-3065

110c, F-3070

3.2.6 Site Specific Inhibitors

As already mentioned in Section 2.4.2, multiple allosteric sites have been revealed for NNIs, particularly two thumb sites (Thumb I and II) and two palm sites (Palm I and II). NNIs have the characteristics that correlate with those of their binding site. Since the features of allosteric binding sites vary across and within diverse HCV genotypes, NNIs have been found to have a distinct spectrum of activity against the various genotypes and subtypes. Chemically distinct NNIs targeting different allosteric sites have been reviewed, many of which are in different phases of trials (Deore and Chern, 2010; Legrand-Abravanel et al., 2010; Li and Shi, 2010; Watkins et al., 2010). Barreca et al. (2011) have nicely summarized information about each allosteric site and related known NNIs limited to those where the X-ray ligand-enzyme complex were available (Table 7.7). Note in Table 7.7 that a mention of an additional palm site (Palm III) has also been made.

In a later communication, on the basis of data literature and in silico analysis, Barreca et al. (2013) pointed out the following chemical features common to diverse chemotypes of NNIs interacting particularly with palm I site:

1. a hydrophobic group occupying a pocket mainly defined by residues Met414 and Gly410
2. a second hydrophobic group mainly buried in a hydrophobic pocket Tyr448, Cys366, Pro197, Tyr415, Arg200, and Leu384
3. a hydrogen-bond acceptor group, which establishes a direct interaction with the backbone amide NH of Tyr448

In addition, in most of the cases (75%) the same acceptor group was found to interact with Gln446 and/or Gly449 residue through a water molecule. This water molecule is called WB, and such inhibitors that interact through WB are called water-mediated inhibitors (WMIs). The inhibitors that do not make WB interactions (25%) are called nonwater-mediated inhibitors (NWMIs). The interactions of both WMIs and NWMIs are shown in Fig. 7.25. Six conserved molecules (W1−W6) were found to be present in palm I site, of which in most of the cases, W2 was observed to have a three-point interaction between the inhibitors and the residues Ser556 and/or Ser288, but in few cases, W5 mediated a hydrogen-bond interaction between the ligands and the residues Tyr415 and Arg836 (Fig. 7.26). The remaining four identified structural water molecules (W1, W3, W4, and W6) do not interact directly with the ligands but fill the space of the site and might play an important role in defining the volume and shape of the binding site.

Table 7.7 NS5B allosteric sites and correlated nonnucleoside inhibitors

Site	Location	Key residues for ligand–protein interaction	Nonnucleoside inhibitor chemotypes	PDB code of nonnucleoside inhibitor/NS5B complex
Thumb I	Upper section of thumb domain, ≈ 30 Å from the active site	Val-37, Leu-392, Ala-393, Ala-395, Ala-396, Thr-399, Ile-424, Leu-425, His-428, Phe-429, Leu-492, GIy-493, Val-494, Pro-495, Trp-500, Arg-503	Indoles	2BRK, 2BRL, 2DXS, 2WCX, 3MWW, 2XWY
Thumb II	At the base of the thumb domain, 35 Å from the active site	Leu-419, Arg-422, Met-423, HIs-475, Ser-476, Tyr-477, Ile-482, Val-485, Leu-497, Leu489, ARg-501, Trp-528, Lys-533	Phenylalanines	1NHU, 1NHV
			Dihydropyranones	1OS5, 2HAI, 3FRZ
			Thiophene carboxylic acids	1YVX, 1YVZ, 2GIR, 3MF5, 2D3U, 2D3Z, 2D41
			Thiazolones	2HWH, 2HWI, 2I1R, 2O5D
			Small fragments and bromopheny1 methanones	3CIZ, 3CJ0, 3CJ2, 3CJ3, 3CJ4, 3CJ5
			Benzisoquinolines -dione	2WHO, 3HVO
			Hexanoic acids	2WRM
			Quinolones	3PHE
Palm I	At the junction of the thumb and palm domains and in close proximity to	Phe-193, Pro-197, Arg-200, Ser-228, Asn-291, Asn-316,	Acrylic acids	1YVF, 1Z4U
			Rhodanines	2AWZ, 2AXO, 2AX1
			Benzothiadiazines	

	the active site. It partially overlaps with palm II and III	Gly-317, Asp-318, Cys-366, Ser-368, Leu-384, Met-414, Tyr-415, Gln-446, Ile-447, Tyr-448, GLy-449, Ser-556		2FVC, 2GIQ, 3HHK, 3BSA, 3BSC, 3CDE, 3BR9,3E51, 3CO9, 3CVK, 3H2L, 3H98, 3GYN, 3IGV
			Benzothiazines	3CWJ, 3G86, 3H59
			Benzoisothiazoles dioxide	3D28, 3D5M, 3H5U, 3H5S,
			Proline sulfonamides	2GC8
			Acylpyrrolidines	2JC0, 2JC1
			Anthranilic acids	2QE2, 2QE5
			Benzodiazepines	3GOL, 3CSO, 3GNV, 3GNW, 3HKW, 3HKY
Palm II	Next to the active site and to the junction between palm and thumb domains, partially overlapping with palm I and III	Phe-193, Pro-197, Arg-200, Leu-204, Leu-314, Leu-360, Ile-363, Val-321, Ser-365, Cys-366, Val-370, Met-414, Tyr-415, Tyr-448	Benzofurans	3FQK, 3FQL
Palm III	Near the top of the palm subdomain, partially overlapping with palm I and II	Tyr-195, Pro-197, Arg-200, Leu-384, Met-414, Tyr-415, Asn-316, Ile447, Tyr-452, Leu-446, Trp-550, Phe-551	Benzamides	3LKH

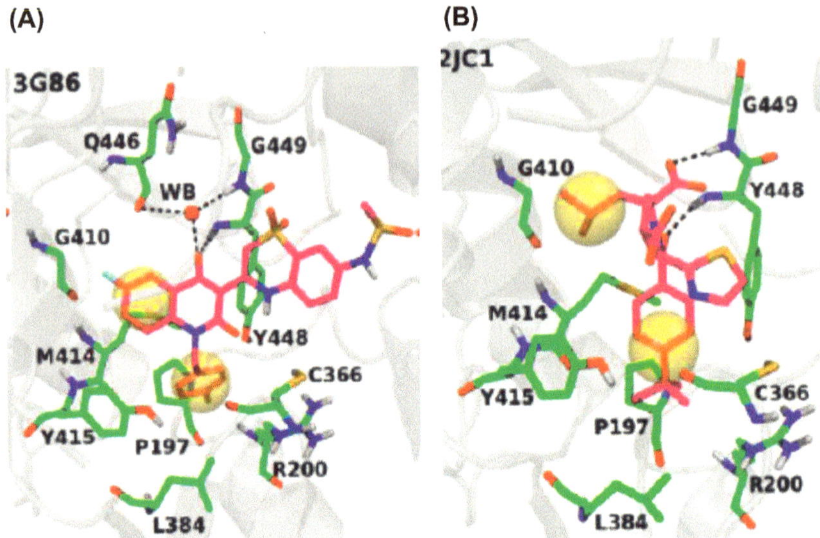

Figure 7.25 Crystallographic positions of a water-mediated inhibitor (a benzothiazine) (A) and a non-WMI (an acylpyrrolodine) (B) at the hepatitis C virus NS5B polymerase palm I site. Hydrogen bonds are represented as *black dashed lines* and the two common hydrophobic regions by *yellow spheres*. *(Reprinted with permission from Barreca, M.L., Manfroni, G., Leyssen, P., Winquist, J., Kaushik-Basu, N., Paeshuyse, J., Krishnan, R., Iraci, N., Sabatini, S., Tabarrini, O., Basu, A., Danielson, U.H., Neyts, J., Cecchetti, V., 2013. Structure-based discovery of pyrazolobenzothiazine derivatives as inhibitors of hepatitis C virus replication. J. Med. Chem. 56, 2270–2282. Copyright 2013 American Chemical Society.)*

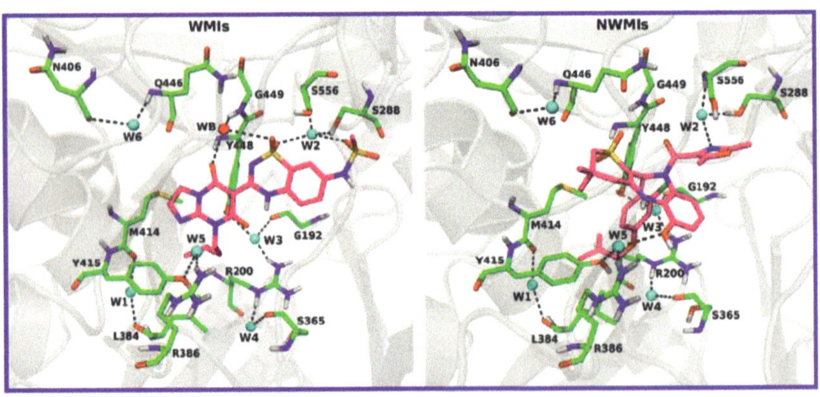

Figure 7.26 Diagram showing, on the left side, conserved water molecules (W1–W6, *cyan*) and a water bridge (WB in *red*) in palm I site in the presence of a water-mediated inhibitor (WMI) (PDB ID: 3H98) and on the right side, with no WB in the presence of a nonwater-mediated inhibitor (NWMI) (PDB ID: 3HKY). Residues forming the H-bonds with the water molecules are shown in *green*. WB, W2, and W5 interact with both the proteins and ligands. WB and ligand are shown in *red*. *(Reprinted with permission from Barreca, M.L., Manfroni, G., Leyssen, P., Winquist, J., Kaushik-Basu, N., Paeshuyse, J., Krishnan, R., Iraci, N., Sabatini, S., Tabarrini, O., Basu, A., Danielson, U.H., Neyts, J., Cecchetti, V., 2013. Structure-based discovery of pyrazolobenzothiazine derivatives as inhibitors of hepatitis C virus replication. J. Med. Chem. 56, 2270–2282. Copyright 2013 American Chemical Society.)*

Based on this knowledge of the roles of water in drug-reception interaction, Barreca et al. tried to exploit the available X-ray structures of the NS5B-ligand complexes to build different models of protein—water combinations that could be useful to study the effects of solvent sites and protein conformations on docking studies. It was found that docking results of WMIs by inclusion of explicit water improved vis-à-vis those of NWMIs. These authors found that the parallel use of different protein structures—water sets outperformed the use of a single target structure, so a two-protein model, 3H98/2W—2FVC/7W, resulted in the best performing ensemble for WMIs and so did a two-protein model, 3HKY/NoW-3SKE/NoW model, for NWMIs. The information gathered from this work confirmed the primary role of water molecules and protein flexibility in docking studies and can be exploited to facilitate the discovery of NS5B inhibitors.

4. CONCLUSIONS

We conclude that anti-HCV agents mainly act through the inhibition of NS proteins of HCV, which are NS3/4A, NS4B, NS5A, and NS5B, among which the most widely studied are those that inhibit the NS5B polymerase. NS5B polymerase as a target has been the most widely exploited because the crystal structure of this polymerase is extensively studied. However, there are also anti-HCV drugs that inhibit the entry of the virus in human cells or inhibit HCV replication. The most important class of NS5B inhibitors have been nucleosides and nucleotides, which function as chain terminators, but they also interfere with the cellular proliferative machine, causing undesirable side effects. There are also NNIs of NS5B polymerase that belong to different classes of chemicals. They exert their effects through allosteric inhibition of the polymerase, i.e., they bind to the different allosteric pockets of the enzyme, leading to the conformational changes of enzyme-catalytic active site, while NIs directly bind to polymerase active site. The most important NNIs studied have been benzimidazole derivatives, aryl diketoacids, acyl pyrrolidines, proline sulfonamides, indole derivatives, thiouracil derivatives, pteridine derivatives, thiozolone derivatives, imidazopyridine derivatives, isatin derivatives, sesquiterpene lactones, etc. Some miscellaneous classes of compounds were also studied for NS5B polymerase inhibition to find that active inhibitors in them tend to be larger, more complex and flexible, more lipophilic, more capable for H-bonding, more reactive, and more charged than inactive inhibitors. For NNIs, there are multiple allosteric sites, particularly two thumb sites (Thumb I and II) and three palm sites (Palm I, II, and III), and the NNIs bind to a site whose characteristics match their characteristics. In most of the cases, the water molecules have been found to play important roles in the interactions of NNIs with the receptors. NNIs interacting particularly with palm I site possess at least two hydrophobic groups and a hydrogen-bond acceptor group.

REFERENCES

Anilkumar, G.N., Lesburg, C.A., Selyutin, O., Rosenblum, S.B., Zeng, Q., Jaing, Y., Chan, T.-Y., Pu, H., Vaccaro, H., Duca, J., Gavalas, S., Huang, Y., Pinto, P., Sannigrahi, M., Velazquez, F., Kozlowski, J.A., 2011. I. Novel HCV NS5B polymerase inhibitors: discovery of indole 2-carboxylic acids with C3-heterocycles. Bioorg. Med. Chem. Lett. 21, 5336–5341.

Anilkumar, G.N., Selyutin, O., Rosenblum, S.B., Zeng, Q., Jaing, Y., Chan, T.-Y., Pu, H., Wang, L., Bennet, F., Chen, K.X., Lesberg, C.A., Duca, J., Gavalas, S., Huang, Y., Pinto, P., Sannigrahi, M., Velazquez, F., Venkatraman, S., Kozlowski, J.A., 2012. II. Novel HCV NS5B polymerase inhibitors: discovery of indole C2 acylsulfonamides. Bioorg. Med. Chem. Lett. 22, 713–717.

Arasappan, A., Bennett, F., Girijavallabhan, V., Huang, Y., Huelgas, R., Alvarez, C., Chen, L., Gavalas, S., Kim, S.-H., Kosinski, A., Pinto, P., Rizvi, R., Rossman, R., Shankar, B., Tong, L., Velazquez, F., Venkatraman, S., Verma, V.A., Kozlowski, J., Shih, N.-Y., Piwinski, J.J., MacCoss, M., Kwong, C.D., Clark, J.L., Fowler, A.T., Geng, F., Kezar III, H.S., Roychowdhury, A., Reynolds, R.C., Maddry, J.A., Ananthan, S., Secrist III, J.A., Li, C., Chase, R., Curry, S., Huang, H.-C., Tong, X., Njoroge, F.G., 2012. 5-Benzothiazole substituted pyrimidine derivatives as HCV replication (replicase) inhibitors. Bioorg. Med. Chem. Lett. 22, 3229–3234.

Asswlah, T., 2011. NS5A inhibitors: a new breakthrough for the treatment of chronic hepatitis C. J. Hepatol. 54, 1069–1072.

Barreca, M.L., Iraci, N., Manfroni, G., Cecchetti, V., 2011. Allosteric inhibition of the hepatitis C virus NS5B polymerase: in silico strategies for drug discovery and development. Future Med. Chem. 3, 1027–1055.

Barreca, M.L., Manfroni, G., Leyssen, P., Winquist, J., Kaushik-Basu, N., Paeshuyse, J., Krishnan, R., Iraci, N., Sabatini, S., Tabarrini, O., Basu, A., Danielson, U.H., Neyts, J., Cecchetti, V., 2013. Structure-based discovery of pyrazolobenzothiazine derivatives as inhibitors of hepatitis C virus replication. J. Med. Chem. 56, 2270–2282.

Bartenschlager, R., Lohmann, V., 2000. Replication of hepatitis C virus. J. Gen. Virol. 81, 1631–1648.

Belema, M., Nguyen, V.N., Bachand, C., Deon, D.H., Goodrich, J.T., James, C.A., Lavoie, R., Lopez, O.D., Martel, A., Romine, J.L., Ruediger, E.H., Snyder, L.B., St Laurent, D.R., Yang, F., Zhu, J., Wong, H.S., Langley, D.R., Adams, S.P., Cantor, G.H., Chimalakonda, A., Fura, A., Johnson, B.M., Knipe, J.O., Parker, D.D., Santone, K.S., Fridell, R.A., Lemm, J.A., O'Boyle II, D.R., Colonno, R.J., Gao, M., Meanwell, N.A., Hamann, L.G., 2014. Hepatitis C Virus NS5A replication complex inhibitors: the discovery of Daclatasvir. J. Med. Chem. 57, 2013–2032.

Bennett, F., Kezar III, H.S., Girijavallabhan, V., Huang, Y., Huelgas, R., Rossman, R., Shih, N.-Y., Piwinski, J.J., MacCoss, M., Kwong, C.D., Clark, J.L., Fowler, A.T., Geng, F., Roychowdhury, A., Reynolds, R.C., Maddry, J.A., Ananthan, S., Secrist III, J.A., Li, C., Chase, R., Curry, S., Huang, H.-C., Tong, X., Njoroge, F.G., Arasappan, A., 2012. Pyridofuran substituted pyrimidine derivatives as HCV replication (replicase) inhibitors. Bioorg. Med. Chem. Lett. 22, 5144–5149.

Berenguer, M., Palau, A., Aguilera, V., Rayon, J.M., Juan, F.S., Preito, M., 2008. Clinical benefits of antiviral therapy in patients with recurrent hepatitis C following liver transplantation. Am. J. Transplant. 8, 679–687.

Bernuth, S., Grim, D., Vollmar, J., Darstein, F., Mittler, J., Heise, M., Hoppe-Lotichius, M., Gale, P.R., Lang, H., Zimmermann, T., 2017. Efficacy and safety of direct-acting antiviral therapy in previous hard-to-treat patients with recurrent hepatitis C virus infection after liver transplantation: a real-word cohort. Drug Des. Develop. Ther. 11, 2131–2138.

Bhatt, A., Gurukumar, K.R., Basu, A., Patel, M.R., Kaushik-Basu, N., Talele, T.T., 2011. Synthesis and SAR optimization of diketo acid pharmacophore for HCV NS5B polymerase inhibition. Eur. J. Med. Chem. 46, 5138–5145.

Boumendjel, A., 2003. Aurones: a subclass of flavones with promising biological potential. Curr. Med. Chem. 10, 2621–2630.

Bäck, M., Johansson, P.-O., Wangsell, F., Thorstensson, F., Kvarnstrom, I., Ayesa, S., Wahling, H., Mikael, P., Jansson, K., Lindstrom, S., Wallberg, H., Classon, B., Rydergard, C., Vrang, L., Hamelink, E., Hallberg, A., Rosenquist, A., Samsuelsson, B., 2007. Novel potent macrocyclic inhibitors of hepatitis C virus NS3 protease: use of cyclopentane and cyclopentene P2-motifs. Bioorg. Med. Chem. 15, 7184–7202.

Cakir, G., Kuçukguzel, I., Guhamazumder, R., Tatar, E., Manvar, D., Basu, A., Patel, B.A., Zia, J., Talele, T.T., Kaushik-Basu, N., 2015. Novel 4-thiazolidinones as non-nucleoside inhibitors of hepatitis C virus NS5B RNA-dependent RNA polymerase. Arch. Pharm. 348, 10–22.

Cannalire, R., Barreca, M.L., Manfroni, G., Cecchetti, V., 2016. A journey around the medicinal chemistry of hepatits C virus inhibitors targeting NS4B: from target to preclinical drug candidates. J. Med. Chem. 59 (1), 16–41.

Carbone, M., Lenci, I., Baiocchi, L., 2012. Prevention of hepatitis C recurrence after liver transplantation: an update. World J Gastrointest. Pharmacol. Ther. 3, 36–48.

Chen, K.X., Lesburg, C.A., Vibulbhan, B., Yang, W., Tin-Yau Chan, T.-Y., Venkatraman, S., Velazquez, F., Zeng, Q., Bennett, F., Anilkumar, G.N., Duca, J., Jiang, Y., Pinto, P., Wang, L., Yuhua Huang, Y., Selyutin, O., Gavalas, S., Pu, H., Agrawal, S., Feld, B., Huang, H.-C., Li, C., Cheng, K.-C., Shih, N.-Y., Kozlowski, J.A., Rosenblum, S.B., Njoroge, F.G., 2012a. A novel class of highly potent irreversible hepatitis C virus NS5B polymerase inhibitors. J. Med. Chem. 55, 2089–2101.

Chen, K.X., Vibulbhan, B., Yang, W., Sannigrahi, M., Velazquez, F., Chan, T.-Y., Venkatraman, S., Anilkumar, G.N., Zeng, Q., Bennet, F., Jiang, Y., Lesburg, C.A., Duca, J., Pinto, P., Gavalas, S., Huang, Y., Wu, W., Selyutin, O., Agrawal, S., Feld, B., Huang, H.-C., Li, C., Cheng, K.-C., Shih, N.-Y., Kozlowski, J.A., Rosenblum, S.B., Njoroge, F.G., 2012b. Structure-activity relationship (SAR) development and discovery of potent indole-based inhibitors of the hepatitis C virus (HCV) NS5B polymerase. J. Med. Chem. 55, 754–765.

Choong, I.C., Cory, D., Gleen, J.S., Yang, W., 2010. Method and Composition of Treating a Flaviviridae Family Viral Infection. WO2010/107739.

Choong, I.C., Cory, D., Gleen, J.S., Yang, W., 2012. Method and Composition of Treating a Flaviviridae Family Viral Infection. US/027605A1.

Clark, J.L., Hollecker, L., Mason, J.C., Stuyver, L.J., Tharnish, P.M., Lostia, S., McBrayer, T.R., Schinazi, R.F., Watanabe, K.A., Otto, M.J., Furman, P.A., Stec, W.J., Patterson, S.E., Pankiewicz, K.W., 2005. Design, synthesis, and antiviral activity of 2'-deoxy-2'-fluoro-2'-C- methylcytidine, a potent inhibitor of hepatitis C virus replication. J. Med. Chem. 48, 5504–5508.

Coburn, C.A., Meinke, P.T., Chang, W., Fandozzi, C.M., Graham, D.J., Hu, B., Huang, Q., Kargman, S., Kozlowski, J., Liu, R., McCauley, J.A., Nomeir, A.A., Soll, R.M., Vacca, J.P., Wang, D., Wu, H., Zhong, B., Olsen, D.B., Ludmerer, S.W., 2013. Discovery of MK-8742: an HCV NS5A inhibitor with broad genotype activity. ChemMedChem 8, 1930–1940.

da Cunha, E.F.F., Ramalho, T.C., Taft, C.A., de Alencastro, R.B., 2006. Computer-assisted analysis of the interactions of macrocyclic inhibitors with wild type and mutant D168A hepatitis C virus NS3 serine protease. Lett. Drug Design Discover. 3, 17–28.

Das, D., 2017. Design and development of HCV NS5B polymerase inhibitors. In: Gupta, S.P. (Ed.), Viral Proteases and Their Inhibitors. Elsevier, Inc., pp. 189–219

Das, D., Hong, J., Chen, S.-H., Wang, G., Beigelman, L., Seiwert, S.D., Buckman, B.O., 2011. Recent advances in drug discovery of benzothiadiazine and related analogs as HCV NS5B polymerase inhibitors. Bioorg. Med. Chem. 19, 4690–4703.

De Francesco, R., Migliaccio, G., 2005. Challenges and successes in developing new therapies for hepatitis C. Nature 436, 953–960.

De Francesco, R., Pessi, A., Steinkuhler, C., 1998. The hepatitis C virus NS3 proteinase: structure and function of a zinc-containing serine proteinase. Antiviral Ther. 3, 99–109.

Delang, L., Coelmont, L., Neyts, J., 2010. Antiviral therapy for hepatitis C virus: beyond the standard of care. Viruses 2, 826–866.

Deore, R.R., Chern, J.W., 2010. NS5B RNA dependent RNA polymerase inhibitors: the promising approach to treat hepatitis C virus infections. Curr. Med. Chem. 17, 3806–3826.

Egger, D., Wolk, B., Gosset, R., Bianchi, L., Blim, H.E., Moradpour, D., Bienz, K., 2002. Expression of hepatitis C virus protems induces distinct membrane alterations including a candidate viral replication complex. J. Virol. 76, 5974–5984.

Elazar, M., Liu, P., Rice, C.M., Glenn, J.S., 2004. An N-terminal amphipathic helix in hepatitis C virus (HCV) NS4B mediates membrane association, correct localization of replication complex proteins, and HCV RNA replication. J. Virol. 78, 11393–11400.

Eldrup, A.B., Prhavc, M., Brooks, J., Bhat, B., Prakash, T.P., Song, Q., Bera, S., Bhat, N., Dande, P., Cook, P.D., Bennet, C.F., Carroll, S.S., Ball, R.G., Bosserman, M., Burlein, C., Colwell, L.F., Fay, J.F., Flores, O.A., Getty, K., LaFemina, R.L., Leone, J., MacCoss, M., McMasters, D.R., Tomassini, J.E., Langen, D., Wolanski, B., Olsen, D.B., 2004. Structure–activity relationship of heterobase-modified 2′-C-methyl ribonucleosides as inhibitors of hepatitis C virus RNA replication. J. Med. Chem. 47, 5284–5297.

Fridell, R.A., Qiu, D., Valera, L., Wang, C., Rose, R.E., Gao, M., 2011. Distinct functions of NS5A in hepatitis C virus RNA replication uncovered by studies with the NS5A inhibitor BMS-790052. J. Virol. 85, 7312–7320.

Furuta, A., Tsubuki, M., Endoh, M., Miyamoto, T., Tanaka, J., Salam, K.A., Akimitsu, A., Tani, H., Yamashita, A., Moriishi, K., Nakakoshi, M., Sekiguchi, Y., Tsuneda, S., Noda, N., 2015. Identification of hydroxyanthraquinones as novel inhibitors of hepatitis C virus NS3 helicase. Int. J. Mol. Sci. 16, 18439–18453.

Gao, M., Nettles, R.E., Belema, M., Snyder, L.B., Nguyen, V.N., Fridell, R.A., Serrano-Wu, M.H., Langley, D.R., Sun II, J.H., Lemm, J.A., Wang, C., Knipe, J.O., Chien, C., Colonno, R.J., Grasela, D.M., Meanwell, N.A., Hamann, L.G., 2010. Chemical genetics strategy identifies an HCV NS5A inhibitor with a potent clinical effect. Nature 465, 96–100.

Geddawy, A., Ibrahim, Y.F., Elbahie, N.M., Ibrahim, M.A., 2017. Direct acting anti-hepatitis C virus drugs: clinical pharmacology and future direction. J. Transl. Int. Med. 5, 8–17.

Girijavallabhan, V.M., Alvarez, C., Bennett, F., Chen, L., Gavalas, S., Huang, Y., Kim, S.-H., Kosinski, A., Pinto, P., Rizvi, R., Rossman, R., Shankar, B., Tong, L., Velazquez, F., Venkatraman, S., Verma, V.A., Kozlowski, J., Shih, N.-Y., Piwinski, J.J., MacCoss, M., Kwong, C.D., Bansal, N., Clark, J.L., Fowler, A.T., Kezar III, H.S., Valiyaveettil, J., Reynolds, R.C., Maddry, J.A., Ananthan, S., Secrist III, J.A., Li, C., Chase, R., Curry, S., Huang, H.-C., Tong, X., Njoroge, F.G., Arasappan, A., 2012. Synthesis and SAR of pyridothiazole substituted pyrimidine derived HCV replication inhibitors. Bioorg. Med. Chem. Lett. 22, 5652–5657.

Giroux, S., Billimoria, D., Cadilchac, C., Cottrell, K.M., Denis, F., Dietrich, E., Ewing, N., Henderson, J.A., Heureux, L.L., Mani, N., Morris, M., Nicolas, O., Reddy, J.T., Selliah, S., Shawgo, R.S., Xu, J., Chauret, N., Berlioz-Seux, F., Maxwell, J.P., 2015a. Discovery of thienoimidazole-based HCV NS5A inhibitors. Part 1: C_2-Symmetric inhibitors with diyne and biphenyl linkers. Bioorg. Med. Chem. Lett. 25, 936–939.

Giroux, S., Billimoria, D., Cadilchac, C., Cottrell, K.M., Denis, F., Dietrich, E., Ewing, N., Henderson, J.A., Heureux, L.L., Mani, N., Morris, M., Nicolas, O., Reddy, J.T., Selliah, S., Shawgo, R.S., Xu, J., Chauret, N., Berlioz-Seux, F., Maxwell, J.P., 2015b. Discovery of thienoimidazole-based HCV NS5A inhibitors. Part 2: non-symmetric inhibitors with potent activity against genotype 1a and 1b. Bioorg. Med. Chem. Lett. 25, 940–943.

Gosett, R., Egger, D., Lohmann, D., Bartenschlager, R., Blum, H.E., Brenz, K., Moradpour, D., 2003. Identification of the hepatitis C virus RNA replication complex in Huh-7 cells harbouring subgenomic replications. J. Virol. 77, 5487–5492.

Gouttenoire, J., Castet, V., Montserret, R., Arora, N., Raussens, V., Ruysschaert, J.M., Diesis, E., Blum, H.E., Penin, F., Moradpour, D., 2009a. Identification of a novel determinant for membrane association in hepatitis C virus non-structural protein 4B. J. Virol. 83, 6257–6268.

Gouttenoire, J., Monfserret, R., Kennel, A., Penin, F., Moradpour, D., 2009b. An amphiphatic helix at the C terminus of hepatits C virus nonstructural protein 4B mediates membrane association. J. Virol. 83, 11378–11384.

Gouttenoire, J., Montserret, R., Paul, D., Castillo, R., Meister, S., Bartenschlager, R., Penin, F., Moradpour, D., 2014. Aminoterminal amphipathic α-helix AH 1 of hepatitis C virus non-structural protein 4B possesses a dual role in RNA replication and virus production. PLoS Pathog. 10, e10045501.

Gu, G., Johnston, V.K., Lester, L., Gutshall, L.L., Nguyen, T.T., Gontarek, R.R., Darcy, M.G., Tedesco, R., Dhanak, D., Duffy, K.J., Kao, C., Sarisky, R.T., 2003. Arresting initiation of hepatitis C virus RNA synthesis using heterocyclic derivatives. J. Biol. Chem. 278, 16602–16607.

Guedji, J., Dahari, H., Rong, L., Sansone, N.D., Nettles, R.E., Cotler, S.J., Layden, T.J., Uprichard, S.L., Perelson, A.S., 2013. Modeling shows that the NS5A inhibitor daclatasvir has two modes of action and yields a shorter estimate of the hepatitis C virus half-life. Proc. Natl. Acad. Sci. U.S.A. 110, 3991–3996.

Gupta, S.P., Samanta, S., Patil, V.M., 2010. A 3D-qsar study on a series of benzimidazole derivatives acting as hepatitis C virus inhibitors: application of kNN-molecular field analysis. Med. Chem. 6, 87–90.

Gupta, S.P., Samanta, S., Masand, N., Patil, V.M., 2013a. k Nearest Neighbor-Molecular Field Analysis on human HCV NS5B polymerase inhibitors: 2,5-disubstituted imidazoj4,5-c]pyridines. Med. Chem. Res. 22, 330–339.

Gupta, S., Gupta, S.P., Agarwal, N., 2013b. A quantitative structure-activity relationship and molecular docking study on a series of pyrimidines acting as anti-hepatitis C virus agents. J. Prot. Proteom. 4, 191–216.

Han, Z., Liang, X., Wang, Y., Qing, J., Cao, L., Shang, L., Yin, Z., 2016. The discovery of indole derivatives as novel hepatitis C virus inhibitors. Eur. J. Med. Chem. 116, 147–155.

Harper, S., Pacini, B., Avolio, S., Di Filippo, M., Migiiaccio, G., Laufer, R., De Francesco, R., Rowley, M., Narjes, F., 2005. Development and preliminary optimization of indole-N-acetaniide inhibitors of hepatitis C virus NS5B polymerase. J. Med. Chem. 48, 1314–1317.

Haudecoeur, R., Ahmed-Belkacem, A., Yi, W., Fortuné, A., Brillet, R., Belle, C., Nicolle, E., Pallier, C., Pawlotsky, J.-M., Boumendjel, A., 2011. Discovery of naturally occurring aurones that are potent allosteric inhibitors of hepatitis C virus RNA-dependent RNA polymerase. J. Med. Chem. 54, 5395–5402.

Henderson, J.A., Billimoria, D., Bubenik, M., Cadilhac, C., Cottrell, K.M., Dietrich, E., Denis, F., Ewing, N., Flardeau, G., Giroux, S., Grey Jr., R., Heureux, L.L., Liu, B., Mani, N., Morris, M., Nicolas, O., Pereira, O.Z., Poisson, C., Maxwell, J.P., 2015. Benzimidazole-containing HCV NS5A inhibitors: effect of 4-substituted pyrrolidines in balancing genotype 1a and 1b potency. Bioorg. Med. Chem. Lett. 25, 944–947.

Hirashima, S., Suzuki, T., Ishida, T., Noji, S., Yata, S., Ando, I., Komatsu, M., Ikeda, S., Hashimoto, H., 2006. Benzimidazole derivatives bearing substituted biphenyls as hepatitis C virus NS5B RNA-dependent RNA polymerase inhibitors: Structure–activity relationship studies and identification of a potent and highly selective inhibitor JTK-109. J. Med. Chem. 49, 4721–4736.

Howe, A.Y., Cheng, H., Johann, S., Mullen, S., Chunduru, S.K., Young, D.C., Bard, J., Chopra, R., Krishnamurthy, G., Mansour, T., O'Cannell, J., 2008. Molecular mechanism of hepatitis C virus replication variants with reduced suspectibility to a benzofuran inhibitor, HCV-796. Antimicrob. Agents Chemother. 52, 3327–3338.

Hugle, T., Fehrmann, F., Bieck, E., Kohara, M., Krausslich, H.G., Rice, C.M., Blum, H.E., Moradpour, D., 2001. The hepatitis C virus non-structural protein 4B is an integral endoplasmic reticulum membrane protein. Virology 284, 70–81.

Ismail, M.A.H., Ella, D.A.A.E., Abouzid, K.A.M., Mahmoud, A.H., 2012. Integrated structure-based activity prediction model of benzothiadiazines on various genotypes of HCV NS5b polymerase (1a, 1b and 4) and its application in the discovery of new derivatives. Bioorg. Med. Chem. 20, 2455–2478.

Jenkins, S., Scola, P., McPhee, F., Knipe, J., Gesenberg, C., Sinz, M., Arora, V., Pilcher, G., Santone, K., 2014. Preclinical pharmacokinetics and in vitro metabolism of BMS-605339: a novel HCV NS3 protease inhibitor. J. Pharm. Sci. 103, 1891–1902.

Jin, G., Lee, S., Choi, M., Son, S., Kim, C.W., Oh, J.W., Lee, C., Lee, K., 2014. Chemical genetics-based discovery of indole derivatives as HCV polymerase inhibitors. Eur. J. Med. Chem. 15, 413–423.

Johansson, P.-O., Back, M., Kvarnstrom, I., Jansson, K., Vrang, L., Hamelink, E., Hallberg, A., Rosenquist, A., Samuelsson, B., 2006. Potent inhibitors of the hepatitis C virus NS3 protease: use of a novel P2 cyclopentane-derived template. Bioorg. Med. Chem. 14, 5136–5559.

Kalgutkar, A.S., Jones, R., Sawant, A., 2010. Sulfonamide as an essential functional group in drug design. In: Gardner, I.R., Obach, S., Smith, D.A., Miao, Z., Alex, A.A., Beaumont, K., Kalgutkar, A., Walker, D., Dalvie, D., Prakash, C., Alf, V. (Eds.), Metabolism, pharmacokinetics and toxicity of functional groups: Impact of chemical building blocks on ADMET. Royal Society of Chemistry, Cambridge, UK, p. 242.

Kakarla, R., Liu, J., Naduthambi, D., Chang, W., Mosley, R.T., Bao, D., Micolochick Steuer, H.M., Keilman, M., Bansal, S., Lam, A.M., Seibel, W., Neilson, S., Furman, P.A., Sofia, M.J., 2014. Discovery of a novel class of potent HCV NS4B inhibitors: SAR studies on piperazinone derivatives. J. Med. Chem. 57, 2136–2160.

Kati, W., Koev, G., Irvin, M., Beyer, J., Liu, Y., Krishnan, P., Relsch, T., Mondal, R., Wagner, R., Molla, A., Maring, C., Collins, C., 2014. In vitro activity and resistance profile of dasabuvir, a nonnucleoside hepatitis C virus polymerase inhibitor. Antimicrobial Agents Chemother. 59, 1505–1511.

Kettner, C.A., Shenvi, A.B., 1984. Inhibition of the serine proteases leukocyte elastase, pancreatic elastase, cathepsin G, and chymotrypsin by peptide boronic acids. J. Biol. Chem. 259, 15106–15114.

Kier, L.B., Hall, L.H., 1976. Molecular Connectivity in Drug Research. Academic Press, New York.

Kim, J., Kim, K.-S., Lee, H.S., Park, K.-S., Park, S.Y., Kang, S.-Y., Lee, S.J., Park, H.S., Kim, D.-E., Chong, Y., 2008. Effects of the aryl linker and the aromatic substituent on the anti-HCV activities of aryl diketoacid (ADK) analogues. Bioorg. Med. Chem. Lett. 18, 4661–4665.

Kneteman, N.M., Howe, A.Y., Gao, T., Lewis, J., Pevear, D., Lund, G., Douglas, D., Mercer, D.F., Tyrrell, D.L., Immermann, F., Chaudhary, I., Speth, J., Villano, S.A., O'Connell, I., 2009. HCV-796: a selective nonstructural protein 5B polymerase inhibitor with potent anti hepatitis C virus activity in vitro, in mice with chimeric human livers and in humans infected with hepatitis C virus. Hepatology 49, 745–752.

Kolykhalov, A.A., Mihalik, K., Feinstone, S.M., Rice, C.M., 2000. Hepatitis C virus-encoded enzymatic activities and conserved RNA elements in the 3' nontranslated region are essential for virus replication in vivo. J. Virol. 74, 2046–2051.

Kowdley, K.V., Lawitz, E., Poordad, F., Cohen, D.E., Nelson, D.R., Zeuzem, S., Everson, G.T., Kwo, P., Foster, G.R., Sulkowski, M.S., Xie, W., Pilot-Matias, T., Liossis, G., Larsen, L., Khatri, A., Podsadecki, T., Bernstein, B., 2014. Phase 2b trial of interferon-free therapy for hepatitis C virus genotype 1. N. Engl. J. Med. 370, 222–232.

Kumar, D.V., Ria, K., Brameld, K.A., Somoza, J.R., Rajagopalan, R., Janc, J.W., Xia, Y.M., Ton, T.L., Shaghafi, M.B., Hu, H.Y., Lehoux, I., To, N., Young, W.B., Green, M.J., 2011. Quinolones as HCV NS5B polymerase inhibitors. Bioorg. Med. Chem. Lett. 21, 82–87.

Küçükgüzel, I., Satimis, G., Gurukumar, K.R., Basu, A., Tatar, E., Nichols, D.B., Talele, T.T., Kaushik-Basu, N., 2013. 2-Heteroarylimino-5-arylidene-4-thiazolidinones as a new class of nonnucleoside inhibitors of HCV NSSB polymerase. Eur. J. Med. Chem. 69, 931–941.

Kwong, C.D., Clark, J.L., Fowler, A.T., Geng, F., Kezar III, H.S., Roychowdhury, A., Reynolds, R.C., Maddry, J.A., Ananthan, S., Secrist III, J.A., Shih, N.Y., Piwinski, J.J., Li, C., Feld, B., Huang, H.C., Tong, X., Njoroge, F.G., Arasappan.A, 2012. Novel substituted pyrimidines as HCV replication (replicase) inhibitors. Bioorg. Med. Chem. Lett. 22, 1160–1164.

Lamarre, D., Anderson, P.C., Bailey, M., Beaulieu, P., Bolger, G., Bonneau, P., Bös, M., Cameron, D.R., Cartier, M., Cordingley, M.G., Faucher, A.M., Goudreau, N., Kawai, S.H., Kukolj, G., Lagacé, L., LaPlante, S.R., Narjes, H., Poupart, M.A., Rancourt, J., Sentjens, R.E., St. George, R., Simoneau, B., Steinmann, G., Thibeault, D., Tsantrizos, Y.S., Weldon, S.M., Yong, C.L., Llinàs-Brunet, M., 2003. An NS3 protease inhibitor with antiviral effects in humans infected with hepatitis C virus. Nature 426, 186–189.

Lavanchy, D., 2009. The global burden of hepatitis C. Liver Int. 29, 74–81.

Lawitz, E., Poordad, F., Kowclley, K.V., Cohen, D.E., Podsadecki, T., Sigjrelkow, S., Larsen, L., Menon, R., Koev, G., Tripathi, R., Pilot-Matias, T., Bernstein, B., 2013. A phase 2a trial of 12-week interferon-free therapy with two direct-acting antivirals (ABT4 50/r, ABT-072) and ribavirin in 1L28B C/C patients with chronic hepatitis C genotype 1. J. Hepatol. 59, 18–23.

Lee, C., Ma, H., Hang, J.Q., Leveque, V., Sklan, E.H., Elazar, M., Klumpp, K., Glenn, J.S., 2011. The hepatitis C virus NS5A inhibitor (BMS-790052) alters the subcellular localization of the NS5A non-structural viral protein. Virol 414 (1), 10–18.

Lee, S., Yoon, K.D., Lee, M., Cho, Y., Choi, G., Jang, H., Kim, B.S., Oh, J.W., Jeong, Y.J., Kwon, H.J., Bae, S.K., Min, D.H., Windisch, M.P., Heo, T.H., Lee, C., 2016. Identification of a resveratrol tetramer as a potent inhibitor of hepatitis C virus helicase. Brit. J. Pharmacol. 173, 191–211.

Legrand-Abravanel, F., Nicot, F., Izopet, J., 2010. New NS5B polymerase inhibitors for hepatitis C. Expert Opin. Inv. Drugs 19, 963–975.

Levin, J. (Ed.), 2015. Macrocycles in Drug Discovery. The Royal Society of Chemistry, ISBN 978-1-84973-701-2.

Li, H., Shi, S.T., 2010. Non-nucleoside inhibitors of hepatitis C virus polymerase: current progress and future challenges. Future Med. Chem. 2, 121–141.

Li, T., Froeyen, M., Herdewijn, P., 2010. Insight into ligand selectivity in HCV NS5B polymerase: molecular dynamics simulations, free energy decomposition and docking. J. Mol. Model. 16, 49–59.

Liu, Y., Jiang, W.W., Pratt, J., Rockway, T., Harris, K., Vasavanonda, S., Tripathi, R., Pithawalla, R., Kati, W.M., 2006. Mechanistic study of HCV polymerase inhibitors at the individual steps of the polymerization reaction. Biochemistry 45, 11312–11323.

Liverton, N.J., Holloway, M.K., McCauley, J.A., Rudd, M.T., Butcher, J.W., Carroll, S.S., DiMuzio, J., Fandozzi, C., Gilbert, K.F., Mao, S.-S., McIntyre, C.J., Nguyen, K.T., Romano, J.J., Stahlhut, M., Wan, B.L., Olsen, D.B., Vacca, J.P., 2008. Molecular modeling based approach to potent P2–P4 macrocyclic inhibitors of hepatitis C NS3/4A protease. J. Am. Chem. Soc. 130, 4607–4609.

Llinás-Brunet, M., Bailey, M., Ghiro, E., Gorys, V., Halmos, T., Poirier, M., Rancourt, J., Goudreau, N., 2004b. A systematic approach to the optimization of substrate-based inhibitors of the hepatitis C virus NS3 protease: discovery of potent and specific tri-peptide inhibitors. J. Med. Chem. 47, 6584–6594.

Llinás-Brunet, M., Bailey, M., Fazal, G., Goulet, S., Halmos, T., Laplante, S., Maurice, R., Poirier, M.A., Poupart, D., Thibeault, D., Wernic, D., Lamarre, D., 1998. Peptide-based inhibitors of the hepatitis C virus serine protease. Bioorg. Med. Chem. Lett. 8, 1713–1718.

Llinás-Brunet, M., Bailey, M.D., Bolger, G., Brocho, C., Faucher, A.M., Ferland, J.M., Garneau, M., Ghiro, E., Gorys, V., Grand-Maitre, C., Halmos, T., Lapeyre-Paquette, N., Liard, F., Poirier, M., Rheaume, M., Tsantrizos, Y.S., Lamarre, D., 2004a. Structure-activity study on a novel series of macrocyclic inhibitors of hepatitis C virus NS3 protease leading to the discovery of BILN 2061. J. Med. Chem. 47, 1605–1608.

Lohmann, V., Körner, F., Herian, U., Bartenschlager, R., 1997. Biochemical properties of hepatitis C virus NS5B RNA-dependent RNA polymerase and identification of amino acid sequence motifs essential for enzymatic activity. J. Virol. 71, 8416–8428.

Mayhoub, A.S., 2012. Hepatitis C RNA-dependent RNA polymerase inhibitors: a review of structure-activity and resistenace relationships; different scaffolds and mutations. Bioorg. Med. Chem. 20 (10), 3150–3161.

McCauley, J.A., Rudd, M.T., 2016. Hepatitis C virus NS3/4a protease inhibitors. Curr. Opin. Pharmacol. 30, 84–92.

Meguellati, A., Ahmed-Belkacem, A., Yi, W., Haudecoeur, R., Crouillere, M., Briilet, R., Pawlotsky, J.-K., Boumendjel, A., Peuchmaur, M., 2014. B-ring modified aurones as promising allosteric inhibitors of hepatitis C virus RNA-dependent RNA polymerase. Eur. J. Med. Chem. 80, 579–592.

Meguellati, A., Ahmed-Belkacem, A., Nurisso, A., Yi, W., Brillet, R., Berqoucha, N., Chavoutier, L., Fortune, A., Pawlotsky, J.-M., Boumndjel, A., Peuchmaur, M., 2016. New pseudodimeric aurones as palm pocket inhibitors of hepatitis C virus RNA-dependent RNA polymerase. Eur. J. Med. Chem. 115, 217–229.
Mo, H., Lu, L., Piloy-Matias, T., Pittawalla, R., Mondal, R., Masse, S., Dekhtyar, T., Ng, T., Koev, G., Stoll, V., Stewart, K.D., Pratt, J., Donner, P., Rockway, T.W., Maring, C., Molla, A., 2005. Mutations conferring resistance to a hepatitis C virus (HCV) RNA dependent RNA polymerase inhibitor alone or in combination with an HCV serine protease inhibitor in vitro. Antimicrob. Agents Chemother. 49, 4305–4314.
Musmuca, I., Caroli, A., Mai, A., Kaushik-Basu, N., Arora, P., Ragno, R., 2010. Combining 3-D quantitative structure–activity relationship with ligand based and structure based alignment procedures for in silico screening of new hepatitis C virus NS5B polymerase inhibitors. J. Chem. Inf. Model. 50, 662–676.
Nair, A.G., Zeng, Q., Selyutin, O., Rosenblum, S.B., Jiang, Y., Ynag, D.Y., Keertikar, K., Zhou, G., Dwyer, M.P., HeonKim, S., Shankar, B., Yu, W., Tong, L., Chen, L., Mazzola, R., Caldwell, J., Tang, H., Allard, M.L., Kozlowski, J.A., 2016. Discovery of silyl proline containing HCV NS5A inhibitors with pan-genotype activity: SAR development. Bioorg. Med. Chem. Lett. 26, 1475–1479.
Nakayama, T., 2002. Enzymology of aurone biosynthesis. J. Biosci. Bioeng. 94, 487–491.
Nakayama, T., Yonekura-Saikakibara, K., Sato, T., Kikuchi, S., Fukui, Y., Fukuchi-Mizutani, M., Ueda, T., Nakao, M., Tanaka, Y., 2000. Aureusidin synthase: a polyphenol oxidase homolog responsible for Rower coloration. Science 290, 1163–1166.
Nichols, D.B., Fournet, G., Gurukumar, K.R., Basu, A., Lee, J.-C., Sakamoto, N., Kozielski, F., Musmuca, I., Joseph, B., Ragno, R., Kaushik-Basu, N., 2012. Inhibition of hepatitis C virus NS5B polymerase by S-trityl-l-cysteine derivatives. Eur. J. Med. Chem. 49, 191–199.
Nurbo, J., Peterson, S.D., Dahl, G., Danielson, H., Karlen, A., Sandström, A., 2008. β-Amino acid substitutions and structure-based CoMFA modeling of hepatitis C virus NS3 protease inhibitors. Bioorg. Med. Chem. 16, 5599–5605.
Olsen, D.B., Eldrup, A.B., Bartholomew, L., Bhat, B., Bosserman, M.R., Ceccacci, A., Colwell, L.F., Fay, J.F., Flores, O.A., Getty, K.L., Grobler, J.A., LaFemina, R.L., Markel, E.J., Migliaccio, G., Prhavc, M., Stahlhut, M.W., Tomassini, J.E., MacCoss, M., Hazuda, D.J., Carroll, S.S., 2004. A 7-deaza-adenosine analog is a potent and selective inhibitor of hepatitis C virus replication with excellent pharmacokinetic properties. Antimicrob. Agents Chemother. 48, 3944–3953.
Patel, P.D., Patel, M.R., Kaushik-Basu, N., Talele, T.T., 2008. 3D QSAR and molecular modeling studies of benzimidazole derivatives as hepatitis C virus NS5B polymerase inhibitors. J. Chem. Inf. Model. 48, 42–55.
Patel, B.A., Krishnan, R., Khadtare, N., Gurukumar, K.R., Basu, A., Arora, P., Bhatt, A., Patel, M.K., Dana, D., Kumar, S., Kaushik-Basu, N., Talele, T.T., 2013. Design and synthesis of L- and D-phenylalanine derived rhodanines with novel C5-arylidenes as inhibitors of HCV NS5B polymerase. Bioorg. Med. Chem. 21, 3262–3271.
Patil, V.M., Gupta, S.P., Samanta, S., 2010. A QSAR study on some series of anti-hepatitis C virus (HCV) agents. Lett. Drug Design Discov. 7, 139–148.
Patil, V.M., Gupta, S.P., Samantha, S., Masand, N., 2011a. Current perspective of HCV NS5B inhibitors: a review. Curr. Med. Chem. 18, 5564–5597.
Patil, V.M., Gupta, S.P., Samanta, S., Masand, N., 2011b. 3D QSAR kNN-MFA studies on thiouracil derivatives as hepatitis C virus inhibitors. Med. Chem. Res. 20, 1616–2162.
Patil, V.M., Gurukumar, K.R., Chudayeu, M., Gupta, S.P., Samanta, S., Masand, N., Kaushik-Basu, N., 2012a. Synthesis, in vitro and in silico NS5B polymerase inhibitory activity of benzimidazole derivatives. Med. Chem. 8, 629–635.

Patil, V.M., Gupta, S.P., Samanta, S., Masand, N., 2012b. 3D-QSAR and docking studies on a series of benzothiadiazine derivatives as genotype 1 HCV polymerase inhibitors. Med. Chem. 8, 1099–1107.

Patil, V.M., Gupta, S.P., Sarnanta, S., Masand, N., 2013. Virtual screening of imidazole analogs as potential hepatitis C virus NS5B polymerase inhibitors. Chem. Papers 67, 236–244.

Pawlotsky, J.M., 2013. NS5A inhibitors in the treatment of hepatitis C. J. Hepatol. 59, 375–382.

Perrone, P., Luoni, G.M., Kelleher, M.R., Daverio, F., Angell, A., Mulready, S., Congiatu, C., Rajyaguru, S., Martin, J.A., Levêque, V., Pogam, S.L., Najera, I., Klumpp, K., Smith, D.B., McGuigan, C., 2007a. Application of the phosphoramidate protide approach to 4'-azidouridine confers sub-micromolar potency versus hepatitis C virus on an inactive nucleoside. J. Med. Chem. 50 (8), 1840–1849.

Perrone, P., Daverio, F., Valente, R., Rajyaguru, S., Martin, J.A., Leveque, V., Le Pogam, S., Najera, I., Klumpp, K., Smith, B.B., McGuigan, C., 2007b. First example of phosphoramidate approach applied to a 4'-substituted purine nucleoside (4'-azidoadenosine): conversion of an inactive nucleoside to a submicromolar compound versus hepatitis C virus. J. Med. Chem. 50, 5463–5470.

Perz, J.F., Armstrong, G.L., Farrington, L.A., Hutin, Y.J., Bell, B.P., 2006. The contributions of hepatitis B virus, hepatitis C. Virus infections to cirrhosis and primary liver cancer worldwide. J. Hepatol. 45, 529–538.

Phillips, B., Cai, R., Delaney, W., Du, Z., Ji, M., Jin, H., Lee, J., Li, J., Niedziela-Majka, A., Mish, M., Pyun, H.J., Saugier, J., Tirunagari, N., Wang, J., Yang, H., Wu, Q., Sheng, C., Zonte, C., 2014. Highly potent HCV NS4B inhibitors with activity against multiple genotypes. J. Med. Chem. 57, 2161–2166.

Polamreddy, P., Vishwakarma, V., Gundla, R., 2016. A review on anti-HCV agents targeting active site and allosteric sites of non-structural protein 5B (NS5B) Int. J. Pharm. Pharmaceut. Sci. 18, 1–18.

Poliakov, A., Johansson, A., Akerblom, E., Oscarsson, K., Samuelsson, B., Hallberg, A., Danielson, U.H., 2004. Structure-activity relationships for the selectivity of hepatitis C virus NS3 protease inhibitors. Biochim. Biophys. Acta 1672, 51–59.

Powers, J.P., Piper, D.E., Li, Y., Mayorga, V., Anzola, J., Chen, J.M., Jaen, J.C., Lee, G., Liu, J., Peterson, M.G., Tonn, G.R., Ye, Q., Walker, N.P., Wang, Z., 2006. SAR and mode of action of novel non-nucleoside inhibitors of hepatitis C NS5B RNA polymerase. J. Med. Chem. 49, 1034–1046.

Priestley, E.S., Decicco, 2000. 1-Aminocyclopropaneboronic acid: synthesis and incorporation into an inhibitor of hepatitis C virus NS3 Protease. Org. Lett. 2, 3095–3097.

Rawal, R.K., Katti, S.B., Kaushik-Basu, N., Arora, P., Pan, Z., 2008. Non-nucleoside inhibitors of the hepatitis C virus NS5B RNA-dependant RNA polymerase: 2-aryl-3-heteroaryl-1,3-thiazolidin-4-one derivatives. Bioorg. Med. Chem. Lett. 18, 6110–6114.

Romero-Brey, I., Merz, A., Chiramel, A., Lee, J.Y., Chlanda, P., Haselman, U., Santarella-Mellwig, R., Habermann, A., Hoppe, S., Kallis, S., Walther, P., Antony, C., Krijnse-Locker, J., Bartenschlager, R., 2012. Three-dimensional architecture and biogenesis of membrane structures associated with hepatitis C virus replication. PLoS Pathog. 8, e1003056.

Rönn, R., Sabnis, Y.A., Gossas, T., Akerblom, E., Danielson, U.H., Hallberg, A., Johansson, A., 2006. Exploration of acyl sulfonamides as carboxylic acid replacements in protease inhibitors of hepatitis C virus full length NS3. Bioorg. Med. Chem. 14, 544–559.

Ruebsam, F., Murphy, D.E., Tran, C.V., Li, L.-S., Zhao, J., Dragovich, P.S., McGuire, H.M., Xiang, A.X., Sun, Z., Ayida, B.K., Blazel, J.K., Kim, S.H., Zhou, Y., Han, Q., Kissinger, C.R., Webber, S.E., Showalter, R.E., Shah, A.M., Tsan, M., Patel, R.A., Thompson, P.A., LeBrun, L.A., Hou, H.J., Kamran, R., Sergee, M.V., Bartkowski, D.M., Nolan, T.G., Norris, D.A., Khandurina, J., Brooks, J., Okamoto, E., Kirkovsky, L., 2009. Discovery of tricyclic 5,6-dihydro-IH-pyridin-2-ones as novel, potent, and orally bioavailable inhibitors of HCVNS5B polymerase. Bioorg. Med. Chem. Lett. 19, 6404–6412.

Samanta, S., Gupta, S.P., Masand, N., Basu, N.K., Patil, V.M., 2012. Docking directed kNN MFA study of pyrimidine nucleosodes as HCV NS5B inhibitors. Int. J. Drug Design Discov. 3, 809–813.

Sharma, A., Gupta, S.P., Siddiqui, A.A., 2013. A QSAR study on a series of thiourea derivatives acting as anti-hepatitis C virus agents. Indian J. Biochem. Biophys. 50, 278–283.

Sharma, A., Gupta, S.P., Siddiqui, A.A., Sharma, N., 2017. HCV NS3/4A protease and its emerging inhibitors. J. Anal. Pharm. Res. https://doi.org/10.15406/japlr.2017.04.00108.

Shotwell, J.B., Baskaran, S., Chong, P., Creech, K.L., Crosby, R.M., Dickson, H., Fang, J., Garrido, D., Mathis, A., Maung, J., Parks, D.J., Pouliot, J.J., Price, D.J., Rai, R., Seal III, J.W., Schmitz, U., Tai, V.W., Thomson, M., Xie, M., Xiong, Z.Z., Peat, A.J., 2012. Imidazo[1,2-a]pyridines that directly interact with hepatitis C NS4B initial preclinical characterization. ACS Med. Chem. Lett. 3, 565–569.

Sofia, M.J., Chang, W., Furman, P.A., Mosley, R.T., Ross, B.S., 2012. Nucleoside, nucleotide, and non-nucleoside inhibitors of hepatitis C virus NS5B RNA-dependent RNA-polymerase. J. Med. Chem. 55 (6), 2481–2531.

Talele, T.T., Arora, P., Kulkarni, S.S., Patel, M.R., Singh, S., Chudayeu, M., Kaushik-Basu, N., 2010. Structure-based virtual screening, synthesis and SAR of novel inhibitors of hepatitis C virus NS 5B polymerase. Bioorg. Med. Chem. 18, 4630–4638.

Tan, S.L., He, Y., 2011. Hepatitis C Antiviral Drug Discovery and Development. Caister Academic Press, Norfolk, UK.

Targett-Adams, P., Graham, E.J., Middleton, J., Palmer, A., Shaw, S.M., Lavender, H., Brain, P., Tran, T.D., Jones, L.H., Wakenhut, F., Stammen, B., Pryde, D., Pickford, C., Westby, M., 2011. Small molecules targeting hepatitis C virus-encoded NS5A cause subcellular redistribution of their target: insights into compound modes of action. J. Virol. 85, 6353–6568.

Tong, L., Yu, W., Coburn, C.A., Meinke, P.T., Nair, A.K.G., Dwyer, M.P., Chen, L., Selyutin, O., Rosenblum, S.B., Jiang, Y., Fells, J., Hu, B., Zhong, B., Soll, R.M., Liu, R., Agarwal, S., Xia, E., Zhai, Y., Kozlowski, J.A., 2016a. Alternative core development around the tetracyclic indole class of HCV NS5A inhibitors. Bioorg. Med. Chem. Lett. 26, 5132–5137.

Tong, L., Zeng, Q., Kozlowski, J.A., 2016b. Discovery of potent macrocyclic HCV NS5A inhibitors. Bioorg. Med. Chem. Lett. 26, 3793–3799.

Tong, L., Yu, W., Chen, L., Selyutin, O., Dwyer, M.P., Nair, A.G., Mazzola, R., Kim, J.H., Sha, D., Yin, J., Ruck, R.T., Davies, I.W., Hu, B., Zhong, B., Hao, J., Ji, T., Zan, S., Liu, R., Agrawal, S., Xia, E., Curry, S., McMonagle, P., Bystol, K., Lahser, F., Carr, D., Rokosz, L., Ingravallo, P., Chen, S., Feng, K.I., Cartwright, M., Asante-Appiah, E., Kozlowski, J.A., 2017. Discovery of Ruzasvir (MK-8408): a potent, pan-genotype HCV NS5A inhibitor with optimized activity against common resistance- associated polymorphisms. J. Med. Chem. 60, 290–306.

Tsantrizos, Y.S., Bolger, G., Bonneau, P., Cameron, D.R., Goudreau, N., Kukolj, G., LaPlante, S.R., Llinas-Brunet, M., Nar, H., Lamarre, D., 2003. Macrocyclic inhibitors of the NS3 protease as potential therapeutic agents of hepatitis C virus infection. Angew. Chem. Int. Ed. 42, 1355–1360.

Varaprasad, C.V.N.S., Ramasamy, K.S., Girardet, J.L., Gunic, E., Lai, V., Zhong, W., An, H., Hong, Z., 2007. Synthesis of pyrrolo[2,3-*d*]pyrimidine nucleoside derivatives as potential anti-HCV agents. Biorg. Chem. 35 (1), 25–34.
Varshney, J., Sharma, A., Gupta, S.P., 2012a. A QSAR study on some thiophene derivatives acting as anti-HCV agents. Lett. Drug Des. Discov. 9, 389–396.
Varshney, J., Sharma, A., Gupta, S.P., 2012b. A quantitative structure-activity relationship study on a few series of anti-hepatitis C virus agents. Med. Chem. 8, 491–504.
Varshney, J., Sharma, P.K., Sharma, A., 2012ca. A review on an update of NS5B polymerase hepatitis C virus inhibitors. Eur. Rev. Med. Phar. Sci. 16, 667–671.
Wang, X., Yang, W., Xu, X., Zhang, H., Li, Y., Wang, Y., 2010. Studies of benzothiadiazine derivatives as hepatitis C virus NS5B polymerase inhibitors using 3D-QSAR, molecular docking and molecular dynamics. Curr. Med. Chem. 17, 2788–2803.
Wang, H., Guo, C., Chen, B.-Z., Ji, M., 2017. Computational study on the drug resistance mechanism of HCV NS3 protease to BMS-605339. Biotechnology 64, 153–164.
Watkins, W.J., Ray, A.S., Chong, L.S., 2010. HCV NS5B polymerase inhibitors. Curr. Opin. Drug Discov. Devel. 13, 441–465.
Wei, H.-Y., Lu, C.-S., Lin, T.-H., 2008. Exploring the P2 and P3 ligand binding features for Hepatitis C virus NS3 protease using some 3D QSAR techniques. J. Mol. Graphics Model 26, 1131–1144.
Weidlich, I.E., Filippov, I.V., Brown, J., Kaushik-Basu, N., Krishnan, R., Nicklaus, M.C., Thorpe, I.F., 2013. Inhibitors for the hepatitis C virus RNA polymerase explored by SAR with advanced machine learning methods. Bioorg. Med. Chem. 21, 3127–3137.
Worachartcheewan, A., Prachayasittikul, V., Anuwongcharoen, N., Shoombuatong, W., Prachayasittikul, V., Nantasenamat, C., 2015. On the origins of hepatitis C virus NS5B polymerase inhibitory activity using machine learning approaches. Curr. Top. Med. Chem. 15, 1814–1826.
Yu, W., Coburn, C.A., Yang, D.Y., Meinke, P.T., Wong, M., Rosenblum, S.B., Chen, K.X., Njoroge, G.F., Chen, L., Dwyer, M.P., Jiang, Y., Nair, A.G., Selyutin, O., Tong, L., Zeng, Q., Zhong, B., Ji, T., Hu, B., Kozlowski, J.A., 2016. Discovery of fused tricyclic core containing HCV NS5A inhibitors with pan-genotype activity. Bioorg. Med. Chem. Lett. 26, 3158–3162.
Zeng, Q., Nair, A.C., Roseriblum, S.B., Huang, H.-C., Lesbui, C.A., Jiang, Y., Selyutin, O., Chan, T.-Y., Bennett, F., Chen, K.X., Venkatraman, S., Sannigrahi, M., Velaquez, F., Duca, J.S., Gavalas, S., Huang, Y., Pu, H., Wang, L., Pinto, P., Vibulbhan, P., Agrawal, S., Ferrari, E., Jiang, C.-K., Li, C., Hesk, D., Gessel, J., Sorota, S., Shih, N.-Y., Njoroge, F.G., Kozlowski, J.A., 2013. Discovery of an irreversible HCV NS5B polymerase inhibitor. Bioorg. Med. Chem. Lett. 23, 6585–6587.
Zhu, T., Fawzi, M.B., Flint, M., Kong, F., Szeliga, J., Tsau, R., Howe, A.Y., Pan, W., 2010. Design and synthesis of HCV agents with sequential triple inhibitory potentials. Bioorg. Med. Chem. Lett. 20, 5212–5216.

CHAPTER 8

Status and Global Strategy of Prevention of Hepatitis Virus Infection

1. INTRODUCTION

Among all virus infections, the hepatitis virus infection has been the most challenging problem across the world, and until recently, it has been largely ignored as a health and development priority, but now, there is a call for specific action to combat it globally. Consequently, a strategy has been planned to completely eradicate hepatitis viral infection by 2030 (Document EB138/29). A 2030 agenda proclaims the goal to reduce by 2030 the incidence of chronic hepatitis infection from the current 6–10 million to 0.9 million and consequently to bring down annual death from 1.4 million to 0.5 million, thus eliminating viral hepatitis as a major public health threat. The agenda envisages a world where viral hepatitis transmission is completely halted and everyone living with viral hepatitis has access to safe, affordable, and effective care and treatment. A detail of the present status of viral hepatitis and the 2030 agenda is given next.

2. PRESENT STATUS OF VIRAL HEPATITIS

Presently, viral hepatitis is a major global health problem. In 2013, it was the seventh highest cause of global mortality. The number of deaths per year attributed to acute hepatitis virus infection and hepatitis-related liver cancer and cirrhosis is 1.4 million, a toll comparable to HIV and tuberculosis. Of the total hepatitis deaths, approximately 47% are estimated to be due to hepatitis B virus (HBV), 48% due to hepatitis C virus (HCV) and the remainder due to hepatitis A virus (HAV) and hepatitis E virus (HEV). Viral hepatitis is also a growing cause of mortality among people living with HIV. About 2.9 million people living with HIV are coinfected with HCV and 2.6 million with HBV (Document EB138/29).

Studies on Hepatitis Viruses
ISBN 978-0-12-813330-9
https://doi.org/10.1016/B978-0-12-813330-9.00008-9

Worldwide, HBV infection dominates over HCV infection, as approximately 240 million people are suffering from this, compared to only 13–150 million people suffering from HCV infection. The number of people suffering from HCV infection is also increasing despite the existence of an effective cure. All five hepatitis viruses widely differ from each other in respect of their structures, modes of infection, populations affected, and health outcomes. However, HBV and HCV infections are blood-borne, occurring significantly in early life and through unsafe injections and medical procedures. They are less common through sexual contacts. As can be seen in Fig. 8.1, HBV infection is highly prevalent in sub-Saharan Africa and East Asia, where between 5% and 10% of the adult population is chronically infected. Mother-to-child transmission of HBV infection is a major mode of transmission in high prevalence areas. The Amazon region of South America and the southern parts of eastern and central Europe also have high rates of HBV chronic infection. In the Middle East and the

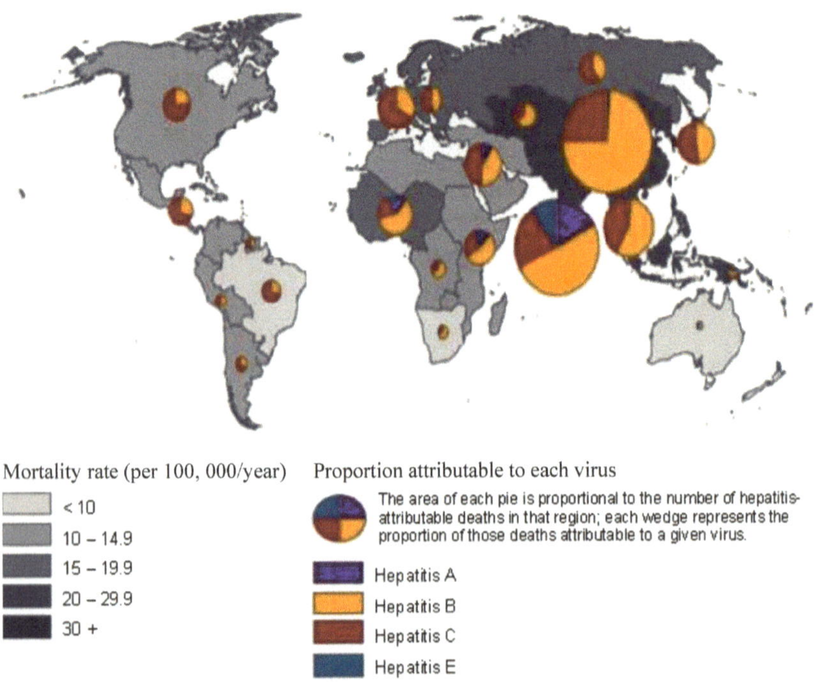

Figure 8.1 Regional distribution of viral hepatitis deaths. *(Global Burden of Disease (http://vizhub.healthdata.org/gbd-compare/#settings=f53a9677cb8e4ff596bc37f980aa0276322ecc39).)*

Indian subcontinent, around 2%—5% of the population is estimated to have chronic HBV infection. Immunization is supposed to be the most effective strategy for prevention of HBV infection.

HCV is found worldwide, but it is most prevalent in central and east Asia and north and west Africa due to unsafe medical injections and other medical procedures. Safe blood products, safe injection practices, harm reduction services for people who inject drugs, and promotion of safe sex are required for the prevention of both HBV and HCV infections. Hepatitis D virus infection occurs only in those people who are already infected with HBV, but HAV and HEV infections are food- and water-borne infections and occur due to unsafe water and poor sanitation, and they never turn to chronic.

3. GLOBAL STRATEGY OF PREVENTION

Hepatitis epidemics can be brought to an end if the attention is paid to five core intervention strategies, namely use of effective vaccines, ways of prevention of mother-to-child transmission of HBV, injection, blood, and surgical safety, use of sterile injecting equipment, and use of new oral, well-tolerated medicines. However, in complete elimination of viral hepatitis, there are certain barriers, as given next, that need to be addressed:

1. uneven leadership and commitment
2. nonavailability of adequate data
3. limited coverage of prevention programs
4. people's ignorance of their hepatitis status
5. inaccessibility to treatment and care services
6. unaffordable medicines and diagnostics for most
7. lack of a public health approach to hepatitis
8. widespread stigma and discrimination, preventing equitable access to services

However, to achieve the 2030 Agenda for Sustainable Development, five "Strategic Directions" were formulated, which were desired to be followed by WHO and partners and countries. These strategic directions have the following roles (Document EB138/31).

Strategic Direction 1 is to focus on the need to understand the viral hepatitis epidemic and response as a basis for advocacy, political commitment, national planning, resource mobilization and allocation, implementation, and program improvement.

Strategic Direction 2 is concerned with the first dimension of universal health coverage and the services that should be delivered globally.

Strategic Direction 3 provides directions as to how these services should be delivered.

Strategic Direction 4 is concerned with the universal health coverage by identifying sustainable and innovative models for financing the hepatitis responses and approaches for reducing costs to a limit affordable by common people.

Strategic Direction 5 refers to the necessity to promote research and innovation that provide opportunities to change the trajectory of the global hepatitis response, improve efficiency and quality of services, and maximize impact, which can be fulfilled by promoting operational research and collaboration between researchers and policy makers to ensure that research findings are accepted and utilized on a scale sufficient to have the desired impact.

The strategy planned on hepatitis is the first global health strategy that contributes to the achievements of 2030 Agenda for Sustainable Development. The agenda covers the first 6 years of the post-2015 health agenda, 2016–21, building on the *Prevention and Control of Viral Hepatitis Infection: Framework for Global Action* (Information on the consultation process and draft strategy documents), and on two resolutions on viral hepatitis adopted by World Health Assembly in 2010 and 2014 (United Nations General Assembly Resolution A/RES/70/1, 2030). The strategy addresses all five hepatitis viruses, with a particular focus on HBV and HCV. The strategy primarily encompasses vision, goal, targets, guiding principles, directions, and priority actions. It also outlines key elements of strategy implementation, including strategic partnerships, monitoring, evaluation, and costing.

4. CONCLUSIONS

In conclusion, viral hepatitis is a major global health problem, and the major infection is due to HBV and HCV, among which the former is dominant. To safeguard from hepatitis virus infection, there are five core intervention studies, namely use of effective vaccines, ways of prevention of mother-to-child transmission of HBV, injection, blood, and surgical safety, use of sterile injecting equipment, and use of new oral, well-tolerated medicines. A global strategy has been planned to eradicate viral hepatitis from the world. The strategy primarily encompasses vision, goal, targets, guiding principles, directions, and priority actions.

SOURCE

World Health Organization, Sixty-Ninth World Health Assembly, Provisional Agenda Item, A69/32, 22 April 2016.

REFERENCES

Document EB138/29.
Document EB138/31.
Information on the consultation process and draft strategy documents is available at: http://www.who.int/hiv/pub/hep-strategy-background/en/.
United Nations General Assembly Resolution A/RES/70/1. Transforming our world: the 2030 Agenda for Sustainable Development http://www.un.org/ga/search/view_doc.asp?symbol=A/RES/70/1&Lang=E.

INDEX

'*Note*: Page numbers followed by "f" indicate figures, "t" indicate tables.'

B

I

MIX
Papier | Fördert gute Waldnutzung
FSC® C083411

Kolibri 360 GmbH
Ferdinand-Jühlke-Straße 7
99095 Erfurt, Deutschland
produktsicherheit@kolibri360.de